LINEAR ALGEBRA WITH APPLICATIONS

Steven J. Leon

LINEAR ALGEBRA WITH APPLICATIONS

MACMILLAN PUBLISHING CO., INC.
New York
COLLIER MACMILLAN PUBLISHERS
London

Printed in the United States of America

Macmillan Publishing Co., Inc.
866 Third Avenue, New York, New York 10022

Collier Macmillan Canada, Ltd.

Library of Congress Cataloging in Publication Data

Leon, Steven J
Linear algebra with applications.

Bibliography: p.
Includes indexes.
1. Algebras, Linear. I. Title.
QA251.L46 512'.5 79-15486
ISBN 0-02-369870-5

Printing: 1 2 3 4 5 6 7 8 Year: 0 1 2 3 4 5 6

To
Judith

Preface

This textbook is suitable for either a sophomore level course or for a junior- or senior-level course. The only prerequisite is calculus. If the text is to be taught at the sophomore level, one should probably spend more time on the earlier chapters and omit most of the more difficult optional sections. On the other hand, if the course is to be taught on the junior or senior level, the instructor should probably spend less time on the early chapters and cover more of the optional material. The explanations in the text are given in sufficient detail so that students at either level will have little trouble reading and understanding. To further aid the student, a large number of examples have been worked out completely. Applications have been scattered throughout the text rather than put together in a chapter at the end. In this way they can be used to motivate new material and to illustrate the relevance of the material that has just been presented. When applications are included at the end of a text, they are more likely to be omitted because of lack of time.

The text tries to give fairly complete coverage to a very broad subject. Consequently, there is probably more material included than can possibly be covered in a one-quarter or one-semester course. The instructor then has some freedom in the choice of topics, and consequently the instructor may design the course to meet the needs of the class. Some instructors may decide to emphasize the mathematical theory in Chapters 3 and 4 and others may decide to skip over these chapters and spend more time on the applied topics in Chapters 5 and 6. As a general rule, if you ask n mathematicians what should go into a course, you will get n different answers. Even if many of the topics in the text are omitted, the students should get a feeling for the overall scope of the subject matter. Furthermore, many of the students may use their text later as a reference and consequently may end up learning many of the optional topics on their own; optional material is preceded by an asterisk. The following is a guide to the various chapters in the text.

Chapter 1. The first two sections deal with systems of equations. Sections 3 and 4 are concerned with matrices and matrix algebra. Most of Section 5 is optional. The instructor should cover the material at the beginning of this section, which includes an explanation of the notation used for column vectors. The material on block multiplication may be omitted at the discretion of the instructor. Block multiplication is used later in the text in some of the optional sections in Chapters 6 and 7, but it has been avoided in those sections that form the core of the text.

Chapter 2. This is a short chapter on determinants. Determinants will be used later in the text for introducing such topics as linear independence and eigenvalues.

Chapter 3. The basic theory of vector spaces is presented in this chapter. All five sections should be covered.

Chapter 4. Chapter 4 is devoted to linear transformations. Instructors wishing to present a more applied course may omit all or part of this chapter.

Chapter 5. This is a long chapter on orthogonality and its applications. Most of Sections 1, 2, 3, 6, and 7 should be covered if at all possible. Section 4 on matrix norms is optional, but it is a prerequisite for Section 7 of Chapter 6 and the last five sections of Chapter 7. Section 5, on least squares problems, is also listed as optional. This is one of the most important applications of linear algebra and is well worth covering if time permits. The last section is an optional section on orthogonal polynomials. Orthogonal polynomials play an important role in many areas of mathematics, but they never seem to get the attention they deserve in the standard courses.

Chapter 6. This chapter treats one of the most important subjects in linear algebra, eigenvalues. Sections 1 and 3 should definitely be covered. Section 2 presents one of the main applications of eigenvalues. If the instructor does not wish to cover the entire section, we recommend that the material through Example 1 be presented. Section 4 deals with matrices with complex entries. This section is optional, but it is recommended that the material in the beginning of the section be covered so that the student gets some exposure to the complex case. In this chapter the really impressive nature of the applications of linear algebra should become apparent. Sections 5 and 6 are optional sections which present some of these applications. A discussion of the Perron–Frobenius theory of nonnegative matrices is given in Section 7. The proofs are omitted. These are powerful theorems and the proofs would be way beyond the scope of a first course.

However, the theory is necessary for an understanding of the Leontief input–output models. This is one of the nicest applications of linear algebra.

Chapter 7. Although this chapter is optional, it may well be the most important chapter for students who are going to work in industry. Instructors who want to incorporate numerics into the course may consider teaching the first three sections of this chapter immediately after completing Chapter 1. Section 4 could then be taught with the section on matrix norms in Chapter 5. The section on the singular value decomposition is highly recommended. Only recently has this subject been given the recognition it deserves. This section is included in Chapter 7 because of its importance to numerical linear algebra, but it could just as well be covered in Chapter 6. The most important algorithms presented in the last two sections are more advanced in nature, and consequently they are only outlined rather than presented in detail. Part of Section 9 is theoretical and could be covered earlier in the text along with Section 7.

ACKNOWLEDGMENTS

Most of this text was written while the author was visiting at Stanford University. The author would like to thank Gene Golub for making that visit not only possible, but also very enjoyable. While at Stanford the author was privileged to attend lectures on numerical linear algebra given by Gene Golub and J. H. Wilkinson. These lectures have undoubtedly had a great deal of influence on this text. The author would also like to thank the students and other visitors at Serra House for their many helpful suggestions. Special thanks are also due to Walter Gander and G. D. Taylor for the help and encouragement they have given. The author would also like to express his gratitude to Robert P. Bulles, Lillian Gough, Joseph Malkevitch, Marion E. Moore, Alan Weinstein, and all of the other reviewers for their helpful comments. The major part of the manuscript was typed by Charlotte Austin. I would like to thank her for a very professional job. Finally, I would like to express my gratitude to all of the students who have worked through parts of this manuscript.

S. L.

Contents

5
ORTHOGONALITY 127

6
EIGENVALUES 187

7*
NUMERICAL LINEAR ALGEBRA 242

LINEAR ALGEBRA WITH APPLICATIONS

1

Matrices and Systems of Equations

INTRODUCTION

Probably the most important problem in mathematics is that of solving a system of linear equations. It would not be conservative to estimate that well over 75 percent of all mathematical problems encountered in scientific or industrial applications involve solving a linear system at some stage. Using the methods of modern mathematics, it is often possible to take a sophisticated problem and reduce it to a single system of linear equations. Linear systems arise in applications to such areas as business, economics, sociology, ecology, demography, genetics, electronics, engineering, and physics. It seems appropriate then that this text should begin with a section on linear systems.

1. SYSTEMS OF LINEAR EQUATIONS

A *linear equation in n unknowns* is an equation of the form

$$a_1x_1 + a_2x_2 + \cdots + a_nx_n = b$$

where $a_1, a_2, \ldots, a_n$ and b are real numbers and $x_1, x_2, \ldots, x_n$ are variables. A linear system of m equations in n unknowns is then a system of the form

$$\begin{aligned} a_{11}x_1 + a_{12}x_2 + \cdots + a_{1n}x_n &= b_1 \\ a_{21}x_1 + a_{22}x_2 + \cdots + a_{2n}x_n &= b_2 \\ &\vdots \\ a_{m1}x_1 + a_{m2}x_2 + \cdots + a_{mn}x_n &= b_m \end{aligned} \tag{1}$$

where a_{ij}'s and the b_i's are all real numbers. We will refer to systems of the form (1) as $m \times n$ linear systems. The following are examples of linear systems:

$$\text{(a)}\quad \begin{aligned} x_1 + 2x_2 &= 5 \\ 2x_1 + 3x_2 &= 8 \end{aligned} \qquad \text{(b)}\quad \begin{aligned} x_1 - x_2 + x_3 &= 2 \\ 2x_1 + x_2 - x_3 &= 4 \end{aligned} \qquad \text{(c)}\quad \begin{aligned} x_1 + x_2 &= 2 \\ x_1 - x_2 &= 1 \\ x_1 &= 4 \end{aligned}$$

System (a) is a 2×2 system, (b) is a 2×3 system, and (c) is a 3×2 system.

By a solution to an $m \times n$ system, we mean an ordered n-tuple of numbers $(x_1, x_2, \ldots, x_n)$ that satisfies all the equations of the system. For example, the ordered pair (1, 2) is a solution to system (a), since

$$\begin{aligned} 1 \cdot (1) + 2 \cdot (2) &= 5 \\ 2 \cdot (1) + 3 \cdot (2) &= 8 \end{aligned}$$

The ordered triple (2, 0, 0) is a solution to system (b), since

$$\begin{aligned} 1 \cdot (2) - 1 \cdot (0) + 1 \cdot (0) &= 2 \\ 2 \cdot (2) + 1 \cdot (0) - 1 \cdot (0) &= 4 \end{aligned}$$

Actually, system (b) has many solutions. If α is any real number, it is easily seen that the ordered triple $(2, \alpha, \alpha)$ is a solution. However, system (c) has no solution. It follows from the third equation that the first coordinate of any solution would have to be 4. Using $x_1 = 4$ in the first two equations, we see that the second coordinate must satisfy

$$\begin{aligned} 4 + x_2 &= 2 \\ 4 - x_2 &= 1 \end{aligned}$$

Since there are no real numbers that satisfy both of these equations, the system has no solution. If a linear system has no solution, we say that the system is *inconsistent.* Thus system (c) is inconsistent, while systems (a) and (b) are both consistent.

The set of all solutions to a linear system is called the *solution set* of the system. If a system is inconsistent, its solution set is empty. A consistent system will have a nonempty solution set. To solve a consistent system, one must find its solution set.

2 × 2 Systems

Let us examine geometrically a system of the form

$$\begin{aligned} a_{11}x_1 + a_{12}x_2 &= b_1 \\ a_{21}x_1 + a_{22}x_2 &= b_2 \end{aligned}$$

Each equation can be represented graphically as a line in the plane. The ordered pair (x_1, x_2) will be a solution to the system if and only if it lies on both lines. For example, consider the three systems

(i) $\begin{aligned} x_1 + x_2 &= 2 \\ x_1 - x_2 &= 2 \end{aligned}$ (ii) $\begin{aligned} x_1 + x_2 &= 2 \\ x_1 + x_2 &= 1 \end{aligned}$ (iii) $\begin{aligned} x_1 + x_2 &= 2 \\ -x_1 - x_2 &= -2 \end{aligned}$

The two lines in system (i) intersect at the point (2, 0). Thus {(2, 0)} is the solution set to (i). In system (ii) the two lines are parallel. Therefore, system (ii) is inconsistent and hence the solution set is ∅. The two equations in system (iii) both represent the same line. Any point on that line will be a solution to the system (see Figure 1.1.1).

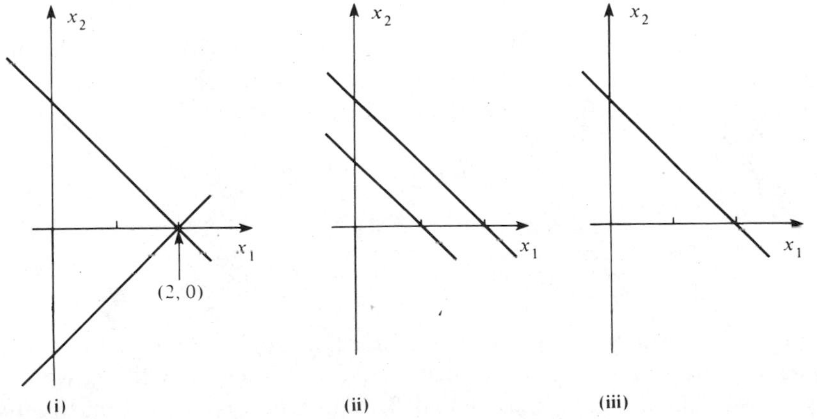

Figure 1.1.1

In general, there are three possibilities: the lines intersect at a point, they are parallel, or both equations represent the same line. The solution set then contains either one, zero, or infinitely many points.

The situation is similar for $m \times n$ systems. If a consistent system has exactly one solution, it is said to be *independent;* otherwise, it is *dependent.* We will see in the next section that if a linear system has more than one solution, it must have infinitely many solutions. The following table summarizes the three possibilities.

Type of System	*Number of Solutions*
Inconsistent	0
Consistent and independent	1
Consistent and dependent	Infinitely many

Equivalent Systems

Consider the two systems

$$\text{(a)}\quad \begin{aligned} 3x_1 + 2x_2 - x_3 &= -2 \\ x_2 \qquad &= 3 \\ 2x_3 &= 4 \end{aligned} \qquad\qquad \text{(b)}\quad \begin{aligned} 3x_1 + 2x_2 - x_3 &= -2 \\ -3x_1 - x_2 + x_3 &= 5 \\ 3x_1 + 2x_2 + x_3 &= 2 \end{aligned}$$

System (a) is easy to solve because it is clear from the last two equations that $x_2 = 3$ and $x_3 = 2$. Using these values in the first equation, we get

$$\begin{aligned} 3x_1 + 2 \cdot 3 - 2 &= -2 \\ x_1 &= -2 \end{aligned}$$

Thus the solution to the system is $(-2, 3, 2)$. System (b) seems to be more difficult to solve. Actually, system (b) has the same solution as system (a). To see this, add the first two equations of the system

$$\begin{aligned} 3x_1 + 2x_2 - x_3 &= -2 \\ -3x_1 - x_2 + x_3 &= 5 \\ \hline x_2 \qquad &= 3 \end{aligned}$$

If (x_1, x_2, x_3) is any solution to (b), it must satisfy all the equations of the system. Thus it must satisfy any new equation formed by adding two of its equations. Therefore, x_2 must equal 3. Similarly, (x_1, x_2, x_3) must satisfy the new equation formed by subtracting the first equation from the third:

$$\begin{array}{r} 3x_1 + 2x_2 + x_3 = 2 \\ 3x_1 + 2x_2 - x_3 = -2 \\ \hline 2x_3 = 4 \end{array}$$

Therefore, any solution to system (b) must also be a solution to system (a). By a similar argument, it can be shown that any solution to (a) is also a solution to (b). This can be done by subtracting the first equation from the second:

$$\begin{array}{r} x_2 \qquad = 3 \\ 3x_1 + 2x_2 - x_3 = -2 \\ \hline -3x_1 - x_2 + x_3 = 5 \end{array}$$

and by adding the first and third equations:

$$\begin{array}{r} 3x_1 + 2x_2 - x_3 = -2 \\ 2x_3 = 4 \\ \hline 3x_1 + 2x_2 + x_3 = 2 \end{array}$$

Thus (x_1, x_2, x_3) is a solution to system (b) if and only if it is a solution to system (a). Therefore, both systems have the same solution set, $\{(-2, 3, 2)\}$.

Definition. Two systems of equations involving the same variables are said to be **equivalent** if they have the same solution set.

Clearly, if we interchange the order in which two equations of a system are written, this will have no effect on the solution set. The reordered system will be equivalent to the original system. For example, the systems

$$\begin{array}{r} x_1 + 2x_2 = 4 \\ 3x_1 - x_2 = 2 \\ 4x_1 + x_2 = 6 \end{array} \qquad \text{and} \qquad \begin{array}{r} 4x_1 + x_2 = 6 \\ 3x_1 - x_2 = 2 \\ x_1 + 2x_2 = 4 \end{array}$$

clearly have the same solution set.

If one of the equations of a system is multiplied through by a nonzero real number, this will have no effect on the solution set and the new system will be equivalent to the original system. For example, the systems

$$\begin{aligned} x_1 + x_2 + x_3 &= 3 \\ -2x_1 - x_2 + 4x_3 &= 1 \end{aligned} \quad \text{and} \quad \begin{aligned} 2x_1 + 2x_2 + 2x_3 &= 6 \\ -2x_1 - x_2 + 4x_3 &= 1 \end{aligned}$$

are equivalent.

If a multiple of one equation is added to another equation, the new system will be equivalent to the original system. This follows since the n-tuple $(x_1, \ldots, x_n)$ will satisfy the two equations

$$\begin{aligned} a_{i1}x_1 + \cdots + a_{in}x_n &= b_i \\ a_{j1}x_1 + \cdots + a_{jn}x_n &= b_j \end{aligned}$$

if and only if it satisfies the equations

$$\begin{aligned} a_{i1}x_1 + \cdots + a_{in}x_n &= b_i \\ (a_{j1} + \alpha a_{i1})x_1 + \cdots + (a_{jn} + \alpha a_{in})x_n &= b_j + \alpha b_i \end{aligned}$$

To summarize, there are three operations that can be used on a system to obtain an equivalent system:

I. The order in which any two equations are written may be interchanged.
II. Both sides of an equation may be multiplied by the same nonzero real number.
III. A multiple of one equation may be added to another.

Given a system of equations, one can use these operations to obtain an equivalent system that is easier to solve.

$n \times n$ Systems

Let us restrict ourselves to $n \times n$ systems for the remainder of this section. We will show that if an $n \times n$ system is independent, then operations I and III can be used to obtain an equivalent "triangular system."

Definition. A system is said to be in **triangular form** if in the kth equation the coefficients of the first $k - 1$ variables are all zero and the coefficient of x_k is nonzero ($k = 1, \ldots, n$).

EXAMPLE 1. The system

$$\begin{aligned} 3x_1 + 2x_2 + x_3 &= 1 \\ x_2 - x_3 &= 2 \\ 2x_3 &= 4 \end{aligned}$$

is in triangular form, since in the second equation the coefficients are 0, 1, -1, respectively, and in the third equation the coefficients are 0, 0, 2, respectively. Because of the triangular form, this system is easy to solve. It follows from the third equation that $x_3 = 2$. Using this value in the second equation, we obtain

$$x_2 - 2 = 2 \qquad \text{or} \qquad x_2 = 4$$

Using $x_2 = 4$, $x_3 = 2$ in the first equation, we end up with

$$\begin{aligned} 3x_1 + 2 \cdot 4 + 2 &= 1 \\ x_1 &= -3 \end{aligned}$$

Thus the solution to the system is $(-3, 4, 2)$.

Any $n \times n$ triangular system can be solved in the same manner as the last example. First, the nth equation is solved for the value of x_n. This value is used in the $(n-1)$st equation to solve for x_{n-1}. The values x_n and x_{n-1} are used in the $(n-2)$nd equation to solve for x_{n-2}, and so on. We will refer to this method of solving a triangular system as *back substitution.*

EXAMPLE 2. Solve the system

$$\begin{aligned} 2x_1 - x_2 + 3x_3 - 2x_4 &= 1 \\ x_2 - 2x_3 + 3x_4 &= 2 \\ 4x_3 + 3x_4 &= 3 \\ 4x_4 &= 4 \end{aligned}$$

SOLUTION. Using back substitution, we obtain

$$\begin{aligned} 4x_4 &= 4 & x_4 &= 1 \\ 4x_3 + 3 \cdot 1 &= 3 & x_3 &= 0 \\ x_2 - 2 \cdot 0 + 3 \cdot 1 &= 2 & x_2 &= -1 \\ 2x_1 - (-1) + 3 \cdot 0 - 2 \cdot 1 &= 1 & x_1 &= 1 \end{aligned}$$

Thus the solution is $(1, -1, 0, 1)$.

If a system of equations is not triangular, we will use operations I and III to try to obtain an equivalent system that is in triangular form.

EXAMPLE 3. Solve the system

$$\begin{aligned} x_1 + 2x_2 + x_3 &= 3 \\ 3x_1 - x_2 - 3x_3 &= -1 \\ 2x_1 + 3x_2 + x_3 &= 4 \end{aligned}$$

SOLUTION. Adding -3 times the first row to the second yields

$$-7x_2 - 6x_3 = -10$$

Adding -2 times the first row to the third row yields

$$-x_1 - x_3 = -2$$

If the second and third equations of our system, respectively, are replaced by these new equations, we obtain the equivalent system

$$\begin{aligned} x_1 + 2x_2 + x_3 &= 3 \\ -7x_2 - 6x_3 &= -10 \\ -x_2 - x_3 &= -2 \end{aligned}$$

If the third equation of this system is replaced by the sum of the third equation and $-\frac{1}{7}$ times the second equation, we end up with the triangular system

$$\begin{aligned} x_1 + 2x_2 + x_3 &= 3 \\ -7x_2 - 6x_3 &= -10 \\ -\tfrac{1}{7}x_3 &= -\tfrac{4}{7} \end{aligned}$$

Using back substitution, we get

$$x_3 = 4, \qquad x_2 = -2, \qquad x_1 = 3$$

Let us look back at the system of equations in the last example. We can associate with that system a 3×3 array of numbers whose entries are the coefficients of the x_i's.

$$\begin{pmatrix} 1 & 2 & 1 \\ 3 & -1 & -3 \\ 2 & 3 & 1 \end{pmatrix}$$

We will refer to this array as the *coefficient matrix* of the system. The term *matrix* means simply a rectangular array of numbers. If we attach to the coefficient matrix an additional column whose entries are the numbers on the right-hand side of the system, we obtain the new matrix

$$\left(\begin{array}{rrr|r} 1 & 2 & 1 & 3 \\ 3 & -1 & -3 & -1 \\ 2 & 3 & 1 & 4 \end{array}\right)$$

We will refer to this new matrix as the *augmented matrix.* In general, when an $m \times r$ matrix B is attached to an $m \times n$ matrix A in this way the augmented matrix is denoted by $(A|B)$. Thus, if

$$A = \begin{pmatrix} a_{11} & a_{12} \cdots a_{1n} \\ a_{21} & a_{22} \cdots a_{2n} \\ \vdots & \\ a_{m1} & a_{m2} \cdots a_{mn} \end{pmatrix} \qquad B = \begin{pmatrix} b_{11} & b_{12} \cdots b_{1r} \\ b_{21} & b_{22} \cdots b_{2r} \\ \vdots & \\ b_{m1} & b_{m2} \cdots b_{mr} \end{pmatrix}$$

then

$$(A|B) = \left(\begin{array}{c|c} a_{11} \cdots a_{1n} & b_{11} \cdots b_{1r} \\ \vdots & \vdots \\ a_{m1} \cdots a_{mn} & b_{m1} \cdots b_{mr} \end{array}\right)$$

With each system of equations we may associate an augmented matrix of the form

$$\left(\begin{array}{c|c} a_{11} \cdots a_{1n} & b_1 \\ \vdots & \vdots \\ a_{m1} \cdots a_{mn} & b_n \end{array}\right)$$

The system can be solved by performing operations on the augmented matrix. The x_i's are placeholders that can be omitted until the end of the computation. Corresponding to the three operations used to obtain equivalent systems, the following row operations may be applied to the augmented matrix.

Elementary Row Operations

I. Interchange two rows.
II. Multiply a row by a nonzero real number.
III. Add a multiple of one row to another row.

Returning to the example, we find that the first row is used to eliminate the elements in the first column of the remaining rows. We refer to the first row as the *pivotal row* and the entry 1 circled in the first row as the *pivot*.

$$\begin{array}{r} \text{pivot} \rightarrow \\ \text{elements to} \rightarrow \\ \text{be eliminated} \rightarrow \end{array} \left(\begin{array}{rrr|r} \textcircled{1} & 2 & 1 & 3 \\ \boxed{3} & -1 & -3 & -1 \\ \boxed{2} & 3 & 1 & 4 \end{array}\right) \begin{array}{l} \leftarrow \text{pivotal row} \\ \\ \\ \end{array}$$

Using row operation III, -3 times the first row is added to the second row and -2 times the first row is added to the third. When this is done, we end up with the matrix

$$\left(\begin{array}{rrr|r} 1 & 2 & 1 & 3 \\ 0 & \textcircled{-7} & -6 & -10 \\ 0 & \boxed{-1} & -1 & -2 \end{array}\right) \begin{array}{l} \\ \leftarrow \text{pivotal row} \\ \\ \end{array}$$

At this step we choose the second row as our new pivotal row and apply row operation III to eliminate the last element in the second column. We end up with the matrix

$$\left(\begin{array}{rrr|r} 1 & 2 & 1 & 3 \\ 0 & -7 & -6 & -10 \\ 0 & 0 & -\frac{1}{7} & -\frac{4}{7} \end{array}\right)$$

This is the augmented matrix for the triangular system, which is equivalent to the original system.

EXAMPLE 4. Solve the system

$$\begin{aligned} -x_2 - x_3 + x_4 &= 0 \\ x_1 + x_2 + x_3 + x_4 &= 6 \\ 2x_1 + 4x_2 + x_3 - 2x_4 &= -1 \\ 3x_1 + x_2 - 2x_3 + 2x_4 &= 3 \end{aligned}$$

SOLUTION. The augmented matrix for this system is

$$\left(\begin{array}{rrrr|r} 0 & -1 & -1 & 1 & 0 \\ 1 & 1 & 1 & 1 & 6 \\ 2 & 4 & 1 & -2 & -1 \\ 3 & 1 & -2 & 2 & 3 \end{array}\right)$$

Since it is not possible to eliminate any entries using 0 as a pivotal element, we will use row operation I to interchange the first two rows of the augmented matrix. The new first row will be the pivotal row and the pivot element will be 1.

$$\text{pivot element} \rightarrow \left[\begin{array}{cccc|c} \textcircled{1} & 1 & 1 & 1 & 6 \\ 0 & -1 & -1 & 1 & 0 \\ \boxed{2} & 4 & 1 & -2 & -1 \\ \boxed{3} & 1 & -2 & 2 & 3 \end{array}\right] \leftarrow \text{pivotal row}$$

Row operation III is then used twice to eliminate the two nonzero entries in the first column.

$$\left[\begin{array}{cccc|c} 1 & 1 & 1 & 1 & 6 \\ 0 & \textcircled{-1} & -1 & 1 & 0 \\ 0 & \boxed{2} & -1 & -4 & -13 \\ 0 & \boxed{-2} & -5 & -1 & -15 \end{array}\right]$$

Next the second row is used as the pivotal row to eliminate the entries in the second column below the pivot element -1.

$$\left[\begin{array}{cccc|c} 1 & 1 & 1 & 1 & 6 \\ 0 & -1 & -1 & 1 & 0 \\ 0 & 0 & \textcircled{-3} & -2 & -13 \\ 0 & 0 & \boxed{-3} & -3 & -15 \end{array}\right]$$

Finally, the third row is used as the pivotal row to eliminate the last element in the third column.

$$\left[\begin{array}{cccc|c} 1 & 1 & 1 & 1 & 6 \\ 0 & -1 & -1 & 1 & 0 \\ 0 & 0 & -3 & -2 & -13 \\ 0 & 0 & 0 & -1 & -2 \end{array}\right]$$

This augmented matrix represents a triangular system. The solution $(2, -1, 3, 2)$ is easily obtained using back substitution.

The process of using row operations to obtain an equivalent system is called *Gaussian elimination.* If the $m \times n$ system can be reduced to triangular form, it will have a unique solution that can be obtained by performing back substitution on the triangular system. In the next section we will see that it is not always possible to reduce an $n \times n$ system to triangular form. If the system is inconsistent or consistent and dependent, we will reduce it to a special echelon or staircase-shaped form. This form will also be used for $m \times n$ systems where $m \neq n$.

EXERCISES

1. Solve each of the following systems of equations:

(a) $$\begin{aligned} x_1 - 3x_2 &= 2 \\ 2x_2 &= 6 \end{aligned}$$

(b) $$\begin{aligned} x_1 + x_2 + x_3 &= 8 \\ 2x_2 + x_3 &= 5 \\ 3x_3 &= 9 \end{aligned}$$

(c) $$\begin{aligned} x_1 + 2x_2 + 2x_3 + x_4 &= 5 \\ 3x_2 + x_3 - 2x_4 &= 1 \\ -x_3 + 2x_4 &= -1 \\ 4x_4 &= 4 \end{aligned}$$

(d) $$\begin{aligned} x_1 + x_2 + x_3 + x_4 + x_5 &= 5 \\ 2x_2 + x_3 - 2x_4 + x_5 &= 1 \\ 4x_3 + x_4 - 2x_5 &= 1 \\ x_4 - 3x_5 &= 0 \\ 2x_5 &= 2 \end{aligned}$$

2. Write out the coefficient matrix for each of the systems in Exercise 1.

3. Determine the solution set of each of the following systems.

(a) $$\begin{aligned} x_1 + 2x_2 + x_3 &= 0 \\ 2x_2 + x_3 &= 1 \end{aligned}$$

(b) $$\begin{aligned} 2x_1 - x_2 &= 0 \\ -4x_1 + 2x_2 &= 1 \end{aligned}$$

(c) $$\begin{aligned} 6x_1 - 3x_2 &= 3 \\ -2x_1 - 6x_2 &= -4 \\ 4x_1 - 9x_2 &= -1 \end{aligned}$$

(d) $$\begin{aligned} x_1 + x_2 &= 2 \\ x_1 - x_2 &= 0 \\ 3x_1 - x_2 &= 1 \end{aligned}$$

4. Which of the systems in Exercise 3 are consistent? Which of the consistent systems are independent? Dependent?

5. Use Gaussian elimination to solve each of the following systems.

(a) $$\begin{aligned} 4x_1 + 3x_2 &= 4 \\ \tfrac{2}{3}x_1 + 4x_2 &= 3 \end{aligned}$$

(b) $$\begin{aligned} x_1 + 2x_2 - x_3 &= 1 \\ 2x_1 - x_2 + x_3 &= 3 \\ -x_1 + 2x_2 + 3x_3 &= 7 \end{aligned}$$

(c) $$\begin{aligned} \tfrac{1}{3}x_1 + \tfrac{2}{3}x_2 + 2x_3 &= -1 \\ x_1 + 2x_2 + \tfrac{3}{2}x_3 &= \tfrac{3}{2} \\ \tfrac{1}{2}x_1 + 2x_2 + \tfrac{12}{5}x_3 &= \tfrac{1}{10} \end{aligned}$$

(d) $$\begin{aligned} x_2 + x_3 + x_4 &= 0 \\ 3x_1 + 3x_3 - 4x_4 &= 7 \\ x_1 + x_2 + x_3 + 2x_4 &= 6 \\ 2x_1 + 3x_2 + x_3 + 3x_4 &= 6 \end{aligned}$$

6. Give a geometrical interpretation of a linear equation in three unknowns. Give a geometrical description of the possible solution sets for a 3×3 linear system.

2. ROW ECHELON FORM

In Section 1 we learned a method for reducing an $n \times n$ linear system to triangular form. This method, however, will fail if at any stage of the reduction process, all of the possible choices for a pivot element in a given column are 0.

EXAMPLE 1. Consider the system represented by the augmented matrix

$$\left[\begin{array}{rrrrr|r} \textcircled{1} & 1 & 1 & 1 & 1 & 1 \\ \boxed{-1} & -1 & 0 & 0 & 1 & -1 \\ -2 & -2 & 0 & 0 & 1 & 1 \\ 0 & 0 & 1 & 1 & 1 & -1 \\ 1 & 1 & 2 & 2 & 2 & 1 \end{array}\right] \leftarrow \text{pivotal row}$$

If row operation III is used to eliminate the last four elements in the first column, the resulting matrix will be

$$\left[\begin{array}{rrrrr|r} 1 & 1 & 1 & 1 & 1 & 1 \\ 0 & 0 & \textcircled{1} & 1 & 2 & 0 \\ 0 & 0 & \boxed{2} & 2 & 3 & 3 \\ 0 & 0 & 1 & 1 & 1 & -1 \\ 0 & 0 & 1 & 1 & 1 & 0 \end{array}\right] \leftarrow \text{pivotal row}$$

At this stage the reduction to triangular form breaks down. All four possible choices for the pivotal element in the second column are 0. How do we proceed from here? Since our goal is to simplify the system as much as possible, it seems natural to move over to the third column and eliminate the last three entries.

$$\left[\begin{array}{rrrrr|r} 1 & 1 & 1 & 1 & 1 & 1 \\ 0 & 0 & 1 & 1 & 2 & 0 \\ 0 & 0 & 0 & 0 & \textcircled{-1} & 3 \\ 0 & 0 & 0 & 0 & \boxed{-1} & -1 \\ 0 & 0 & 0 & 0 & -1 & 0 \end{array}\right]$$

In the fourth column all the choices for a pivotal element are 0; so again we move on to the next column. Using the third row as the pivotal row, the last two entries in the fifth column are eliminated.

$$\left[\begin{array}{rrrrr|r} 1 & 1 & 1 & 1 & 1 & 1 \\ 0 & 0 & 1 & 1 & 2 & 0 \\ 0 & 0 & 0 & 0 & -1 & 3 \\ 0 & 0 & 0 & 0 & 0 & -4 \\ 0 & 0 & 0 & 0 & 0 & -3 \end{array}\right]$$

The equations represented by the last two rows are

$$\begin{aligned} 0x_1 + 0x_2 + 0x_3 + 0x_4 + 0x_5 &= -4 \\ 0x_1 + 0x_2 + 0x_3 + 0x_4 + 0x_5 &= -3 \end{aligned}$$

Since there are no 5-tuples that could possibly satisfy these equations, the system is inconsistent. Note that the coefficient matrix we end up with is not in triangular form; it is in staircase or echelon form.

Suppose now that we change the right-hand side of the system in the last example so as to obtain a consistent system. For example, if we start with

$$\left(\begin{array}{rrrrr|r} 1 & 1 & 1 & 1 & 1 & 1 \\ -1 & -1 & 0 & 0 & 1 & -1 \\ -2 & -2 & 0 & 0 & 1 & 1 \\ 0 & 0 & 1 & 1 & 1 & 3 \\ 1 & 1 & 2 & 2 & 2 & 4 \end{array}\right)$$

then the reduction process will yield the augmented matrix

$$\left(\begin{array}{rrrrr|r} 1 & 1 & 1 & 1 & 1 & 1 \\ 0 & 0 & 1 & 1 & 2 & 0 \\ 0 & 0 & 0 & 0 & -1 & 3 \\ 0 & 0 & 0 & 0 & 0 & 0 \\ 0 & 0 & 0 & 0 & 0 & 0 \end{array}\right)$$

Clearly, the last two equations of the reduced system will be satisfied for any 5-tuple. Thus the solution set will be the set of all 5-tuples satisfying the first three equations.

(1)
$$\begin{aligned} x_1 + x_2 + x_3 + x_4 + x_5 &= 1 \\ x_3 + x_4 + 2x_5 &= 0 \\ -x_5 &= 3 \end{aligned}$$

The variables corresponding to the first nonzero elements in each row of the augmented matrix will be referred to as *lead variables.* Thus x_1, x_3, and x_5 are the lead variables. The remaining variables corresponding to the columns skipped in the reduction process will be referred to as *free variables.* Thus x_2 and x_4 are the free variables. If we transfer the free variables over to the right-hand side in (1), we obtain the system

(2)
$$\begin{aligned} x_1 + x_3 + x_5 &= 1 - x_2 - x_4 \\ x_3 + 2x_5 &= -x_4 \\ -x_5 &= 3 \end{aligned}$$

System (2) is triangular in the unknowns x_1, x_3, x_5. Thus for each pair of values assigned to x_2 and x_4, there will be a unique solution. For example, if $x_2 = x_4 = 0$, then $x_5 = -3$, $x_3 = 6$, $x_1 = -2$, and hence $(-2, 0, 6, 0, -3)$ is a solution to the system.

Definition. A matrix is said to be in **row echelon form** if

(1) The first nonzero entry in each row is 1.

(2) If row k does not consist entirely of zeros, the number of leading zero entries in row $k+1$ is greater than the number of leading zero entries in row k.

(3) If there are rows whose entries are all zero, they are below the rows having nonzero entries.

EXAMPLE 2. The following matrices are in row echelon form.

$$\begin{bmatrix} 1 & 4 & 2 \\ 0 & 1 & 3 \\ 0 & 0 & 1 \end{bmatrix}, \quad \begin{bmatrix} 1 & 2 & 3 \\ 0 & 0 & 1 \\ 0 & 0 & 0 \end{bmatrix}, \quad \begin{bmatrix} 1 & 3 & 1 & 0 \\ 0 & 0 & 1 & 3 \\ 0 & 0 & 0 & 0 \end{bmatrix}$$

EXAMPLE 3. The following matrices are not in row echelon form.

$$\begin{bmatrix} 2 & 4 & 6 \\ 0 & 3 & 5 \\ 0 & 0 & 4 \end{bmatrix}, \quad \begin{pmatrix} 0 & 0 & 0 \\ 0 & 1 & 0 \end{pmatrix}, \quad \begin{pmatrix} 0 & 1 \\ 1 & 0 \end{pmatrix}$$

The first matrix does not satisfy condition 1. The second matrix fails to satisfy condition 3, and the third matrix fails to satisfy the second condition.

To solve an $m \times n$ linear system, the general strategy is to use Gaussian elimination to transform the system into one whose augmented matrix is in row echelon form. This can be done using row operations I, II, and III. (Row operation II is used to scale the rows so that the lead coefficients are all 1.) If the row echelon matrix contains a row of the form

$$(0 \; 0 \cdots 0|1)$$

then the system is inconsistent. Otherwise, the system will be consistent. A consistent system will be independent if the nonzero rows of the row echelon matrix represent a triangular system and dependent otherwise.

Overdetermined Systems

A linear system is said to be overdetermined if there are more equations than unknowns ($m > n$). Overdetermined systems are usually (but not always) inconsistent.

EXAMPLE 4

(a)
$$\begin{aligned} x_1 + x_2 &= 1 \\ x_1 - x_2 &= 3 \\ -x_1 + 2x_2 &= -2 \end{aligned}$$

(b)
$$\begin{aligned} x_1 + 2x_2 + x_3 &= 1 \\ 2x_1 - x_2 + x_3 &= 2 \\ 4x_1 + 3x_2 + 3x_3 &= 4 \\ 2x_1 - x_2 + 3x_3 &= 5 \end{aligned}$$

(c)
$$\begin{aligned} x_1 + 2x_2 + x_3 &= 1 \\ 2x_1 - x_2 + x_3 &= 2 \\ 4x_1 + 3x_2 + 3x_3 &= 4 \\ 3x_1 + x_2 + 2x_3 &= 3 \end{aligned}$$

We will assume that, by now, the reader is familiar with the elimination process, and hence we will omit all the intermediate steps in reducing each of these systems.

System (a)

$$\left(\begin{array}{rr|r} 1 & 1 & 1 \\ 1 & -1 & 3 \\ -1 & 2 & -2 \end{array}\right) \rightarrow \left(\begin{array}{rr|r} 1 & 1 & 1 \\ 0 & 1 & -1 \\ 0 & 0 & 1 \end{array}\right)$$

It follows from the last row of the reduced matrix that the system is inconsistent. The three equations in system (a) represent lines in the plane. The first two lines intersect at the point $(2, -1)$. However, the third line does not pass through this point. Thus there are no points that lie on all three lines (see Figure 1.2.1).

System (b)

$$\left(\begin{array}{rrr|r} 1 & 2 & 1 & 1 \\ 2 & -1 & 1 & 2 \\ 4 & 3 & 3 & 4 \\ 2 & -1 & 3 & 5 \end{array}\right) \rightarrow \left(\begin{array}{rrr|r} 1 & 2 & 1 & 1 \\ 0 & 1 & \frac{1}{5} & 0 \\ 0 & 0 & 1 & \frac{3}{2} \\ 0 & 0 & 0 & 0 \end{array}\right)$$

Thus system (b) is independent. The solution $(0.1, -0.3, 1.5)$ is easily

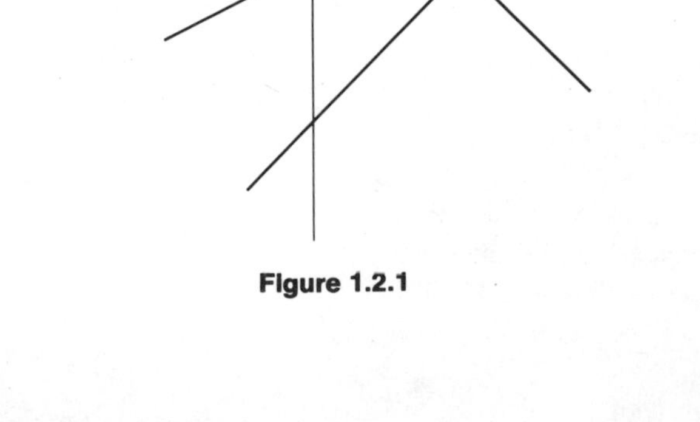

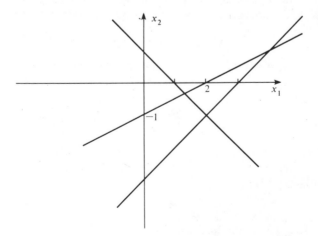

Figure 1.2.1

obtained using back substitution.

System (c)

$$\left[\begin{array}{rrr|r} 1 & 2 & 1 & 1 \\ 2 & -1 & 1 & 2 \\ 4 & 3 & 3 & 4 \\ 3 & 1 & 2 & 3 \end{array}\right] \to \left[\begin{array}{rrr|r} 1 & 2 & 1 & 1 \\ 0 & 1 & \frac{1}{5} & 0 \\ 0 & 0 & 0 & 0 \\ 0 & 0 & 0 & 0 \end{array}\right]$$

The system is consistent and dependent. There is one free variable, x_3. Since

$$x_2 = -0.2x_3$$

and

$$x_1 = 1 - 2x_2 - x_3 = 1 - 0.6x_3$$

it follows that the solution set is the set of all ordered triples of the form $(1 - 0.6\alpha, -0.2\alpha, \alpha)$, where α is a real number.

Underdetermined Systems

A linear system is said to be *underdetermined* if there are less equations than unknowns ($m < n$). Underdetermined systems are usually consistent and dependent, although they may turn out to be inconsistent. *An underdetermined system cannot be independent.* The reason for this is that any row echelon form of the coefficient matrix will involve $r \leqslant m$ nonzero rows. Thus there will be r lead variables and $n - r$ free variables, where $n - r \geqslant n - m > 0$. If the system is consistent, we can assign the free variables arbitrary values and solve for the lead variables. Therefore, a consistent underdetermined system must be dependent.

EXAMPLE 5

(a)
$$\begin{aligned} x_1 + 2x_2 + x_3 &= -1 \\ 2x_1 + 4x_4 + 2x_3 &= 3 \end{aligned}$$

(b)
$$\begin{aligned} x_1 + x_2 + x_3 + x_4 + x_5 &= 2 \\ x_1 + x_2 + x_3 + 2x_4 + 2x_5 &= 3 \\ x_1 + x_2 + x_3 + 2x_4 + 3x_5 &= 2 \end{aligned}$$

System (a)

$$\left(\begin{array}{rrr|r} 1 & 2 & 1 & -1 \\ 2 & 4 & 2 & 3 \end{array}\right) \to \left(\begin{array}{rrr|r} 1 & 2 & 1 & -1 \\ 0 & 0 & 0 & 1 \end{array}\right)$$

Clearly, system (a) is inconsistent. We can think of the two equations in system (a) as representing planes in 3-space. Usually, two planes intersect in a line; however, in this case the planes are parallel.

System (b)

$$\left[\begin{array}{ccccc|c} 1 & 1 & 1 & 1 & 1 & 2 \\ 1 & 1 & 1 & 2 & 2 & 3 \\ 1 & 1 & 1 & 2 & 3 & 2 \end{array}\right] \rightarrow \left[\begin{array}{ccccc|c} 1 & 1 & 1 & \boxed{1} & \boxed{1} & 2 \\ 0 & 0 & 0 & 1 & \boxed{1} & 1 \\ 0 & 0 & 0 & 0 & 1 & -1 \end{array}\right]$$

Clearly, system (b) is consistent and dependent. Often with systems like this it is convenient to continue with the elimination process until all the terms above each leading 1 are eliminated. Thus for system (b) we will continue and eliminate the first two entries in the fifth column and then the first element in the fourth column.

$$\left[\begin{array}{ccccc|c} 1 & 1 & 1 & 1 & 1 & 2 \\ 0 & 0 & 0 & 1 & 1 & 1 \\ 0 & 0 & 0 & 0 & 1 & -1 \end{array}\right] \rightarrow \left[\begin{array}{ccccc|c} 1 & 1 & 1 & 1 & 0 & 3 \\ 0 & 0 & 0 & 1 & 0 & 2 \\ 0 & 0 & 0 & 0 & 1 & -1 \end{array}\right]$$

$$\rightarrow \left[\begin{array}{ccccc|c} 1 & 1 & 1 & 0 & 0 & 1 \\ 0 & 0 & 0 & 1 & 0 & 2 \\ 0 & 0 & 0 & 0 & 1 & -1 \end{array}\right]$$

If we put the free variables over on the right-hand side, it follows that

$$\begin{aligned} x_1 &= 1 - x_2 - x_3 \\ x_4 &= 2 \\ x_5 &= -1 \end{aligned}$$

Thus, for any real numbers α and β, the 5-tuple $(1 - \alpha - \beta, \alpha, \beta, 2, -1)$ is a solution to the system.

Definition. A matrix is said to be in **reduced row echelon form** if:

(1) The matrix is in row echelon form.

(2) The first nonzero entry in each row is the only nonzero entry in its column.

The following matrices are in reduced row echelon form:

$$\begin{pmatrix} 1 & 0 \\ 0 & 1 \end{pmatrix}, \quad \begin{bmatrix} 1 & 0 & 0 & 3 \\ 0 & 1 & 0 & 2 \\ 0 & 0 & 1 & 1 \end{bmatrix}, \quad \begin{bmatrix} 0 & 1 & 2 & 0 \\ 0 & 0 & 0 & 1 \\ 0 & 0 & 0 & 0 \end{bmatrix}, \quad \begin{bmatrix} 1 & 2 & 0 & 1 \\ 0 & 0 & 1 & 3 \\ 0 & 0 & 0 & 0 \end{bmatrix}$$

The process of using elementary row operations to transform a matrix into reduced row echelon form is called *Gauss–Jordan reduction.*

EXAMPLE 6. Use Gauss–Jordan reduction to solve the system

$$\begin{aligned} -2x_1 + x_2 - x_3 + 3x_4 &= 0 \\ 2x_1 + 3x_2 + x_3 + x_4 &= 0 \\ 4x_1 - x_2 - 3x_3 + 2x_4 &= 0 \end{aligned}$$

SOLUTION

$$\left[\begin{array}{cccc|c} \textcircled{-2} & 1 & -1 & 3 & 0 \\ \boxed{2} & 3 & 1 & 1 & 0 \\ \boxed{4} & -1 & -3 & 2 & 0 \end{array}\right] \rightarrow \left[\begin{array}{cccc|c} -2 & 1 & -1 & 3 & 0 \\ 0 & \textcircled{4} & 0 & 4 & 0 \\ 0 & \boxed{1} & -5 & 8 & 0 \end{array}\right] \rightarrow \left[\begin{array}{cccc|c} -2 & 1 & -1 & 3 & 0 \\ 0 & 4 & 0 & 4 & 0 \\ 0 & 0 & -5 & 7 & 0 \end{array}\right]$$

$$\rightarrow \left[\begin{array}{cccc|c} 1 & -0.5 & \boxed{0.5} & -1.5 & 0 \\ 0 & 1 & 0 & 1 & 0 \\ 0 & 0 & \textcircled{1} & -1.4 & 0 \end{array}\right] \text{ row echelon form}$$

$$\rightarrow \left[\begin{array}{cccc|c} 1 & \boxed{-0.5} & 0 & -0.8 & 0 \\ 0 & \textcircled{1} & 0 & 1.0 & 0 \\ 0 & 0 & 1 & -1.4 & 0 \end{array}\right] \rightarrow \left[\begin{array}{cccc|c} 1 & 0 & 0 & -0.3 & 0 \\ 0 & 1 & 0 & 1.0 & 0 \\ 0 & 0 & 1 & -1.4 & 0 \end{array}\right] \text{ reduced row echelon form}$$

If we set x_4 equal to any real number α, then $x_1 = 0.3\alpha$, $x_2 = -\alpha$, and $x_3 = 1.4\alpha$. Thus all ordered 4-tuples of the form $(0.3\alpha, -\alpha, 1.4\alpha, \alpha)$ are solutions to the system.

A system of linear equations is said to be *homogeneous* if the constants on the right-hand side are all zero. Homogeneous systems are always consistent. It is a trivial matter to find a solution; just set all the variables equal to zero. Thus an $m \times n$ homogeneous system is either independent, in which case $(0, \ldots, 0)$ is the only solution, or it is dependent. Furthermore, if $n > m$, the system must be dependent. In particular, if $n > m$, we can conclude that the homogeneous system has nontrivial solutions. This result has essentially been proved in our discussion of underdetermined systems, but because of its importance we state it as a theorem.

Theorem 1.2.1. *An $m \times n$ homogeneous system of linear equations has a nontrivial solution if $n > m$.*

PROOF. A homogeneous system is always consistent. A consistent underdetermined system must be dependent. Therefore, an $m \times n$ homogeneous system will have infinitely many solutions if $n > m$.

Application

In an electrical network it is possible to determine the amount of current in each branch in terms of the resistances and the voltages. In Figure 1.2.2 the —| |— represents a source of power (a battery or a generator), which is measured in volts. The symbol —WW— represents a resistor. The resistances are measured in ohms. The capital letters represent nodes and the i's represent the currents between the nodes. The currents are measured in amperes. The arrows show the direction of the currents. If, for example, i_3 turns out to be negative, this means that the current along the branch is in

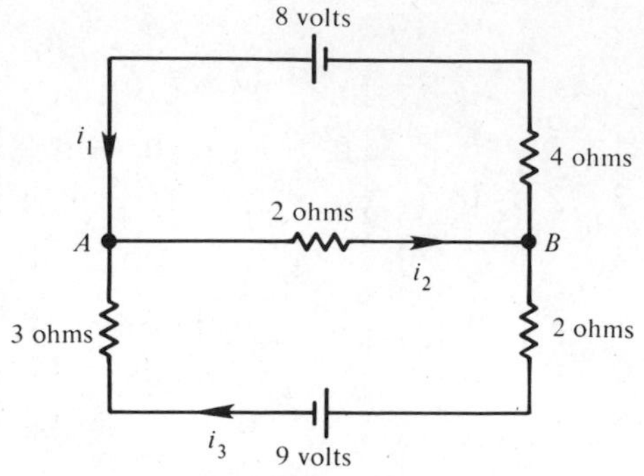

Figure 1.2.2

the opposite direction of the arrow. In order to determine the currents, *Kirchhoff's laws* are used:

1. At every node the sum of the incoming currents equals the sum of the outgoing currents.
2. Around every closed loop the algebraic sum of the voltage must equal the algebraic sum of the voltage drops.

The voltage drops E for each resistor are given by *Ohm's law:*

$$E = iR$$

where i represents the current in amperes and R the resistance in ohms.

Let us find the currents in the network pictured in Figure 1.2.2. From the first law, we have

$$\begin{aligned} i_1 - i_2 + i_3 &= 0 \qquad \text{(node A)} \\ -i_1 + i_2 - i_3 &= 0 \qquad \text{(node B)} \end{aligned}$$

By the second law,

$$\begin{aligned} 4i_1 + 2i_2 \qquad &= 8 \qquad \text{(top loop)} \\ 2i_2 + 5i_3 &= 9 \qquad \text{(bottom loop)} \end{aligned}$$

The network can be represented by the augmented matrix

$$\left(\begin{array}{rrr|r} 1 & -1 & 1 & 0 \\ -1 & 1 & -1 & 0 \\ 4 & 2 & 0 & 8 \\ 0 & 2 & 5 & 9 \end{array}\right)$$

This matrix is easily reduced to row echelon form:

$$\left(\begin{array}{rrr|r} 1 & -1 & 1 & 0 \\ 0 & 1 & -\frac{2}{3} & \frac{4}{3} \\ 0 & 0 & 1 & 1 \\ 0 & 0 & 0 & 0 \end{array}\right)$$

Solving by back substitution, we see that $i_1 = 1$, $i_2 = 2$, and $i_3 = 1$.

EXERCISES

1. Which of the following matrices are in row echelon form? Which are in reduced row echelon form?

(a) $\begin{pmatrix} 1 & 2 & 3 & 4 \\ 0 & 0 & 1 & 2 \end{pmatrix}$ (b) $\begin{bmatrix} 1 & 0 & 0 \\ 0 & 0 & 0 \\ 0 & 0 & 1 \end{bmatrix}$

(c) $\begin{bmatrix} 1 & 3 & 0 \\ 0 & 0 & 1 \\ 0 & 0 & 0 \end{bmatrix}$ (d) $\begin{bmatrix} 0 & 1 \\ 0 & 0 \\ 0 & 0 \end{bmatrix}$

(e) $\begin{bmatrix} 1 & 1 & 1 \\ 0 & 1 & 2 \\ 0 & 0 & 3 \end{bmatrix}$ (f) $\begin{bmatrix} 1 & 4 & 6 \\ 0 & 0 & 1 \\ 0 & 1 & 3 \end{bmatrix}$

(g) $\begin{bmatrix} 1 & 0 & 0 & 1 & 2 \\ 0 & 1 & 0 & 2 & 4 \\ 0 & 0 & 1 & 3 & 6 \end{bmatrix}$ (h) $\begin{bmatrix} 0 & 1 & 3 & 4 \\ 0 & 0 & 1 & 3 \\ 0 & 0 & 0 & 0 \end{bmatrix}$

2. For each of the following systems of equations, use Gaussian elimination to obtain an equivalent system whose coefficient matrix is in row echelon form. Determine whether the system is inconsistent, consistent and dependent, or consistent and independent, and find the solution set.

(a) $\begin{aligned} x_1 - 2x_2 &= 3 \\ 2x_1 - x_2 &= 9 \end{aligned}$

(b) $\begin{aligned} 2x_1 - 3x_2 &= 5 \\ -4x_1 + 6x_2 &= 8 \end{aligned}$

(c) $\begin{aligned} x_1 + x_2 &= 0 \\ 2x_1 + 3x_2 &= 0 \\ 3x_1 - 2x_2 &= 0 \end{aligned}$

(d) $\begin{aligned} 3x_1 + 2x_2 - x_3 &= 4 \\ x_1 - 2x_2 + 2x_3 &= 1 \\ 11x_1 + 2x_2 + x_3 &= 14 \end{aligned}$

(e) $\begin{aligned} 2x_1 + 3x_2 + x_3 &= 1 \\ x_1 + x_2 + x_3 &= 3 \\ 3x_1 + 4x_2 + 2x_3 &= 4 \end{aligned}$

(f) $\begin{aligned} x_1 - x_2 + 2x_3 &= 4 \\ 2x_1 + 3x_2 - x_3 &= 1 \\ 7x_1 + 3x_2 + 4x_3 &= 7 \end{aligned}$

(g) $\begin{aligned} x_1 + x_2 + x_3 + x_4 &= 0 \\ 2x_1 + 3x_2 - x_3 - x_4 &= 2 \\ x_1 - x_2 + 2x_3 + 2x_4 &= 3 \\ 2x_1 + 5x_2 - 2x_3 - 2x_4 &= 4 \end{aligned}$

(h) $\begin{aligned} x_1 - 2x_2 &= 3 \\ 2x_1 + x_2 &= 1 \\ -5x_1 + 8x_2 &= 4 \end{aligned}$

(i) $\begin{aligned} -x_1 + 2x_2 - x_3 &= 2 \\ -2x_1 + 2x_2 + x_3 &= 4 \\ 3x_1 + 2x_2 + 2x_3 &= 5 \\ -3x_1 + 8x_2 + 5x_3 &= 17 \end{aligned}$

(j) $\begin{aligned} x_1 + 2x_2 - 3x_3 + x_4 &= -2 \\ -x_1 - x_2 + 4x_3 - x_4 &= 6 \\ -2x_1 - 4x_2 + 7x_3 - 2x_4 &= 5 \end{aligned}$

(k) $\begin{aligned} x_1 + 3x_2 + x_3 + x_4 &= 3 \\ 2x_1 - 2x_2 + x_3 + 2x_4 &= 8 \\ x_1 + 11x_2 + 2x_3 + x_4 &= 1 \end{aligned}$

(l) $\begin{aligned} x_1 - 3x_2 + x_3 &= 1 \\ 2x_1 + x_2 - x_3 &= 2 \\ x_1 + 4x_2 - 2x_3 &= 1 \\ 5x_1 - 8x_2 + 2x_3 &= 5 \end{aligned}$

3. Use Gauss–Jordan reduction to solve each of the following systems.

(a) $\begin{aligned} x_1 + x_2 &= -1 \\ 4x_1 - 3x_2 &= 3 \end{aligned}$

(b) $\begin{aligned} x_1 + 3x_2 + x_3 + x_4 &= 3 \\ 2x_1 - 2x_2 + x_3 + 2x_4 &= 8 \\ 3x_1 + x_2 + 2x_3 - x_4 &= -1 \end{aligned}$

(c) $\begin{aligned} x_1 + x_2 + x_3 &= 0 \\ x_1 - x_2 - x_3 &= 0 \end{aligned}$

(d) $\begin{aligned} x_1 + x_2 + x_3 + x_4 &= 0 \\ 2x_1 + x_2 - x_3 + 3x_3 &= 0 \\ x_1 - 2x_2 + x_3 + x_4 &= 0 \end{aligned}$

4. Determine the amount of each of the currents for the network pictured.

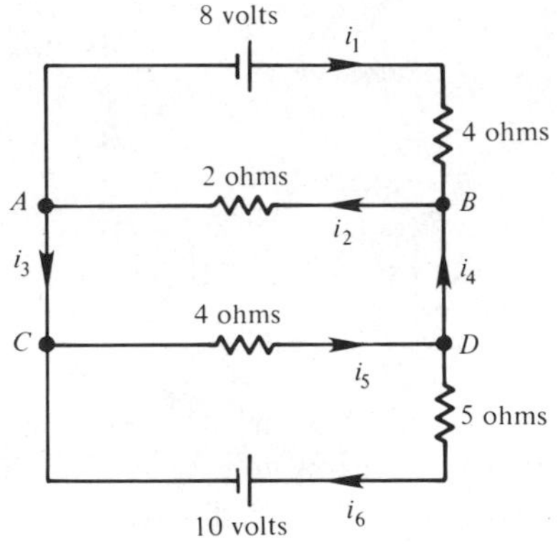

3. MATRIX ALGEBRA

In this section we will define arithmetic operations with matrices and look at some of their algebraic properties. Matrices are one of the most powerful tools in mathematics. To use matrices effectively, we must be adept at matrix arithmetic.

The entries of a matrix are called *scalars.* They are usually either real or complex numbers. For the most part we will be working with matrices whose entries are real numbers. Unless otherwise stated, the reader may assume that the term *scalar* refers to a real number. There will, however, be some occasions (mainly in Chapter 6) when we will use the set of complex numbers as our scalar field.

If we wish to refer to matrices without specifically writing out all their entries, we will use capital letters A, B, C, and so on. In general, a_{ij} will denote the entry of the matrix A that is in the ith row and the jth column. Thus, if A is an $m \times n$ matrix, then

$$A = \begin{bmatrix} a_{11} & a_{12} \cdots & a_{1n} \\ a_{21} & a_{22} \cdots & a_{2n} \\ \vdots & & \\ a_{m1} & a_{m2} \cdots & a_{mn} \end{bmatrix}$$

We will sometimes shorten this to $A = (a_{ij})$. Similarly, a matrix B may be referred to as (b_{ij}), a matrix C as (c_{ij}), and so on.

Definition. Two $m \times n$ matrices A and B are said to be **equal** if $a_{ij} = b_{ij}$ for each i and j.

Scalar Multiplication

If A is a matrix and α is a scalar, then αA is the matrix formed by multiplying each of the entries of A by α. For example, if

$$A = \begin{pmatrix} 4 & 8 & 2 \\ 6 & 8 & 10 \end{pmatrix}$$

then

$$\tfrac{1}{2}A = \begin{pmatrix} 2 & 4 & 1 \\ 3 & 4 & 5 \end{pmatrix} \quad \text{and} \quad 3A = \begin{pmatrix} 12 & 24 & 6 \\ 18 & 24 & 30 \end{pmatrix}$$

Matrix Addition

If $A = (a_{ij})$ and $B = (b_{ij})$ are both $m \times n$ matrices, then the sum $A + B$ is the $m \times n$ matrix whose ijth entry is $a_{ij} + b_{ij}$ for each ordered pair (i, j). For example,

$$\begin{pmatrix} 3 & 2 & 1 \\ 4 & 5 & 6 \end{pmatrix} + \begin{pmatrix} 2 & 2 & 2 \\ 1 & 2 & 3 \end{pmatrix} = \begin{pmatrix} 5 & 4 & 3 \\ 5 & 7 & 9 \end{pmatrix}$$

$$\begin{bmatrix} 2 \\ 1 \\ 8 \end{bmatrix} + \begin{bmatrix} -8 \\ 3 \\ 2 \end{bmatrix} = \begin{bmatrix} -6 \\ 4 \\ 10 \end{bmatrix}$$

If we define $A - B$ to be $A + (-1)B$, then it turns out that $A - B$ is formed by subtracting the corresponding entry of B from each entry of A. Thus

$$\begin{aligned} \begin{pmatrix} 2 & 4 \\ 3 & 1 \end{pmatrix} - \begin{pmatrix} 4 & 5 \\ 2 & 3 \end{pmatrix} &= \begin{pmatrix} 2 & 4 \\ 3 & 1 \end{pmatrix} + (-1)\begin{pmatrix} 4 & 5 \\ 2 & 3 \end{pmatrix} \\ &= \begin{pmatrix} 2 & 4 \\ 3 & 1 \end{pmatrix} + \begin{pmatrix} -4 & -5 \\ -2 & -3 \end{pmatrix} \\ &= \begin{pmatrix} 2-4 & 4-5 \\ 3-2 & 1-3 \end{pmatrix} \\ &= \begin{pmatrix} -2 & -1 \\ 1 & -2 \end{pmatrix} \end{aligned}$$

If O represents a matrix, with the same dimensions as A, whose entries are all 0, then

$$A + O = O + A = A$$

That is, the zero matrix acts as an additive identity on the set of all $m \times n$ matrices. Furthermore, each $m \times n$ matrix A has an additive inverse. Indeed,

$$A + (-1)A = O = (-1)A + A$$

It is customary to denote the additive inverse by $-A$. Thus

$$-A = (-1)A$$

We have yet to define the most important operation, the multiplication of two matrices. Before doing so, it may be helpful to look at the following applied problem.

Application 1

A company manufactures three products. Its production expenses are divided into three categories. In each of these categories, an estimate is given for the cost of producing a single item of each of the products. An estimate is also made of the amount of each product to be produced per quarter. These estimates are given in the following tables.

Production Costs per Item (dollars)

	Product		
Expenses	A	B	C
Raw materials	0.10	0.30	0.15
Labor	0.30	0.40	0.25
Overhead and miscellaneous	0.10	0.20	0.15

Amount Produced per Quarter

	Season			
Product	*Summer*	*Fall*	*Winter*	*Spring*
A	4000	4500	4500	4000
B	2000	2600	2400	2200
C	5800	6200	6000	6000

The company would like to present at their stockholders meeting a single table showing the total costs for each quarter in each of the three categories: raw materials, labor, and overhead.

Let us consider the problem in terms of matrices. Each of the two tables can be represented by a matrix.

$$M = \begin{bmatrix} 0.10 & 0.30 & 0.15 \\ 0.30 & 0.40 & 0.25 \\ 0.10 & 0.20 & 0.15 \end{bmatrix} \quad \text{and} \quad P = \begin{bmatrix} 4000 & 4500 & 4500 & 4000 \\ 2000 & 2600 & 2400 & 2200 \\ 5800 & 6200 & 6000 & 6000 \end{bmatrix}$$

Let us form a new matrix MP with four columns corresponding to the quarters of the year. The entries in a given column are to represent the total costs during that quarter.

The costs for the summer quarter will be

Raw materials: $(0.10)(4000) + (0.30)(2000) + (0.15)(5800) = 1870$

Labor: $(0.30)(4000) + (0.40)(2000) + (0.25)(5800) = 3450$

Overhead and miscellaneous: $(0.10)(4000) + (0.20)(2000) + (0.15)(5800) = 1670$

The first column of MP will be

$$\begin{bmatrix} 1870 \\ 3450 \\ 1670 \end{bmatrix}$$

The costs for the fall quarter are computed in the same manner using the entries in the second column of P.

Raw materials: $(0.10)(4500) + (0.30)(2600) + (0.15)(6200) = 2160$

Labor: $(0.30)(4500) + (0.40)(2600) + (0.25)(6200) = 3940$

Overhead and miscellaneous: $(0.10)(4500) + (0.20)(2600) + (0.15)(6200) = 1900$

The costs for winter and spring quarter are calculated in a similar manner using the entries from the third and fourth columns of P. The four columns are then combined to form

$$MP = \begin{bmatrix} 1870 & 2160 & 2070 & 1960 \\ 3450 & 3940 & 3810 & 3580 \\ 1670 & 1900 & 1830 & 1740 \end{bmatrix}$$

The new matrix MP has the desired properties. The entries in row 1 represent the total cost of raw materials for each of the four quarters. The entries in rows 2 and 3 represent total cost for labor and overhead, respectively, for each of the four quarters. The yearly expenses in each category may be obtained by adding the entries in each row. The numbers in each of the columns may be added to obtain the total production costs for each quarter. The following table summarizes the total production costs.

	Season				
	Summer	Fall	Winter	Spring	Year
Raw materials	1,870	2,160	2,070	1,960	8,060
Labor	3,450	3,940	3,810	3,580	14,780
Overhead and miscellaneous	1,670	1,900	1,830	1,740	7,140
Total production costs	6,990	8,000	7,710	7,280	29,980

In Application 1, two matrices were combined to produce a third matrix. The operation involved may seem unusual, but it is extremely useful. We will call this operation *matrix multiplication.*

Definition. If $A = (a_{ij})$ is an $m \times n$ matrix and $B = (b_{ij})$ is an $n \times r$ matrix, then the product $AB = C = (c_{ij})$ is the $m \times r$ matrix whose entries are defined by

$$c_{ij} = \sum_{k=1}^{n} a_{ik} b_{kj}$$

What this definition says is that to find the ijth element of the product, you take the ith row of A and the jth column of B, multiply the corresponding elements pairwise, and add the resulting numbers.

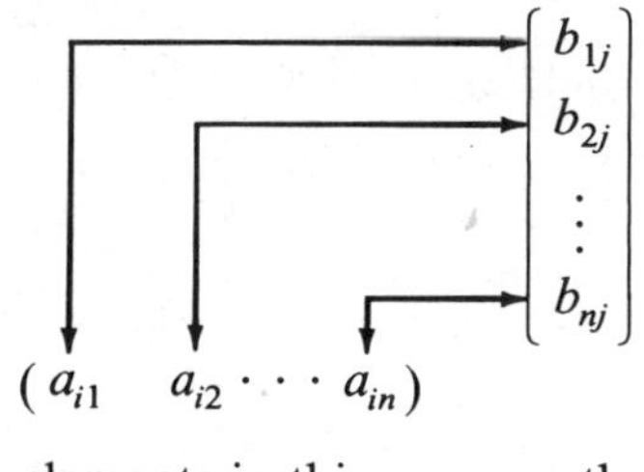

In order to pair off the elements in this manner, the number of columns of A must equal the number of rows of B. If this does not occur, the multiplication is impossible.

EXAMPLE 1. If

$$A = \begin{pmatrix} -2 & 1 & 3 \\ 4 & 1 & 6 \end{pmatrix} \quad \text{and} \quad B = \begin{bmatrix} 3 & -2 \\ 2 & 4 \\ 1 & -3 \end{bmatrix}$$

then

$$\begin{aligned} AB &= \begin{pmatrix} -2\cdot 3 + 1\cdot 2 + 3\cdot 1 & -2\cdot(-2) + 1\cdot 4 + 3\cdot(-3) \\ 4\cdot 3 + 1\cdot 2 + 6\cdot 1 & 4\cdot(-2) + 1\cdot 4 + 6\cdot(-3) \end{pmatrix} \\ &= \begin{pmatrix} -1 & -1 \\ 20 & -22 \end{pmatrix} \end{aligned}$$

and

$$BA = \begin{bmatrix} 3\cdot(-2)+(-2)\cdot 4 & 3\cdot 1-2\cdot 1 & 3\cdot 3-2\cdot 6 \\ 2\cdot(-2)+\ \ 4\cdot 4 & 2\cdot 1+4\cdot 1 & 2\cdot 3+4\cdot 6 \\ 1\cdot(-2)-\ \ 3\cdot 4 & 1\cdot 1-3\cdot 1 & 1\cdot 3-3\cdot 6 \end{bmatrix}$$

$$= \begin{bmatrix} -14 & 1 & -3 \\ 12 & 6 & 30 \\ -14 & -2 & -15 \end{bmatrix}$$

EXAMPLE 2. If

$$A = \begin{pmatrix} 3 & 4 \\ 1 & 2 \end{pmatrix} \quad \text{and} \quad B = \begin{bmatrix} 1 & 2 \\ 4 & 5 \\ 3 & 6 \end{bmatrix}$$

then it is impossible to multiply A times B, since the number of columns of A does not equal the number of rows of B. However, it is possible to multiply B times A.

$$BA = \begin{bmatrix} 1 & 2 \\ 4 & 5 \\ 3 & 6 \end{bmatrix}\begin{pmatrix} 3 & 4 \\ 1 & 2 \end{pmatrix} = \begin{bmatrix} 5 & 8 \\ 17 & 26 \\ 15 & 24 \end{bmatrix}$$

If A and B are both $n \times n$ matrices, then AB and BA will also be $n \times n$ matrices, but in general they will not be equal. *Multiplication of matrices is not commutative.*

EXAMPLE 3. If

$$A = \begin{pmatrix} 1 & 1 \\ 0 & 0 \end{pmatrix} \quad \text{and} \quad B = \begin{pmatrix} 1 & 1 \\ 2 & 2 \end{pmatrix}$$

then

$$AB = \begin{pmatrix} 1 & 1 \\ 0 & 0 \end{pmatrix}\begin{pmatrix} 1 & 1 \\ 2 & 2 \end{pmatrix} = \begin{pmatrix} 3 & 3 \\ 0 & 0 \end{pmatrix}$$

and

$$BA = \begin{pmatrix} 1 & 1 \\ 2 & 2 \end{pmatrix}\begin{pmatrix} 1 & 1 \\ 0 & 0 \end{pmatrix} = \begin{pmatrix} 1 & 1 \\ 2 & 2 \end{pmatrix}$$

and hence $AB \neq BA$.

Application 2

Bob weighs 178 pounds. He wishes to lose weight through a program of dieting and exercise. After consulting Table 1 he sets up the exercise schedule in Table 2. How many calories will he burn up exercising each day if he follows this program?

TABLE 1
Calories Burned per Hour

	Exercise Activity			
Weight	*Walking 2 mph*	*Running 5.5 mph*	*Bicycling 5.5 mph*	*Tennis (Moderate)*
152	213	651	304	420
161	225	688	321	441
170	237	726	338	468
178	249	764	356	492

TABLE 2
Hours per Day for Each Activity

	Exercise Schedule			
	Walking	*Running*	*Bicycling*	*Tennis*
Monday	1.0	0.0	1.0	0.0
Tuesday	0.0	0.0	0.0	2.0
Wednesday	0.4	0.5	0.0	0.0
Thursday	0.0	0.0	0.5	2.0
Friday	0.4	0.5	0.0	0.0

SOLUTION. The information pertaining to Bob is located in row four of Table 1. This information can be represented by a 4×1 matrix X. The information in Table 2 can be represented by a 5×4 matrix A. To answer the question we simply calculate AX.

$$\begin{bmatrix} 1.0 & 0.0 & 1.0 & 0.0 \\ 0.0 & 0.0 & 0.0 & 2.0 \\ 0.4 & 0.5 & 0.0 & 0.0 \\ 0.0 & 0.0 & 0.5 & 2.0 \\ 0.4 & 0.5 & 0.0 & 0.0 \end{bmatrix} \begin{bmatrix} 249 \\ 764 \\ 356 \\ 492 \end{bmatrix} = \begin{bmatrix} 605.0 \\ 984.0 \\ 481.6 \\ 1162.0 \\ 481.6 \end{bmatrix} \begin{matrix} \textit{Monday} \\ \textit{Tuesday} \\ \textit{Wednesday} \\ \textit{Thursday} \\ \textit{Friday} \end{matrix}$$

Notational Rules

Just as in ordinary algebra, if an expression involves both multiplication and addition and there are no parentheses to indicate the order of the operation, multiplications are carried out before additions. This is true for both scalar and matrix multiplications. For example, if

$$A = \begin{pmatrix} 3 & 4 \\ 1 & 2 \end{pmatrix}, \quad B = \begin{pmatrix} 1 & 3 \\ 2 & 1 \end{pmatrix}, \quad \text{and} \quad C = \begin{pmatrix} -2 & 1 \\ 3 & 2 \end{pmatrix}$$

then

$$A + BC = \begin{pmatrix} 3 & 4 \\ 1 & 2 \end{pmatrix} + \begin{pmatrix} 7 & 7 \\ -1 & 4 \end{pmatrix} = \begin{pmatrix} 10 & 11 \\ 0 & 6 \end{pmatrix}$$

and

$$3A + B = \begin{pmatrix} 9 & 12 \\ 3 & 6 \end{pmatrix} + \begin{pmatrix} 1 & 3 \\ 2 & 1 \end{pmatrix} = \begin{pmatrix} 10 & 15 \\ 5 & 7 \end{pmatrix}$$

Algebraic Rules

The following theorem provides some useful rules for doing matrix arithmetic.

Theorem 1.3.1. *Each of the following statements is valid for any scalars α and β and for any matrices A, B, and C for which the indicated operations are defined.*

(1) $A + B = B + A$
(2) $(A + B) + C = A + (B + C)$
(3) $(AB)C = A(BC)$
(4) $A(B + C) = AB + AC$
(5) $(A + B)C = AC + BC$
(6) $(\alpha\beta)A = \alpha(\beta A)$
(7) $\alpha(AB) = (\alpha A)B = A(\alpha B)$
(8) $(\alpha + \beta)A = \alpha A + \beta A$
(9) $\alpha(A + B) = \alpha A + \alpha B$

We will prove two of the rules and leave the rest for the reader to verify.

PROOF OF (4). Assume that $A = (a_{ij})$ is an $m \times n$ matrix and $B = (b_{ij})$ and $C = (c_{ij})$ are both $n \times r$ matrices. Let $D = A(B + C)$ and $E = AB + AC$. It follows that

$$d_{ij} = \sum_{k=1}^{n} a_{ik}(b_{kj} + c_{kj})$$

and

$$e_{ij} = \sum_{k=1}^{n} a_{ik}b_{kj} + \sum_{k=1}^{n} a_{ik}c_{kj}$$

But

$$\sum_{k=1}^{n} a_{ik}(b_{kj} + c_{kj}) = \sum_{k=1}^{n} a_{ik}b_{kj} + \sum_{k=1}^{n} a_{ik}c_{kj}$$

so that $d_{ij} = e_{ij}$ and hence $A(B + C) = AB + AC$.

PROOF OF (3). Let A be an $m \times n$ matrix, B an $n \times r$ matrix, and C an $r \times s$ matrix. Let $D = AB$ and $E = BC$. We must show that $DC = AE$. By the definition of matrix multiplication,

$$d_{il} = \sum_{k=1}^{n} a_{ik}b_{kl} \qquad \text{and} \qquad e_{kj} = \sum_{l=1}^{r} b_{kl}c_{lj}$$

The ijth term of DC is

$$\sum_{l=1}^{r} d_{il}c_{lj} = \sum_{l=1}^{r} \left(\sum_{k=1}^{n} a_{ik}b_{kl} \right) c_{lj}$$

and the ijth entry of AE is

$$\sum_{k=1}^{n} a_{ik}e_{kj} = \sum_{k=1}^{n} a_{ik} \left(\sum_{l=1}^{r} b_{kl}c_{lj} \right)$$

Since

$$\sum_{l=1}^{r} \left(\sum_{k=1}^{n} a_{ik}b_{kl} \right) c_{lj} = \sum_{l=1}^{r} \sum_{k=1}^{n} a_{ik}b_{kl}c_{lj} = \sum_{k=1}^{n} a_{ik} \left(\sum_{l=1}^{r} b_{kl}c_{lj} \right)$$

it follows that

$$(AB)C = DC = AE = A(BC)$$

EXAMPLE 4. If

$$A = \begin{pmatrix} 1 & 2 \\ 3 & 4 \end{pmatrix}, \qquad B = \begin{pmatrix} 2 & 1 \\ -3 & 2 \end{pmatrix}, \qquad \text{and} \qquad C = \begin{pmatrix} 1 & 0 \\ 2 & 1 \end{pmatrix}$$

verify that $A(BC) = (AB)C$ and $A(B + C) = AB + AC$.

SOLUTION.

$$A(BC) = \begin{pmatrix} 1 & 2 \\ 3 & 4 \end{pmatrix}\begin{pmatrix} 4 & 1 \\ 1 & 2 \end{pmatrix} = \begin{pmatrix} 6 & 5 \\ 16 & 11 \end{pmatrix}$$

$$(AB)C = \begin{pmatrix} -4 & 5 \\ -6 & 11 \end{pmatrix}\begin{pmatrix} 1 & 0 \\ 2 & 1 \end{pmatrix} = \begin{pmatrix} 6 & 5 \\ 16 & 11 \end{pmatrix}$$

Thus

$$A(BC) = \begin{pmatrix} 6 & 5 \\ 16 & 11 \end{pmatrix} = (AB)C$$

$$A(B + C) = \begin{pmatrix} 1 & 2 \\ 3 & 4 \end{pmatrix}\begin{pmatrix} 3 & 1 \\ -1 & 3 \end{pmatrix} = \begin{pmatrix} 1 & 7 \\ 5 & 15 \end{pmatrix}$$

$$AB + AC = \begin{pmatrix} -4 & 5 \\ -6 & 11 \end{pmatrix} + \begin{pmatrix} 5 & 2 \\ 11 & 4 \end{pmatrix} = \begin{pmatrix} 1 & 7 \\ 5 & 15 \end{pmatrix}$$

Therefore,

$$A(B + C) = AB + AC$$

Notation

Since $(AB)C = A(BC)$, one may simply omit the parentheses and write ABC. The same is true for a product of four or more matrices. In the case where an $n \times n$ matrix is multiplied by itself a number of times, it is convenient to use exponential notation. Thus, if k is a positive integer, then

$$A^k = \underbrace{AA \cdots A}_{k \text{ times}}$$

EXAMPLE 5. If

$$A = \begin{pmatrix} 1 & 1 \\ 1 & 1 \end{pmatrix}$$

then

$$A^2 = \begin{pmatrix} 1 & 1 \\ 1 & 1 \end{pmatrix}\begin{pmatrix} 1 & 1 \\ 1 & 1 \end{pmatrix} = \begin{pmatrix} 2 & 2 \\ 2 & 2 \end{pmatrix}$$

$$A^3 = AAA = AA^2 = \begin{pmatrix} 1 & 1 \\ 1 & 1 \end{pmatrix}\begin{pmatrix} 2 & 2 \\ 2 & 2 \end{pmatrix} = \begin{pmatrix} 4 & 4 \\ 4 & 4 \end{pmatrix}$$

and in general

$$A^n = \begin{pmatrix} 2^{n-1} & 2^{n-1} \\ 2^{n-1} & 2^{n-1} \end{pmatrix}$$

Application 3

In a certain town 30 percent of the married women get divorced each year and 20 percent of the single women get married each year. There are 8000 married women and 2000 single women. Assuming that the total population of women remains constant, how many married women and how many single women will there be after 1 year? After 2 years?

SOLUTION. Form a matrix A as follows. The entries in the first row of A will be the percent of married and single women, respectively, that are married after 1 year. The entries in the second row will be the percent of women who are single after 1 year. Thus

$$A = \begin{pmatrix} 0.70 & 0.20 \\ 0.30 & 0.80 \end{pmatrix}$$

If we let $X = \begin{pmatrix} 8000 \\ 2000 \end{pmatrix}$, then the number of married and single women after 1 year can be computed by multiplying A times X.

$$AX = \begin{pmatrix} 0.70 & 0.20 \\ 0.30 & 0.80 \end{pmatrix}\begin{pmatrix} 8000 \\ 2000 \end{pmatrix} = \begin{pmatrix} 6000 \\ 4000 \end{pmatrix}$$

After 1 year there will be 6000 married women and 4000 single women. To find the number of married and single women after 2 years, we compute

$$A^2X = A(AX) = \begin{pmatrix} 0.70 & 0.20 \\ 0.30 & 0.80 \end{pmatrix}\begin{pmatrix} 6000 \\ 4000 \end{pmatrix} = \begin{pmatrix} 5000 \\ 5000 \end{pmatrix}$$

After 2 years half of the women will be married and half will be single.

Given an $m \times n$ matrix A, it is often useful to form a new $n \times m$ matrix whose columns are the rows of A.

Definition. The **transpose** of an $m \times n$ matrix A is the $n \times m$ matrix B defined by

(1) $$b_{ji} = a_{ij}$$

for $j = 1, \ldots, n$ and $i = 1, \ldots, m$. The transpose of A is denoted by A^T.

It follows from (1) that the jth row of A^T has the same entries, respectively, as the jth column of A and the ith column of A^T has the same entries, respectively, as the ith row of A.

EXAMPLE 6.

(i) If $A = \begin{pmatrix} 1 & 2 & 3 \\ 4 & 5 & 6 \end{pmatrix}$, then $A^T = \begin{bmatrix} 1 & 4 \\ 2 & 5 \\ 3 & 6 \end{bmatrix}$.

(ii) If $B = \begin{bmatrix} -3 & 2 & 1 \\ 4 & 3 & 2 \\ 1 & 2 & 5 \end{bmatrix}$, then $B^T = \begin{bmatrix} -3 & 4 & 1 \\ 2 & 3 & 2 \\ 1 & 2 & 5 \end{bmatrix}$.

(iii) If $C = \begin{pmatrix} 1 & 2 \\ 2 & 3 \end{pmatrix}$, then $C^T = \begin{pmatrix} 1 & 2 \\ 2 & 3 \end{pmatrix} = C$.

The matrix C is its own transpose.

Definition. An $n \times n$ matrix A is said to be **symmetric** if $A^T = A$.

The following are examples of symmetric matrices:

$$\begin{pmatrix} 1 & 0 \\ 0 & -4 \end{pmatrix} \quad \begin{bmatrix} 2 & 3 & 4 \\ 3 & 1 & 5 \\ 4 & 5 & 3 \end{bmatrix} \quad \begin{bmatrix} 0 & 1 & 2 \\ 1 & 1 & -2 \\ 2 & -2 & -3 \end{bmatrix}$$

There are four basic algebraic rules involving transposes.

1. $(A^T)^T = A$
2. $(\alpha A)^T = \alpha A^T$
3. If A and B are both $m \times n$ matrices, then $(A + B)^T = A^T + B^T$.
4. If A is an $m \times n$ matrix and B is an $n \times r$ matrix, then $(AB)^T = B^T A^T$.

We will prove the fourth rule and leave the first three for the reader to verify. Let $C = AB$ and denote the ijth entries of A^T, B^T, and C^T by a_{ij}^T, b_{ij}^T, and c_{ij}^T, respectively. Thus

$$c_{ij}^T = c_{ji}, \qquad a_{ij}^T = a_{ji}, \qquad b_{ij}^T = b_{ji}$$

The ijth term of B^TA^T is given by

$$\sum_{k=1}^{n} b_{ik}^T a_{kj}^T = \sum_{k=1}^{n} a_{jk} b_{ki}$$

The ijth term of $C^T = (AB)^T$ is given by

$$c_{ij}^T = c_{ji} = \sum_{k=1}^{n} a_{jk} b_{ki}$$

Therefore, $(AB)^T = B^TA^T$.

EXERCISES

1. If

$$A = \begin{bmatrix} 3 & 1 & 4 \\ -2 & 0 & 1 \\ 1 & 2 & 2 \end{bmatrix} \quad \text{and} \quad B = \begin{bmatrix} 1 & 0 & 2 \\ -3 & 1 & 1 \\ 2 & -4 & 1 \end{bmatrix}$$

compute:
(a) $2A$ (b) $A + B$ (c) $2A - 3B$
(d) $(2A)^T - (3B)^T$ (e) AB (f) BA
(g) A^TB^T (h) $(BA)^T$

2. For each of the following pairs of matrices, determine whether or not it is possible to multiply the first matrix times the second. If it is possible, perform the multiplication.

(a) $\begin{pmatrix} 3 & 5 & 1 \\ -2 & 0 & 2 \end{pmatrix} \begin{bmatrix} 2 & 1 \\ 1 & 3 \\ 4 & 1 \end{bmatrix}$ (b) $\begin{bmatrix} 4 & -2 \\ 6 & -4 \\ 8 & -6 \end{bmatrix} (1 \quad 2 \quad 3)$

(c) $\begin{bmatrix} 1 & 4 & 3 \\ 0 & 1 & 4 \\ 0 & 0 & 2 \end{bmatrix} \begin{bmatrix} 3 & 2 \\ 1 & 1 \\ 4 & 5 \end{bmatrix}$ (d) $\begin{pmatrix} 4 & 6 \\ 2 & 1 \end{pmatrix} \begin{pmatrix} 3 & 1 & 5 \\ 4 & 1 & 6 \end{pmatrix}$

(e) $\begin{pmatrix} 4 & 6 & 1 \\ 2 & 1 & 1 \end{pmatrix} \begin{pmatrix} 3 & 1 & 5 \\ 4 & 1 & 6 \end{pmatrix}$

3. For which of the pairs in Exercise 2 is it possible to multiply the second matrix times the first, and what would the dimension of the product be?

4. If $A = \begin{bmatrix} 3 & 4 \\ 1 & 1 \\ 2 & 7 \end{bmatrix}$, verify that

(a) $5A = 3A + 2A$ (b) $6A = 3(2A)$ (c) $(A^T)^T = A$

5. If

$$A = \begin{pmatrix} 4 & 1 & 6 \\ 2 & 3 & 5 \end{pmatrix} \quad \text{and} \quad B = \begin{pmatrix} 1 & 3 & 0 \\ -2 & 2 & -4 \end{pmatrix}$$

verify that:
(a) $A + B = B + A$
(b) $3(A + B) = 3A + 3B$
(c) $(A + B)^T = A^T + B^T$

6. If

$$A = \begin{bmatrix} 2 & 1 \\ 6 & 3 \\ -2 & 4 \end{bmatrix} \quad \text{and} \quad B = \begin{pmatrix} 2 & 4 \\ 1 & 6 \end{pmatrix}$$

verify that:
(a) $3(AB) = (3A)B = A(3B)$
(b) $(AB)^T = B^T A^T$

7. If

$$A = \begin{pmatrix} 2 & 4 \\ 1 & 3 \end{pmatrix}, \quad B = \begin{pmatrix} -2 & 1 \\ 0 & 4 \end{pmatrix}, \quad \text{and} \quad C = \begin{pmatrix} 3 & 1 \\ 2 & 1 \end{pmatrix}$$

verify that:
(a) $(A + B) + C = A + (B + C)$
(b) $(AB)C = A(BC)$
(c) $A(B + C) = AB + AC$
(d) $(A + B)C = AC + BC$

8. Let

$$A = \begin{bmatrix} \frac{1}{2} & -\frac{1}{2} \\ -\frac{1}{2} & \frac{1}{2} \end{bmatrix}$$

Compute A^2 and A^3. What will A^n turn out to be?

9. Let

$$A = \begin{bmatrix} \frac{1}{2} & -\frac{1}{2} & -\frac{1}{2} & -\frac{1}{2} \\ -\frac{1}{2} & \frac{1}{2} & -\frac{1}{2} & -\frac{1}{2} \\ -\frac{1}{2} & -\frac{1}{2} & \frac{1}{2} & -\frac{1}{2} \\ -\frac{1}{2} & -\frac{1}{2} & -\frac{1}{2} & \frac{1}{2} \end{bmatrix}$$

Compute A^2 and A^3. What will A^{2n} and A^{2n+1} turn out to be?

10. Let A and B be symmetric $n \times n$ matrices. Prove that $AB = BA$ if and only if AB is also symmetric.

11. Suppose that in Application 2, Bob loses 8 pounds. If he continues the same exercise program, how many calories will he burn up each day?

12. In Application 3, how many married women and how many single women will there be after 3 years?

4. SPECIAL TYPES OF MATRICES

In this section we will learn how to represent a system of linear equations as a single matrix equation. Next, we will look at special types of $n \times n$ matrices, such as triangular matrices, diagonal matrices, and elementary matrices. These special types of matrices play an important role in the solution of matrix equations. We will also discuss the existence and computation of the multiplicative inverse of an $n \times n$ matrix.

The system

$$\begin{aligned} a_{11}x_1 + a_{12}x_2 + \cdots + a_{1n}x_n &= b_1 \\ a_{21}x_1 + a_{22}x_2 + \cdots + a_{2n}x_n &= b_2 \\ \vdots \qquad\qquad & \quad \vdots \\ a_{m1}x_1 + a_{m2}x_2 + \cdots + a_{mn}x_n &= b_m \end{aligned}$$

can be written as the matrix equation

$$\begin{bmatrix} a_{11}x_1 + a_{12}x_2 + \cdots + a_{1n}x_n \\ a_{21}x_1 + a_{22}x_2 + \cdots + a_{2n}x_n \\ \vdots \\ a_{m1}x_1 + a_{m2}x_2 + \cdots + a_{mn}x_n \end{bmatrix} = \begin{bmatrix} b_1 \\ b_2 \\ \vdots \\ b_m \end{bmatrix}$$

which in turn can be written in the form

$$\begin{bmatrix} a_{11} & a_{12} \cdots a_{1n} \\ a_{21} & a_{22} \cdots a_{2n} \\ \vdots & \\ a_{m1} & a_{m2} \cdots a_{mn} \end{bmatrix} \begin{bmatrix} x_1 \\ x_2 \\ \vdots \\ x_n \end{bmatrix} = \begin{bmatrix} b_1 \\ b_2 \\ \vdots \\ b_m \end{bmatrix}$$

If we let

$$\mathbf{x} = \begin{bmatrix} x_1 \\ x_2 \\ \vdots \\ x_n \end{bmatrix} \quad \text{and} \quad \mathbf{b} = \begin{bmatrix} b_1 \\ b_2 \\ \vdots \\ b_m \end{bmatrix}$$

the system can then be represented by the matrix equation $A\mathbf{x} = \mathbf{b}$, where A is the $m \times n$ coefficient matrix of the system.

EXAMPLE 1. The system

$$\begin{aligned} 3x_1 + 2x_2 + x_3 &= 5 \\ x_1 - 2x_2 + 5x_3 &= -2 \\ 2x_1 + x_2 - 3x_3 &= 1 \end{aligned}$$

can be written as the matrix equation

$$\begin{bmatrix} 3 & 2 & 1 \\ 1 & -2 & 5 \\ 2 & 1 & -3 \end{bmatrix} \begin{bmatrix} x_1 \\ x_2 \\ x_3 \end{bmatrix} = \begin{bmatrix} 5 \\ -2 \\ 1 \end{bmatrix}$$

The set of all n-tuples of real numbers is called *Euclidean n-space* and is usually denoted by R^n. The elements of R^n are called *vectors*. Note, however, that the solution to the matrix equation $A\mathbf{x} = \mathbf{b}$ will be an $n \times 1$ matrix rather than an n-tuple. In general, when working with matrix equations it is more convenient to think of R^n as consisting of column vectors ($n \times 1$ matrices) rather than row vectors ($1 \times n$ matrices). The standard notation for a column vector is a boldface lowercase letter.

$$\mathbf{x} = \begin{bmatrix} x_1 \\ x_2 \\ \vdots \\ x_n \end{bmatrix} \qquad \mathbf{x}^T = (x_1, \ldots, x_n)$$

column vector row vector

In order to solve an $m \times n$ system $A\mathbf{x} = \mathbf{b}$, we will apply matrix operations to both sides to obtain an equivalent system $U\mathbf{x} = \mathbf{c}$, where the matrix

U is in a simpler form. In particular, if A is $n \times n$, the new system will be easy to solve if U is "triangular."

Triangular Matrices

An $n \times n$ matrix A is said to be *upper triangular* if $a_{ij} = 0$ for $i > j$ and *lower triangular* if $a_{ij} = 0$ for $i < j$. Also, A is said to be *triangular* if it is either upper triangular or lower triangular. For example, the 3×3 matrices

$$\begin{bmatrix} 3 & 2 & 1 \\ 0 & 2 & 1 \\ 0 & 0 & 5 \end{bmatrix} \quad \text{and} \quad \begin{bmatrix} 1 & 0 & 0 \\ 6 & 2 & 0 \\ 1 & 4 & 3 \end{bmatrix}$$

are both triangular. The first is upper triangular and the second is lower triangular.

Diagonal Matrices

An $n \times n$ matrix A is *diagonal* if $a_{ij} = 0$ whenever $i \neq j$. The matrices

$$\begin{pmatrix} 1 & 0 \\ 0 & 2 \end{pmatrix} \quad \begin{bmatrix} 1 & 0 & 0 \\ 0 & 3 & 0 \\ 0 & 0 & 1 \end{bmatrix} \quad \begin{bmatrix} 0 & 0 & 0 \\ 0 & 2 & 0 \\ 0 & 0 & 0 \end{bmatrix}$$

are diagonal. Of particular importance is the diagonal matrix $I = (\delta_{ij})$, where

$$\delta_{ij} = \begin{cases} 1 & \text{if } i = j \\ 0 & \text{if } i \neq j \end{cases}$$

I is a diagonal matrix with 1's along the diagonal. If A is an $n \times n$ matrix and I is $n \times n$, then $AI = IA = A$. Matrix I acts as an identity for the multiplication of $n \times n$ matrices and consequently is referred to as the *identity matrix.* For example,

$$\begin{bmatrix} 1 & 0 & 0 \\ 0 & 1 & 0 \\ 0 & 0 & 1 \end{bmatrix} \begin{bmatrix} 3 & 4 & 1 \\ 2 & 6 & 3 \\ 0 & 1 & 8 \end{bmatrix} = \begin{bmatrix} 3 & 4 & 1 \\ 2 & 6 & 3 \\ 0 & 1 & 8 \end{bmatrix}$$

and

$$\begin{bmatrix} 3 & 4 & 1 \\ 2 & 6 & 3 \\ 0 & 1 & 8 \end{bmatrix} \begin{bmatrix} 1 & 0 & 0 \\ 0 & 1 & 0 \\ 0 & 0 & 1 \end{bmatrix} = \begin{bmatrix} 3 & 4 & 1 \\ 2 & 6 & 3 \\ 0 & 1 & 8 \end{bmatrix}$$

Definition. A matrix A is said to be **nonsingular** or **invertible** if there exists a matrix B such that $AB = BA = I$. The matrix B is said to be a multiplicative inverse of A.

If B and C are both multiplicative inverses of A, then

$$B = BI = B(AC) = (BA)C = IC = C$$

Thus a matrix can have at most one multiplicative inverse. We will refer to the multiplicative inverse of a nonsingular matrix A as simply the **inverse of A** and denote it by A^{-1}.

EXAMPLE 2. The matrices

$$\begin{pmatrix} 2 & 4 \\ 3 & 1 \end{pmatrix} \quad \text{and} \quad \begin{pmatrix} -\frac{1}{10} & \frac{2}{5} \\ \frac{3}{10} & -\frac{1}{5} \end{pmatrix}$$

are inverses of each other, since

$$\begin{pmatrix} 2 & 4 \\ 3 & 1 \end{pmatrix}\begin{pmatrix} -\frac{1}{10} & \frac{2}{5} \\ \frac{3}{10} & -\frac{1}{5} \end{pmatrix} = \begin{pmatrix} 1 & 0 \\ 0 & 1 \end{pmatrix}$$

and

$$\begin{pmatrix} -\frac{1}{10} & \frac{2}{5} \\ \frac{3}{10} & -\frac{1}{5} \end{pmatrix}\begin{pmatrix} 2 & 4 \\ 3 & 1 \end{pmatrix} = \begin{pmatrix} 1 & 0 \\ 0 & 1 \end{pmatrix}$$

EXAMPLE 3. The triangular matrices

$$\begin{pmatrix} 1 & 2 & 3 \\ 0 & 1 & 4 \\ 0 & 0 & 1 \end{pmatrix} \quad \text{and} \quad \begin{pmatrix} 1 & -2 & 5 \\ 0 & 1 & -4 \\ 0 & 0 & 1 \end{pmatrix}$$

are inverses, since

$$\begin{pmatrix} 1 & 2 & 3 \\ 0 & 1 & 4 \\ 0 & 0 & 1 \end{pmatrix}\begin{pmatrix} 1 & -2 & 5 \\ 0 & 1 & -4 \\ 0 & 0 & 1 \end{pmatrix} = \begin{pmatrix} 1 & 0 & 0 \\ 0 & 1 & 0 \\ 0 & 0 & 1 \end{pmatrix}$$

and

$$\begin{pmatrix} 1 & -2 & 5 \\ 0 & 1 & -4 \\ 0 & 0 & 1 \end{pmatrix}\begin{pmatrix} 1 & 2 & 3 \\ 0 & 1 & 4 \\ 0 & 0 & 1 \end{pmatrix} = \begin{pmatrix} 1 & 0 & 0 \\ 0 & 1 & 0 \\ 0 & 0 & 1 \end{pmatrix}$$

Equivalent Systems

Given an $m \times n$ linear system $A\mathbf{x} = \mathbf{b}$, we can obtain an equivalent system by multiplying both sides of the equation by a nonsingular $m \times m$ matrix M.

(1) $$A\mathbf{x} = \mathbf{b}$$

(2) $$MA\mathbf{x} = M\mathbf{b}$$

Clearly, any solution to (1) will also be a solution to (2). On the other hand, if $\hat{\mathbf{x}}$ is a solution to (2), then

$$M^{-1}(MA\hat{\mathbf{x}}) = M^{-1}(M\mathbf{b})$$
$$A\hat{\mathbf{x}} = \mathbf{b}$$

so the two systems are equivalent.

To obtain an equivalent system that is easier to solve, we can apply a sequence of nonsingular matrices $E_1, \ldots, E_k$ to both sides of the equation $A\mathbf{x} = \mathbf{b}$ to obtain a simpler system

$$U\mathbf{x} = \mathbf{c}$$

where $U = E_k \cdots E_1 A$ and $\mathbf{c} = E_k \cdots E_2 E_1 \mathbf{b}$. The new system will be equivalent to the original provided that $M = E_k \cdots E_1$ is nonsingular. However, M is the product of nonsingular matrices. The following theorem shows that any product of nonsingular matrices is nonsingular.

Theorem 1.4.1. *If A and B are nonsingular $n \times n$ matrices, then AB is also nonsingular and $(AB)^{-1} = B^{-1}A^{-1}$.*

PROOF

$$(B^{-1}A^{-1})AB = B^{-1}(A^{-1}A)B = B^{-1}B = I$$
$$(AB)(B^{-1}A^{-1}) = A(BB^{-1})A^{-1} = AA^{-1} = I$$

It follows by induction that if $E_1, \ldots, E_k$ are all nonsingular, then the product $E_1E_2 \cdots E_k$ is nonsingular and

$$(E_1E_2 \cdots E_k)^{-1} = E_k^{-1} \cdots E_2^{-1}E_1^{-1}$$

We will show next that any of the three elementary row operations can be accomplished by multiplying A on the left by a nonsingular matrix.

Elementary Matrices

A matrix obtained from the identity matrix I by the performance of one elementary row operation is called an *elementary* matrix.

There are three types of elementary matrices corresponding to the three types of elementary row operations.

TYPE I. An elementary matrix of type I is a matrix obtained by interchanging two rows of I.

EXAMPLE 4. Let

$$E_1 = \begin{pmatrix} 0 & 1 & 0 \\ 1 & 0 & 0 \\ 0 & 0 & 1 \end{pmatrix}$$

E_1 is an elementary matrix of type I, since it was obtained by interchanging the first two rows of I. Let A be a 3×3 matrix.

$$E_1A = \begin{bmatrix} 0 & 1 & 0 \\ 1 & 0 & 0 \\ 0 & 0 & 1 \end{bmatrix} \begin{bmatrix} a_{11} & a_{12} & a_{13} \\ a_{21} & a_{22} & a_{23} \\ a_{31} & a_{32} & a_{33} \end{bmatrix} = \begin{bmatrix} a_{21} & a_{22} & a_{23} \\ a_{11} & a_{12} & a_{13} \\ a_{31} & a_{32} & a_{33} \end{bmatrix}$$

$$AE_1 = \begin{bmatrix} a_{11} & a_{12} & a_{13} \\ a_{21} & a_{22} & a_{23} \\ a_{31} & a_{32} & a_{33} \end{bmatrix} \begin{bmatrix} 0 & 1 & 0 \\ 1 & 0 & 0 \\ 0 & 0 & 1 \end{bmatrix} = \begin{bmatrix} a_{12} & a_{11} & a_{13} \\ a_{22} & a_{21} & a_{23} \\ a_{32} & a_{31} & a_{33} \end{bmatrix}$$

Multiplying A on the left by E_1 interchanges the first and second rows of A. Right multiplication of A by E_1 is equivalent to the elementary column operation of interchanging the first and second columns.

Type II. An elementary matrix of type II is a matrix obtained by multiplying a row of I by a nonzero constant.

EXAMPLE 5

$$E_2 = \begin{bmatrix} 1 & 0 & 0 \\ 0 & 1 & 0 \\ 0 & 0 & 3 \end{bmatrix}$$

is an elementary matrix of type II.

$$\begin{bmatrix} 1 & 0 & 0 \\ 0 & 1 & 0 \\ 0 & 0 & 3 \end{bmatrix} \begin{bmatrix} a_{11} & a_{12} & a_{13} \\ a_{21} & a_{22} & a_{23} \\ a_{31} & a_{32} & a_{33} \end{bmatrix} = \begin{bmatrix} a_{11} & a_{12} & a_{13} \\ a_{21} & a_{22} & a_{23} \\ 3a_{31} & 3a_{32} & 3a_{33} \end{bmatrix}$$

$$\begin{bmatrix} a_{11} & a_{12} & a_{13} \\ a_{21} & a_{22} & a_{23} \\ a_{31} & a_{32} & a_{33} \end{bmatrix} \begin{bmatrix} 1 & 0 & 0 \\ 0 & 1 & 0 \\ 0 & 0 & 3 \end{bmatrix} = \begin{bmatrix} a_{11} & a_{12} & 3a_{13} \\ a_{21} & a_{22} & 3a_{23} \\ a_{31} & a_{32} & 3a_{33} \end{bmatrix}$$

Multiplication on the left by E_2 performs the elementary row operation of multiplying the third row by 3, while multiplication on the right by E_2 performs the elementary column operation of multiplying the third column by 3.

Type III. An elementary matrix of type III is a matrix obtained from I by adding a multiple of one row to another row.

EXAMPLE 6

$$E_3 = \begin{bmatrix} 1 & 0 & 3 \\ 0 & 1 & 0 \\ 0 & 0 & 1 \end{bmatrix}$$

is an elementary matrix of type III. If A is a 3×3 matrix, then

$$E_3A = \begin{bmatrix} a_{11}+3a_{31} & a_{12}+3a_{32} & a_{13}+3a_{33} \\ a_{21} & a_{22} & a_{23} \\ a_{31} & a_{32} & a_{33} \end{bmatrix}$$

$$AE_3 = \begin{bmatrix} a_{11} & a_{12} & 3a_{11}+a_{13} \\ a_{21} & a_{22} & 3a_{21}+a_{23} \\ a_{31} & a_{32} & 3a_{31}+a_{33} \end{bmatrix}$$

Multiplication on the left by E_3 adds 3 times the third row to the first row. Multiplication on the right adds 3 times the first column to the third column.

In general, suppose that E is an $n \times n$ elementary matrix. We can think of E as being obtained from I by either a row operation or a column operation. If A is any $n \times r$ matrix, then premultiplying A by E has the effect of performing that same row operation on A. If B is an $m \times n$ matrix, postmultiplying B by E is equivalent to performing that same column operation on B.

Theorem 1.4.2. *If E is an elementary matrix, E is invertible and E^{-1} is an elementary matrix of the same type.*

Proof. If E is the elementary matrix of type I formed from I by interchanging the ith and jth rows, then E can be transformed back into I by interchanging these same rows again. Thus $EE = I$ and hence E is its own inverse. If E is the elementary matrix of type II formed by multiplying the ith row of I by a nonzero scalar α, then E can be transformed into the identity by multiplying either its ith row or its ith column by $1/\alpha$. Thus

$$E^{-1} = \begin{bmatrix} 1 & & & & & & O \\ & \ddots & & & & & \\ & & 1 & & & & \\ & & & 1/\alpha & & & \\ & & & & 1 & & \\ & & & & & \ddots & \\ O & & & & & & 1 \end{bmatrix} \quad i\text{th row}$$

Finally, suppose that E is the elementary matrix of type III formed from I

by adding m times the ith row to the jth row.

$$E = \begin{bmatrix} 1 & & & & & & \\ \vdots & \ddots & & & O & & \\ 0 & \cdots & 1 & & & & \\ \vdots & & & \ddots & & & \\ 0 & \cdots & m & \cdots & 1 & & \\ \vdots & & & & & \ddots & \\ 0 & \cdots & 0 & \cdots & 0 & \cdots & 1 \end{bmatrix} \begin{matrix} \\ \\ i\text{th row} \\ \\ j\text{th row} \\ \\ \\ \end{matrix}$$

E can be transformed back into I by either subtracting m times the ith row from the jth row or by subtracting m times the jth column from the ith column. Thus

$$E^{-1} = \begin{bmatrix} 1 & & & & & & \\ \vdots & \ddots & & & O & & \\ 0 & \cdots & 1 & & & & \\ \vdots & & & \ddots & & & \\ 0 & \cdots & -m & \cdots & 1 & & \\ \vdots & & & & & \ddots & \\ 0 & \cdots & 0 & \cdots & 0 & \cdots & 1 \end{bmatrix}$$

Definition. A matrix B is **row equivalent** to A if there exists a finite sequence $E_1, E_2, \ldots, E_k$ of elementary matrices such that

$$B = E_k E_{k-1} \cdots E_1 A$$

In other words, B is row equivalent to A if B can be obtained from A by a finite number of row operations. In view of Theorem 1.4.2, it follows that:

(i) If A is row equivalent to B, then B is row equivalent to A.
(ii) If A is row equivalent to B, and B is row equivalent to C, then A is row equivalent to C.

The details of the proofs of (i) and (ii) are left as an exercise for the reader.

Theorem 1.4.3. *The following are equivalent*:
(a) *A is nonsingular.*
(b) *$A\mathbf{x} = \mathbf{0}$ has only the trivial solution $\mathbf{0}$.*
(c) *A is row equivalent to I.*

PROOF. We prove first that statement (a) implies statement (b). If A is nonsingular and $\hat{\mathbf{x}}$ is a solution to $A\mathbf{x} = \mathbf{0}$, then

$$\hat{\mathbf{x}} = I\hat{\mathbf{x}} = (A^{-1}A)\hat{\mathbf{x}} = A^{-1}(A\hat{\mathbf{x}}) = A^{-1}\mathbf{0} = \mathbf{0}$$

Thus $A\mathbf{x} = \mathbf{0}$ has only the trivial solution. Next we show that statement (b) implies statement (c). If we use elementary row operations, the system can be transformed into the form $U\mathbf{x} = \mathbf{0}$, where U is in row echelon form. If one of the diagonal elements of U were 0, the last row of U would consist entirely of 0's. But then $A\mathbf{x} = \mathbf{0}$ would be equivalent to a system with more unknowns than equations and hence by Theorem 1.2.1 would have nontrivial solution. Thus U must be a triangular matrix with diagonal elements all equal to 1. It follows then that I is the reduced row echelon form of A and hence A is row equivalent to I.

Finally, we will show that statement (c) implies statement (a). If A is row equivalent to I, there exists elementary matrices $E_1, E_2, \ldots, E_k$ such that

$$A = E_k E_{k-1} \cdots E_1 I = E_k E_{k-1} \cdots E_1$$

But since E_i is invertible, $i = 1, \ldots, k$, the product $E_k E_{k-1} \cdots E_1$ is also invertible. Hence A is nonsingular and

$$A^{-1} = (E_k E_{k-1} \cdots E_1)^{-1} = E_1^{-1} E_2^{-1} \cdots E_k^{-1}$$

Corollary. *The system of equations $A\mathbf{x} = \mathbf{b}$ has a unique solution if and only if A is nonsingular.*

PROOF. If A is nonsingular, then $A^{-1}\mathbf{b}$ is the only solution to $A\mathbf{x} = \mathbf{b}$. Conversely, suppose that $A\mathbf{x} = \mathbf{b}$ has a unique solution $\hat{\mathbf{x}}$. If A is singular, $A\mathbf{x} = \mathbf{0}$ has a solution $\mathbf{z} \neq \mathbf{0}$. Let $\mathbf{y} = \hat{\mathbf{x}} + \mathbf{z}$. Clearly, $\mathbf{y} \neq \hat{\mathbf{x}}$ and

$$A\mathbf{y} = A(\hat{\mathbf{x}} + \mathbf{z}) = A\hat{\mathbf{x}} + A\mathbf{z} = \mathbf{b} + \mathbf{0} = \mathbf{b}$$

Thus $\mathbf{y}$ is also a solution to $A\mathbf{x} = \mathbf{b}$, which is a contradiction. Therefore, if $A\mathbf{x} = \mathbf{b}$ has a unique solution, then A must be nonsingular.

If A is nonsingular, A is row equivalent to I, so there exist elementary matrices $E_1, \ldots, E_k$ such that

$$E_k E_{k-1} \cdots E_1 A = I$$

Multiplying both sides of this equation on the right by A^{-1}, one obtains

$$E_k E_{k-1} \cdots E_1 I = A^{-1}$$

Thus the same series of elementary row operations that transform a nonsingular matrix A into I will transform I into A^{-1}. This gives us a method for computing A^{-1}. If we augment A by I and perform the elementary row operations that transform A into I on the augmented matrix, then I will be transformed into A^{-1}. That is, the reduced row echelon form of the augmented matrix $(A|I)$ will be $(I|A^{-1})$.

EXAMPLE 7. Compute A^{-1} if

$$A = \begin{pmatrix} 1 & 4 & 3 \\ -1 & -2 & 0 \\ 2 & 2 & 3 \end{pmatrix}$$

SOLUTION

$$\left(\begin{array}{rrr|rrr} 1 & 4 & 3 & 1 & 0 & 0 \\ -1 & -2 & 0 & 0 & 1 & 0 \\ 2 & 2 & 3 & 0 & 0 & 1 \end{array}\right) \rightarrow \left(\begin{array}{rrr|rrr} 1 & 4 & 3 & 1 & 0 & 0 \\ 0 & 2 & 3 & 1 & 1 & 0 \\ 0 & -6 & -3 & -2 & 0 & 1 \end{array}\right) \rightarrow \left(\begin{array}{rrr|rrr} 1 & 4 & 3 & 1 & 0 & 0 \\ 0 & 2 & 3 & 1 & 1 & 0 \\ 0 & 0 & 6 & 1 & 3 & 1 \end{array}\right)$$

$$\rightarrow \left(\begin{array}{rrr|rrr} 1 & 4 & 0 & \frac{1}{2} & -\frac{3}{2} & -\frac{1}{2} \\ 0 & 2 & 0 & \frac{1}{2} & -\frac{1}{2} & -\frac{1}{2} \\ 0 & 0 & 6 & 1 & 3 & 1 \end{array}\right) \rightarrow \left(\begin{array}{rrr|rrr} 1 & 0 & 0 & -\frac{1}{2} & -\frac{1}{2} & \frac{1}{2} \\ 0 & 2 & 0 & \frac{1}{2} & -\frac{1}{2} & -\frac{1}{2} \\ 0 & 0 & 6 & 1 & 3 & 1 \end{array}\right) \rightarrow \left(\begin{array}{rrr|rrr} 1 & 0 & 0 & -\frac{1}{2} & -\frac{1}{2} & \frac{1}{2} \\ 0 & 1 & 0 & \frac{1}{4} & -\frac{1}{4} & -\frac{1}{4} \\ 0 & 0 & 1 & \frac{1}{6} & \frac{1}{2} & \frac{1}{6} \end{array}\right)$$

Thus

$$A^{-1} = \begin{pmatrix} -\frac{1}{2} & -\frac{1}{2} & \frac{1}{2} \\ \frac{1}{4} & -\frac{1}{4} & -\frac{1}{4} \\ \frac{1}{6} & \frac{1}{2} & \frac{1}{6} \end{pmatrix}$$

EXAMPLE 8. Solve the system

$$\begin{aligned} x_1 + 4x_2 + 3x_3 &= 12 \\ -x_1 - 2x_2 &= -12 \\ 2x_1 + 2x_2 + 3x_3 &= 8 \end{aligned}$$

The coefficient matrix of this system is the matrix A of the last example. The solution to the system then is

$$\mathbf{x} = A^{-1}\mathbf{b} = \begin{pmatrix} -\frac{1}{2} & -\frac{1}{2} & \frac{1}{2} \\ \frac{1}{4} & -\frac{1}{4} & -\frac{1}{4} \\ \frac{1}{6} & \frac{1}{2} & \frac{1}{6} \end{pmatrix} \begin{pmatrix} 12 \\ -12 \\ 8 \end{pmatrix} = \begin{pmatrix} 4 \\ 4 \\ -\frac{8}{3} \end{pmatrix}$$

EXERCISES

1. Write each of the following systems of equations as a matrix equation.

(a) $$\begin{aligned} 3x_1 + 2x_2 &= 1 \\ 2x_1 - 3x_2 &= 5 \end{aligned}$$

(b) $$\begin{aligned} x_1 + x_2 \qquad &= 5 \\ 2x_1 + x_2 - x_3 &= 6 \\ 3x_1 - 2x_2 + 2x_3 &= 7 \end{aligned}$$

(c) $$\begin{aligned} 2x_1 + x_2 + x_3 &= 4 \\ x_1 - x_2 + 2x_3 &= 2 \\ 3x_1 - 2x_2 - x_3 &= 0 \end{aligned}$$

2. Show that the product of two diagonal matrices is a diagonal matrix. Is the product of two elementary matrices an elementary matrix? Justify your answer.

3. Prove that if A is invertible, then A^T is invertible and $(A^T)^{-1} = (A^{-1})^T$.
[(*HINT*: $(AB)^T = B^T A^T$).]

4. For each of the following pairs of matrices, find an elementary matrix E such that $EA = B$.

(a) $A = \begin{pmatrix} 2 & -1 \\ 5 & 3 \end{pmatrix}$ $\qquad B = \begin{pmatrix} -4 & 2 \\ 5 & 3 \end{pmatrix}$

(b) $A = \begin{bmatrix} 2 & 1 & 3 \\ -2 & 4 & 5 \\ 3 & 1 & 4 \end{bmatrix}$ $\qquad B = \begin{bmatrix} 2 & 1 & 3 \\ 3 & 1 & 4 \\ -2 & 4 & 5 \end{bmatrix}$

(c) $A = \begin{bmatrix} 4 & -2 & 3 \\ 1 & 0 & 2 \\ -2 & 3 & 1 \end{bmatrix}$ $\qquad B = \begin{bmatrix} 4 & -2 & 3 \\ 1 & 0 & 2 \\ 0 & 3 & 5 \end{bmatrix}$

5. For each of the following pairs of matrices, find an elementary matrix E such that $AE = B$.

(a) $A = \begin{bmatrix} 4 & 1 & 3 \\ 2 & 1 & 4 \\ 1 & 3 & 2 \end{bmatrix}$ $\qquad B = \begin{bmatrix} 3 & 1 & 4 \\ 4 & 1 & 2 \\ 2 & 3 & 1 \end{bmatrix}$

(b) $A = \begin{pmatrix} 2 & 4 \\ 3 & 6 \end{pmatrix}$ $\qquad B = \begin{pmatrix} 2 & -2 \\ 3 & -3 \end{pmatrix}$

(c) $A = \begin{bmatrix} 4 & -2 & 3 \\ -2 & 4 & 2 \\ 6 & 1 & -2 \end{bmatrix}$ $\qquad B = \begin{bmatrix} 2 & -2 & 3 \\ -1 & 4 & 2 \\ 3 & 1 & -2 \end{bmatrix}$

6. Show that the transpose of an elementary matrix is an elementary matrix of the same type.

7. Let

$$A = \begin{pmatrix} a_{11} & a_{12} \\ a_{21} & a_{22} \end{pmatrix}$$

Show that if $d = a_{11}a_{22} - a_{21}a_{12} \neq 0$, then

$$A^{-1} = \frac{1}{d}\begin{pmatrix} a_{22} & -a_{12} \\ -a_{21} & a_{11} \end{pmatrix}$$

8. Find the inverse of each of the following matrices.

(a) $\begin{pmatrix} 3 & 2 \\ 1 & 5 \end{pmatrix}$ (b) $\begin{pmatrix} 2 & 6 \\ 3 & -2 \end{pmatrix}$

(c) $\begin{bmatrix} -1 & -3 & 3 \\ 2 & 6 & 1 \\ 3 & 8 & 3 \end{bmatrix}$ (d) $\begin{bmatrix} 1 & 0 & 1 \\ -1 & 1 & 1 \\ -1 & -2 & -3 \end{bmatrix}$

9. Let

$$A = \begin{bmatrix} 1 & 0 & 1 \\ 3 & 3 & 4 \\ 2 & 2 & 3 \end{bmatrix}$$

(a) Verify that

$$A^{-1} = \begin{bmatrix} 1 & 2 & -3 \\ -1 & 1 & -1 \\ 0 & -2 & 3 \end{bmatrix}$$

(b) Use A^{-1} to solve $A\mathbf{x} = \mathbf{b}$ for the following choices of $\mathbf{b}$:
(i) $\mathbf{b} = (1, 1, 1)^T$ (ii) $\mathbf{b} = (1, 2, 3)^T$
(iii) $\mathbf{b} = (-2, 1, 0)^T$

10. Prove that if A is row equivalent to B, then B is row equivalent to A.

11. (a) Prove that if A is row equivalent to B and B is row equivalent to C, then A is row equivalent to C.
(b) Prove that any two nonsingular $n \times n$ matrices are row equivalent.

12. Prove that B is row equivalent to A if and only if there exists a nonsingular matrix M such that $B = MA$.

5. PARTITIONED MATRICES

Often it is useful to think of a matrix as being composed of a number of submatrices. A matrix A can be partitioned into smaller matrices by drawing horizontal lines between the rows and vertical lines between the columns. For example, let

$$A = \begin{pmatrix} 1 & -2 & 4 & 1 & 3 \\ 2 & 1 & 1 & 1 & 1 \\ 3 & 3 & 2 & -1 & 2 \\ 4 & 6 & 2 & 2 & 4 \end{pmatrix}$$

If lines are drawn between the second and third rows and between the third and fourth columns, then A will be divided into four submatrices, A_{11}, A_{12}, A_{21}, and A_{22}.

$$\begin{pmatrix} A_{11} & A_{12} \\ A_{21} & A_{22} \end{pmatrix} = \left(\begin{array}{ccc|cc} 1 & -2 & 4 & 1 & 3 \\ 2 & 1 & 1 & 1 & 1 \\ \hline 3 & 3 & 2 & -1 & 2 \\ 4 & 6 & 2 & 2 & 4 \end{array}\right)$$

One of the most useful ways to partition an $m \times n$ matrix is into n column vectors ($m \times 1$ matrices). In this case we will write

$$A = (\mathbf{a}_1, \mathbf{a}_2, \ldots, \mathbf{a}_n)$$

where

$$\mathbf{a}_j = \begin{pmatrix} a_{1j} \\ a_{2j} \\ \vdots \\ a_{mj} \end{pmatrix} \qquad j = 1, \ldots, n$$

Similarly, if B is an $n \times r$ matrix, then $B = (\mathbf{b}_1, \mathbf{b}_2, \ldots, \mathbf{b}_r)$, where

$$\mathbf{b}_j = (b_{1j}, \ldots, b_{nj})^T \qquad j = 1, \ldots, r$$

The only exception to this notation is in the case of the identity matrix I. The standard notation for the jth column vector of I is $\mathbf{e}_j$ rather than $\mathbf{i}_j$. Thus the $n \times n$ identity matrix can be written

$$I = (\mathbf{e}_1, \ldots, \mathbf{e}_n)$$

EXAMPLE 1. Let

$$A = \begin{pmatrix} 1 & 3 & 1 \\ 2 & 1 & -2 \end{pmatrix} \quad \text{and} \quad B = \begin{pmatrix} -1 & 2 & 1 \\ 2 & 3 & 1 \\ 1 & 4 & 1 \end{pmatrix}$$

Suppose that in some application it is necessary to know the second column of AB. Rather than compute AB, we need only calculate $A\mathbf{b}_2$.

$$A\mathbf{b}_2 = \begin{pmatrix} 1 & 3 & 1 \\ 2 & 1 & -2 \end{pmatrix} \begin{pmatrix} 2 \\ 3 \\ 4 \end{pmatrix} = \begin{pmatrix} 15 \\ -1 \end{pmatrix}$$

In general, if A is an $m \times n$ matrix and $B = (\mathbf{b}_1, \ldots, \mathbf{b}_r)$ is an $n \times r$ matrix, then the jth column of AB will be $A\mathbf{b}_j$. This is a direct consequence of the definition of matrix multiplication. If $C = AB$, then

$$c_{ij} = \sum_{k=1}^{n} a_{ik} b_{kj}$$

and hence

$$\mathbf{c}_j = \begin{bmatrix} \Sigma a_{1k} b_{kj} \\ \Sigma a_{2k} b_{kj} \\ \vdots \\ \Sigma a_{mk} b_{kj} \end{bmatrix} = A\mathbf{b}_j$$

Thus

$$AB = (A\mathbf{b}_1, A\mathbf{b}_2, \ldots, A\mathbf{b}_n)$$

In particular,

$$(\mathbf{a}_1, \ldots, \mathbf{a}_n) = A = AI = (A\mathbf{e}_1, \ldots, A\mathbf{e}_n)$$

Similarly, for A^T we have

(i) $$A^T = (A^T\mathbf{e}_1, \ldots, A^T\mathbf{e}_m)$$

Transposing both sides of (i), we get

$$A = \begin{bmatrix} \mathbf{e}_1^T A \\ \mathbf{e}_2^T A \\ \vdots \\ \mathbf{e}_m^T A \end{bmatrix}$$

Thus the jth row vector of A is $\mathbf{e}_j^T A$.

If B is an $n \times r$ matrix and $C = AB$, then

$$C = \begin{bmatrix} \mathbf{e}_1^T C \\ \mathbf{e}_2^T C \\ \vdots \\ \mathbf{e}_m^T C \end{bmatrix} = \begin{bmatrix} (\mathbf{e}_1^T A)B \\ (\mathbf{e}_2^T A)B \\ \vdots \\ (\mathbf{e}_m^T A)B \end{bmatrix}$$

Thus

$$AB = \begin{bmatrix} \mathbf{e}_1^T A \\ \mathbf{e}_2^T A \\ \vdots \\ \mathbf{e}_m^T A \end{bmatrix} B = \begin{bmatrix} (\mathbf{e}_1^T A)B \\ (\mathbf{e}_2^T A)B \\ \vdots \\ (\mathbf{e}_m^T A)B \end{bmatrix}$$

*Block Multiplication

Let A be an $m \times n$ matrix and B an $n \times r$ matrix. It is often useful to partition A and B and express the product in terms of the submatrices of A and B. Consider the following four cases:

1. $B = (B_1 \quad B_2)$, where B_1 is an $n \times t$ matrix and B_2 is an $n \times (r - t)$ matrix.

$$\begin{aligned} AB &= A(\mathbf{b}_1, \ldots, \mathbf{b}_t, \mathbf{b}_{t+1}, \ldots, \mathbf{b}_r) \\ &= (A\mathbf{b}_1, \ldots, A\mathbf{b}_t, A\mathbf{b}_{t+1}, \ldots, A\mathbf{b}_r) \\ &= (AB_1 \quad AB_2) \end{aligned}$$

Thus

$$A(B_1 \quad B_2) = (AB_1 \quad AB_2)$$

2. $A = \begin{pmatrix} A_1 \\ A_2 \end{pmatrix}$, where A_1 is a $k \times n$ matrix and A_2 is an $(m - k) \times n$ matrix.

$$AB = \begin{bmatrix} \mathbf{e}_1^T A \\ \vdots \\ \mathbf{e}_k^T A \\ \mathbf{e}_{k+1}^T A \\ \vdots \\ \mathbf{e}_m^T A \end{bmatrix} B = \begin{bmatrix} \mathbf{e}_1^T AB \\ \vdots \\ \mathbf{e}_k^T AB \\ \mathbf{e}_{k+1}^T AB \\ \vdots \\ \mathbf{e}_m^T AB \end{bmatrix} = \begin{pmatrix} A_1 B \\ A_2 B \end{pmatrix}$$

Thus

$$\begin{pmatrix} A_1 \\ A_2 \end{pmatrix} B = \begin{pmatrix} A_1 B \\ A_2 B \end{pmatrix}$$

3. $A = (A_1 \quad A_2)$ and $B = \begin{pmatrix} B_1 \\ B_2 \end{pmatrix}$, where A_1 is an $m \times s$ matrix, A_2 is an $m \times (n-s)$ matrix, B_1 is an $s \times r$ matrix, and B_2 is an $(n-s) \times r$ matrix. If $C = AB$, then

$$\begin{aligned} c_{ij} &= \sum_{l=1}^{n} a_{il} b_{lj} \\ &= \sum_{l=1}^{s} a_{il} b_{lj} + \sum_{l=s+1}^{n} a_{il} b_{lj} \end{aligned}$$

Thus c_{ij} is the sum of the ijth entry of A_1B_1 and the ijth entry of A_2B_2.

$$AB = C = A_1B_1 + A_2B_2$$

$$\boxed{(A_1 \quad A_2)\begin{pmatrix} B_1 \\ B_2 \end{pmatrix} = A_1B_1 + A_2B_2}$$

4. Let A and B both be partitioned as follows:

$$A = \begin{matrix} k \\ m-k \end{matrix}\left(\begin{array}{c|c} A_{11} & A_{12} \\ \hline A_{21} & A_{22} \end{array}\right) \qquad B = \left(\begin{array}{c|c} B_{11} & B_{12} \\ \hline B_{21} & B_{22} \end{array}\right)\begin{matrix} s \\ n-s \end{matrix}$$

$$\qquad\quad s \quad n-s \qquad\qquad\qquad\qquad t \quad r-t$$

Let

$$A_1 = \begin{pmatrix} A_{11} \\ A_{21} \end{pmatrix} \qquad A_2 = \begin{pmatrix} A_{12} \\ A_{22} \end{pmatrix}$$

$$B_1 = (B_{11} \quad B_{12}) \qquad B_2 = (B_{21} \quad B_{22})$$

It follows from (3) that

$$AB = (A_1 \quad A_2)\begin{pmatrix} B_1 \\ B_2 \end{pmatrix} = A_1B_1 + A_2B_2$$

It follows from (1) and (2) that

$$A_1B_1 = \begin{pmatrix} A_{11} \\ A_{21} \end{pmatrix} B_1 = \begin{pmatrix} A_{11}B_1 \\ A_{21}B_1 \end{pmatrix} = \begin{pmatrix} A_{11}B_{11} & A_{11}B_{12} \\ A_{21}B_{11} & A_{21}B_{12} \end{pmatrix}$$

$$A_2B_2 = \begin{pmatrix} A_{12} \\ A_{22} \end{pmatrix} B_2 = \begin{pmatrix} A_{12}B_2 \\ A_{22}B_2 \end{pmatrix} = \begin{pmatrix} A_{12}B_{21} & A_{12}B_{22} \\ A_{22}B_{21} & A_{22}B_{22} \end{pmatrix}$$

Therefore,

$$\begin{pmatrix} A_{11} & A_{12} \\ A_{21} & A_{22} \end{pmatrix}\begin{pmatrix} B_{11} & B_{12} \\ B_{21} & B_{22} \end{pmatrix} = \begin{pmatrix} A_{11}B_{11} + A_{12}B_{21} & A_{11}B_{12} + A_{12}B_{22} \\ A_{21}B_{11} + A_{22}B_{21} & A_{21}B_{12} + A_{22}B_{22} \end{pmatrix}$$

In general, if the blocks have the proper dimensions, the block multiplication can be carried out in the same manner as ordinary matrix multiplication. If

$$A = \begin{bmatrix} A_{11} & \cdots & A_{1t} \\ \vdots & & \\ A_{s1} & \cdots & A_{st} \end{bmatrix} \quad \text{and} \quad B = \begin{bmatrix} B_{11} & \cdots & B_{1r} \\ \vdots & & \\ B_{t1} & \cdots & B_{tr} \end{bmatrix}$$

then

$$AB = \begin{bmatrix} C_{11} & \cdots & C_{1r} \\ \vdots & & \vdots \\ C_{s1} & \cdots & C_{sr} \end{bmatrix}$$

where

$$C_{ij} = \sum_{k=1}^{t} A_{ik}B_{kj}$$

The multiplication can be carried out in this manner only if the number of columns of A_{ik} equals the number of rows of B_{kj} for each k.

EXAMPLE 2. Let

$$A = \begin{bmatrix} 1 & 1 & 1 & 1 \\ 2 & 2 & 1 & 1 \\ 3 & 3 & 2 & 2 \end{bmatrix} \quad \text{and} \quad B = \begin{pmatrix} B_{11} & B_{12} \\ B_{21} & B_{22} \end{pmatrix} = \left[\begin{array}{cc|cc} 1 & 1 & 1 & 1 \\ 1 & 2 & 1 & 1 \\ \hline 3 & 1 & 1 & 1 \\ 3 & 2 & 1 & 2 \end{array}\right]$$

Partition A into four blocks and perform the block multiplication.

SOLUTION. Since each B_{kj} has two rows, the A_{ik}'s must each have two columns. Thus there are two possibilities:

(i)
$$\begin{pmatrix} A_{11} & A_{12} \\ A_{21} & A_{22} \end{pmatrix} = \left[\begin{array}{cc|cc} 1 & 1 & 1 & 1 \\ \hline 2 & 2 & 1 & 1 \\ 3 & 3 & 2 & 2 \end{array}\right]$$

in which case

$$\left[\begin{array}{cc|cc} 1 & 1 & 1 & 1 \\ \hline 2 & 2 & 1 & 1 \\ 3 & 3 & 2 & 2 \end{array}\right]\left[\begin{array}{cc|cc} 1 & 1 & 1 & 1 \\ 1 & 2 & 1 & 1 \\ \hline 3 & 1 & 1 & 1 \\ 3 & 2 & 1 & 2 \end{array}\right] = \left[\begin{array}{cc|cc} 8 & 6 & 4 & 5 \\ \hline 10 & 9 & 6 & 7 \\ 18 & 15 & 10 & 12 \end{array}\right]$$

or

(ii) $$\begin{pmatrix} A_{11} & A_{12} \\ A_{21} & A_{22} \end{pmatrix} = \left[\begin{array}{cc|cc} 1 & 1 & 1 & 1 \\ 2 & 2 & 1 & 1 \\ \hline 3 & 3 & 2 & 2 \end{array}\right]$$

in which case

$$\left[\begin{array}{cc|cc} 1 & 1 & 1 & 1 \\ 2 & 2 & 1 & 1 \\ \hline 3 & 3 & 2 & 2 \end{array}\right]\left[\begin{array}{cc|cc} 1 & 1 & 1 & 1 \\ 1 & 2 & 1 & 1 \\ \hline 3 & 1 & 1 & 1 \\ 3 & 2 & 1 & 2 \end{array}\right] = \left[\begin{array}{cc|cc} 8 & 6 & 4 & 5 \\ 10 & 9 & 6 & 7 \\ \hline 18 & 15 & 10 & 12 \end{array}\right]$$

EXAMPLE 3. Let A be an $n \times n$ matrix of the form

$$\begin{pmatrix} A_{11} & O \\ O & A_{22} \end{pmatrix}$$

where A_{11} is a $k \times k$ matrix ($k < n$). Show that A is nonsingular if and only if A_{11} and A_{22} are nonsingular.

SOLUTION. If A_{11} and A_{22} are nonsingular, then

$$\begin{pmatrix} A_{11}^{-1} & O \\ O & A_{22}^{-1} \end{pmatrix}\begin{pmatrix} A_{11} & O \\ O & A_{22} \end{pmatrix} = \begin{pmatrix} I_k & O \\ O & I_{n-k} \end{pmatrix} = I$$

and

$$\begin{pmatrix} A_{11} & O \\ O & A_{22} \end{pmatrix}\begin{pmatrix} A_{11}^{-1} & O \\ O & A_{22}^{-1} \end{pmatrix} = \begin{pmatrix} I_k & O \\ O & I_{n-k} \end{pmatrix} = I$$

so A is nonsingular and

$$A^{-1} = \begin{pmatrix} A_{11}^{-1} & O \\ O & A_{22}^{-1} \end{pmatrix}$$

Conversely, if A is nonsingular, then let $B = A^{-1}$ and partition B in the same manner as A. Since

$$BA = I = AB$$

it follows that

$$\begin{pmatrix} B_{11} & B_{12} \\ B_{21} & B_{22} \end{pmatrix}\begin{pmatrix} A_{11} & O \\ O & A_{22} \end{pmatrix} = \begin{pmatrix} I_k & O \\ O & I_{n-k} \end{pmatrix} = \begin{pmatrix} A_{11} & O \\ O & A_{22} \end{pmatrix}\begin{pmatrix} B_{11} & B_{12} \\ B_{21} & B_{22} \end{pmatrix}$$

$$\begin{pmatrix} B_{11}A_{11} & B_{12}A_{22} \\ B_{21}A_{11} & B_{22}A_{22} \end{pmatrix} = \begin{pmatrix} I_k & O \\ O & I_{n-k} \end{pmatrix} = \begin{pmatrix} A_{11}B_{11} & A_{11}B_{12} \\ A_{22}B_{21} & A_{22}B_{22} \end{pmatrix}$$

Thus

$$B_{11}A_{11} = I_k = A_{11}B_{11}$$
$$B_{22}A_{22} = I_{n-k} = A_{22}B_{22}$$

and hence A_{11} and A_{22} are both nonsingular with inverses B_{11} and B_{22}, respectively.

EXERCISES

1. Let A be a nonsingular $n \times n$ matrix. Perform the following multiplications.

(a) $A^{-1}(A \quad I)$ (b) $\begin{pmatrix} A \\ I \end{pmatrix}A^{-1}$ (c) $(A \quad I)^T(A \quad I)$

(d) $(A \quad I)(A \quad I)^T$ (e) $\begin{pmatrix} A^{-1} \\ I \end{pmatrix}(A \quad I)$

2. Let $B = A^TA$. Show that $b_{ij} = \mathbf{a}_i^T\mathbf{a}_j$.

3. Let

$$A = \begin{pmatrix} 1 & 1 \\ 2 & -1 \end{pmatrix} \quad \text{and} \quad B = \begin{pmatrix} 2 & 1 \\ 1 & 3 \end{pmatrix}$$

(i) Calculate $A\mathbf{b}_1$ and $A\mathbf{b}_2$.
(ii) Calculate $(1 \quad 1)B$ and $(2 \quad -1)B$.
(iii) Multiply AB and verify that its column vectors are the vectors in (i) and its row vectors are the vectors in (ii).

4. Let

$$I = \begin{pmatrix} 1 & 0 \\ 0 & 1 \end{pmatrix} \quad E = \begin{pmatrix} 0 & 1 \\ 1 & 0 \end{pmatrix} \quad O = \begin{pmatrix} 0 & 0 \\ 0 & 0 \end{pmatrix}$$

$$C = \begin{pmatrix} 1 & 0 \\ -1 & 1 \end{pmatrix} \quad D = \begin{pmatrix} 2 & 0 \\ 0 & 2 \end{pmatrix}$$

and

$$B = \begin{pmatrix} B_{11} & B_{12} \\ B_{21} & B_{22} \end{pmatrix} = \left[\begin{array}{cc|cc} 1 & 1 & 1 & 1 \\ 1 & 2 & 1 & 1 \\ \hline 3 & 1 & 1 & 1 \\ 3 & 2 & 1 & 2 \end{array}\right]$$

Perform each of the following block multiplications.

(a) $\begin{pmatrix} O & I \\ I & O \end{pmatrix}\begin{pmatrix} B_{11} & B_{12} \\ B_{21} & B_{22} \end{pmatrix}$ (b) $\begin{pmatrix} C & O \\ O & C \end{pmatrix}\begin{pmatrix} B_{11} & B_{12} \\ B_{21} & B_{22} \end{pmatrix}$

(c) $\begin{pmatrix} D & O \\ O & I \end{pmatrix}\begin{pmatrix} B_{11} & B_{12} \\ B_{21} & B_{22} \end{pmatrix}$ (d) $\begin{pmatrix} E & O \\ O & E \end{pmatrix}\begin{pmatrix} B_{11} & B_{12} \\ B_{21} & B_{22} \end{pmatrix}$

5. Perform each of the following block multiplications.

(a) $\left(\begin{array}{ccc|c} 1 & 1 & 1 & -1 \\ 2 & 1 & 2 & -1 \end{array}\right)\left[\begin{array}{ccc} 4 & -2 & 1 \\ 2 & 3 & 1 \\ 1 & 1 & 2 \\ \hline 1 & 2 & 3 \end{array}\right]$

(b) $\left[\begin{array}{cc} 4 & -2 \\ 2 & 3 \\ 1 & 1 \\ \hline 1 & 2 \end{array}\right]\left(\begin{array}{ccc|c} 1 & 1 & 1 & -1 \\ 2 & 1 & 2 & -1 \end{array}\right)$

(c) $\left[\begin{array}{cc|cc} \frac{3}{5} & -\frac{4}{5} & 0 & 0 \\ \frac{4}{5} & \frac{3}{5} & 0 & 0 \\ \hline 0 & 0 & 1 & 0 \end{array}\right]\left[\begin{array}{cc|c} \frac{3}{5} & \frac{4}{5} & 0 \\ -\frac{4}{5} & \frac{3}{5} & 0 \\ \hline 0 & 0 & 1 \\ 0 & 0 & 0 \end{array}\right]$

(d) $\left[\begin{array}{ccc|cc} 0 & 0 & 1 & 0 & 0 \\ 0 & 1 & 0 & 0 & 0 \\ 1 & 0 & 0 & 0 & 0 \\ \hline 0 & 0 & 0 & 0 & 1 \\ 0 & 0 & 0 & 1 & 0 \end{array}\right]\left[\begin{array}{cc} 1 & -1 \\ 2 & -2 \\ 3 & -3 \\ \hline 4 & -4 \\ 5 & -5 \end{array}\right]$

6. Let

$$A = \begin{pmatrix} A_{11} & A_{12} \\ A_{21} & A_{22} \end{pmatrix} \quad \text{and} \quad A^T = \begin{pmatrix} A_{11}^T & A_{21}^T \\ A_{12}^T & A_{22}^T \end{pmatrix}$$

Is it possible to perform the block multiplications of AA^T and A^TA? Explain.

7. Let

$$A = \begin{pmatrix} A_{11} & A_{12} \\ O & A_{22} \end{pmatrix}$$

where all four blocks are $n \times n$ matrices.

(a) If A_{11} and A_{22} are nonsingular, show that A must also be nonsingular and that A^{-1} must be of the form

$$\left[\begin{array}{c|c} A_{11}^{-1} & C \\ \hline O & A_{22}^{-1} \end{array}\right]$$

(b) Determine C.

2

Determinants

With each square matrix it is possible to associate a real number called the determinant of the matrix. The value of this number will tell us whether or not the matrix is singular.

In Section 1 the definition of the determinant of a matrix is given. Although the value of the determinant can be calculated using the definition, it is generally more convenient to use some other method that involves fewer arithmetic operations. One such method is the method of cofactors. This method is probably the simplest method for evaluating determinants of 3×3 matrices. It also works well for larger matrices that have many zero entries. In Section 2 we will study properties of determinants and derive an elimination method for evaluating determinants. The elimination method is generally the simplest method to use for evaluating the determinant of an $n \times n$ matrix when $n > 3$. In Section 3 we will see how determinants can be applied to solving $n \times n$ linear systems and how they can be used to calculate the inverse of a matrix. Further applications of determinants will be presented in Chapters 3 and 6.

1. THE DETERMINANT OF A MATRIX

Before defining the determinant of an $n \times n$ matrix, it is necessary to introduce some preliminary definitions and notation.

Definition. Let S be a finite set. An ordering of the elements of S is said to be a **permutation** of S.

If a set S has n elements, there are $n!$ permutations of S. Each permutation can be represented by an ordered n-tuple. For example, if $S = \{1, 2, 3\}$, there are six possible permutations.

$$(1, 2, 3), \quad (1, 3, 2), \quad (2, 1, 3), \quad (2, 3, 1), \quad (3, 1, 2), \quad (3, 2, 1)$$

Let π_n denote the set of all permutations of $\{1, 2, \ldots, n\}$. An element $\mathbf{k} = (k_1, k_2, \ldots, k_n)$ of π_n is said to have an *inversion* if $k_i > k_j$ for $i < j$ (i.e., a larger number precedes a smaller number). A permutation is said to be *even* if it has an even number of inversions and it is said to be *odd* if it has an odd number of inversions. In π_3 the permutations (1, 2, 3), (2, 3, 1), and (3, 1, 2) are even and the permutations (1, 3, 2), (2, 1, 3), and (3, 2, 1) are odd. For each permutation $\mathbf{k}$ in π_n, we define

$$\delta(\mathbf{k}) = \begin{cases} 0 & \text{if } \mathbf{k} \text{ is even} \\ 1 & \text{if } \mathbf{k} \text{ is odd} \end{cases}$$

Definition. Let $A = (a_{ij})$ be an $n \times n$ matrix. The **determinant of A**, denoted $|A|$, is defined by

$$|A| = \Sigma(-1)^{\delta(\mathbf{k})} a_{1k_1} a_{2k_2} \cdots a_{nk_n}$$

where the sum is taken over all $\mathbf{k} = (k_1, \ldots, k_n)$ in π_n.

The determinant of a 1×1 matrix (a_{11}) is just a_{11}. If A is a 2×2 matrix, then

$$|A| = (-1)^0 a_{11} a_{22} + (-1)^1 a_{12} a_{21} = a_{11} a_{22} - a_{12} a_{21}$$

This formula is easy to remember because it involves only subtracting the product of the off-diagonal elements from the product of the diagonal elements. If A is a 3×3 matrix, then

$$\text{(1)} \qquad \begin{aligned} |A| = {} & a_{11}a_{22}a_{33} - a_{11}a_{23}a_{32} - a_{12}a_{21}a_{33} \\ & + a_{12}a_{23}a_{31} + a_{13}a_{21}a_{32} - a_{13}a_{22}a_{31} \end{aligned}$$

The formula for the determinant of a 4×4 matrix would involve the sum of 24 products of four elements.

To simplify matters, it is convenient to introduce the notation of minors and cofactors.

Definition. Let $A = (a_{ij})$ be an $n \times n$ matrix. Let M_{ij} be the $n-1 \times n-1$ matrix obtained from A by deleting the row and column containing a_{ij}. The determinant of M_{ij} is called the **minor** of a_{ij}. We define the **cofactor** A_{ij} of a_{ij} by

$$A_{ij} = (-1)^{i+j}|M_{ij}|$$

Returning to the 3×3 case, if we rewrite (1) in the form

$$|A| = a_{11}(a_{22}a_{33} - a_{23}a_{32}) - a_{12}(a_{21}a_{33} - a_{23}a_{31}) + a_{13}(a_{21}a_{32} - a_{22}a_{31})$$

we have

$$|A| = a_{11}|M_{11}| - a_{12}|M_{12}| + a_{13}|M_{13}|$$

or

$$|A| = a_{11}A_{11} + a_{12}A_{12} + a_{13}A_{13} \tag{2}$$

Equation (2) is called the *cofactor expansion of* $|A|$ using the first row of A. Actually, any row of A can be used for the cofactor expansion. We could just as well have rewritten (1) in the form

$$\begin{aligned} |A| &= -a_{21}(a_{12}a_{33} - a_{13}a_{32}) + a_{22}(a_{11}a_{33} - a_{13}a_{31}) - a_{23}(a_{11}a_{32} - a_{12}a_{31}) \\ &= a_{21}A_{21} + a_{22}A_{22} + a_{23}A_{23} \end{aligned}$$

In a similar manner, it can be shown that

$$|A| = a_{31}A_{31} + a_{32}A_{32} + a_{33}A_{33}$$

The determinant of A can also be expanded using the columns of A. Rewriting (1) in the form

$$|A| = a_{11}(a_{22}a_{33} - a_{23}a_{32}) - a_{21}(a_{12}a_{33} - a_{13}a_{32}) + a_{31}(a_{12}a_{23} - a_{13}a_{22})$$

we see that

$$|A| = a_{11}A_{11} + a_{21}A_{21} + a_{31}A_{31}$$

The reader may verify that $|A|$ can also be expressed as a cofactor expansion using the second and third columns.

In general, if A is an $n \times n$ matrix with $n \geqslant 2$, then $|A|$ can be expressed as a cofactor expansion using any row or any column of A.

$$\begin{aligned} |A| &= a_{i1}A_{i1} + a_{i2}A_{i2} + \cdots + a_{in}A_{in} \\ &= a_{1j}A_{1j} + a_{2j}A_{2j} + \cdots + a_{nj}A_{nj} \end{aligned}$$

We will not bother to prove that the cofactor expansion is valid in the general case; however, the proof of this is essentially the same as in the 3×3 case.

EXAMPLE 1. Use cofactors to evaluate

$$\begin{vmatrix} 3 & 2 & 4 \\ 1 & -2 & 3 \\ 2 & 3 & 2 \end{vmatrix}$$

SOLUTION. Expanding along the first row, we get

$$\begin{aligned} \begin{vmatrix} 3 & 2 & 4 \\ 1 & -2 & 3 \\ 2 & 3 & 2 \end{vmatrix} &= 3\begin{vmatrix} -2 & 3 \\ 3 & 2 \end{vmatrix} - 2\begin{vmatrix} 1 & 3 \\ 2 & 2 \end{vmatrix} + 4\begin{vmatrix} 1 & -2 \\ 2 & 3 \end{vmatrix} \\ &= 3 \cdot (-13) - 2 \cdot (-4) + 4 \cdot 7 \\ &= -3 \end{aligned}$$

The cofactor expansion of a 4×4 determinant will involve four 3×3 determinants. One can often save work by expanding along the row or column that contains the most zeros. For example, to evaluate

$$\begin{vmatrix} 0 & 2 & 3 & 0 \\ 0 & 4 & 5 & 0 \\ 0 & 1 & 0 & 3 \\ 2 & 0 & 1 & 3 \end{vmatrix}$$

one would expand down the first column. The first three terms will drop out, leaving

$$-2\begin{vmatrix} 2 & 3 & 0 \\ 4 & 5 & 0 \\ 1 & 0 & 3 \end{vmatrix} = -2 \cdot 3 \cdot \begin{vmatrix} 2 & 3 \\ 4 & 5 \end{vmatrix} = 12$$

Using cofactors to evaluate determinants is generally more efficient than using the definition since the cofactor method involves fewer arithmetic operations. The cofactor method is particularly well suited for sparse matrices (i.e., matrices with many zero entries).

Theorem 2.1.1. *If A is an $n \times n$ matrix, then $|A^T| = |A|$.*

PROOF. The proof is by induction on n. Clearly, the result holds if $n = 1$, since a 1×1 matrix is necessarily symmetric. Assume that the result holds for all $k \times k$ matrices and that A is a $(k+1) \times (k+1)$ matrix. Expanding $|A|$ along the first row of A, we get

$$|A| = a_{11}|M_{11}| - a_{12}|M_{12}| + - \cdots \pm a_{1,k+1}|M_{1,k+1}|$$

Since the M_{ij}'s are all $k \times k$ matrices, it follows from the induction hypothesis that

$$(3) \qquad |A| = a_{11}|M_{11}^T| - a_{12}|M_{12}^T| + - \cdots \pm a_{1,k+1}|M_{1,k+1}^T|$$

The right-hand side of (3) is just the expansion by minors of $|A^T|$ using the first column of A^T. Therefore, $|A^T| = |A|$.

Theorem 2.1.2. *If A is an $n \times n$ triangular matrix, the determinant of A equals the product of the diagonal elements of A.*

PROOF. In view of Theorem 2.1.1, it suffices to prove the theorem for lower triangular matrices. The result follows easily using the cofactor expansion and induction on n. The details of this are left for the reader as an exercise.

In the next section we will look at the effect of row operations on the value of the determinant. This will allow us to make use of Theorem 2.1.2 to derive a third and more efficient method for computing the value of a determinant.

EXERCISES

1. Determine whether the following permutations in π_5 are odd or even.
(a) (5, 1, 2, 3, 4) (b) (5, 2, 3, 4, 1) (c) (5, 4, 3, 2, 1)
(d) (2, 1, 3, 5, 4) (e) (3, 2, 1, 4, 5)

2. Let $\mathbf{k} = (k_1, k_2, \ldots, k_n)$, $\mathbf{j} = (k_n, k_1, k_2, \ldots, k_{n-1})$, and $\mathbf{i} = (k_n, k_2, k_3, \ldots, k_{n-1}, k_1)$ be elements of π_n.
(a) Let $m = k_n$. Show that the number of inversions in $\mathbf{k}$ and $\mathbf{j}$ differ by $2m - n - 1$.
(b) Let $l = k_1$. If $l < m$, show that the number of inversions in $\mathbf{j}$ and $\mathbf{i}$ differ by $n - 2l$.
(c) If $l > m$, show that the number of inversions in $\mathbf{j}$ and $\mathbf{i}$ differ by $n - 2l + 2$.
(d) Show that $\mathbf{k}$ and $\mathbf{i}$ differ in parity (i.e., $\mathbf{k}$ is even if and only if $\mathbf{i}$ is odd). Thus whenever two elements of a permutation are switched, the parity is changed.

3. Evaluate the following determinants.

(a) $\begin{vmatrix} 1 & 3 & 2 \\ 4 & 1 & -2 \\ 2 & 1 & 3 \end{vmatrix}$ (b) $\begin{vmatrix} 2 & -1 & 2 \\ 1 & 3 & 2 \\ 5 & 1 & 6 \end{vmatrix}$

(c) $\begin{vmatrix} 2 & 0 & 0 & 1 \\ 0 & 1 & 0 & 0 \\ 1 & 6 & 2 & 0 \\ 1 & 1 & -2 & 3 \end{vmatrix}$ (d) $\begin{vmatrix} 2 & 1 & 2 & 1 \\ 3 & 0 & 1 & 1 \\ -1 & 2 & -2 & 1 \\ -3 & 2 & 3 & 1 \end{vmatrix}$

4. Evaluate the following determinants by inspection.

(a) $\begin{vmatrix} 3 & 5 \\ 2 & 4 \end{vmatrix}$ (b) $\begin{vmatrix} 2 & 0 & 0 \\ 4 & 1 & 0 \\ 7 & 3 & -2 \end{vmatrix}$

(c) $\begin{vmatrix} 3 & 0 & 0 \\ 2 & 1 & 1 \\ 1 & 2 & 2 \end{vmatrix}$ (d) $\begin{vmatrix} 4 & 0 & 2 & 1 \\ 5 & 0 & 4 & 2 \\ 2 & 0 & 3 & 4 \\ 1 & 0 & 2 & 3 \end{vmatrix}$

5. Write out the details of the proof of Theorem 2.1.2.

6. Does $|A + B| = |A| + |B|$? Justify your answer.

7. Let A be a symmetric tridiagonal matrix (i.e., A is symmetric and $a_{ij} = 0$ whenever $|i - j| > 1$). Let B be the matrix formed from A by deleting the first two rows and columns. Show that

$$|A| = a_{11}|M_{11}| - a_{12}^2|B|$$

8. Prove that if a row or a column of an $n \times n$ matrix consists entirely of zeros, then $|A| = 0$.

9. Use mathematical induction to prove that if A is an $(n + 1) \times (n + 1)$ matrix with two identical rows, then $|A| = 0$.

10. Let A, B, and C be three $n \times n$ matrices such that for some k,

$$a_{kj} = b_{kj} + c_{kj} \qquad j = 1, \ldots, n$$

and

$$a_{ij} = b_{ij} = c_{ij} \qquad \text{whenever } i \neq k$$

Show that

$$|A| = |B| + |C|$$

2. PROPERTIES OF DETERMINANTS

In this section we will consider the effect of row operations on the determinant of a matrix. From this we will see that the value of the determinant actually tells us whether or not the matrix is singular. We will also learn a method for evaluating determinants using row operations. Finally, we will establish a multiplicative property of determinants.

Row Operation I

Two rows of A are interchanged. Let E_{ij} be the elementary matrix of type I formed by interchanging the ith and jth rows of I. $E_{ij}A$ is the matrix formed by interchanging these same rows of A. If $\mathbf{k} = (k_1, k_2, \ldots, k_i, \ldots, k_j, \ldots, k_n)$ is an element of π_n and $\mathbf{k}^*$ is the permutation formed from $\mathbf{k}$ by interchanging k_i and k_j, then $\mathbf{k}$ and $\mathbf{k}^*$ differ in parity (see Exercise 2 of Section 1). It follows that

$$\begin{aligned} |A| &= \sum_{\mathbf{k}\in\pi_n} (-1)^{\delta(\mathbf{k})}(a_{1k_1}a_{2k_2}\cdots a_{ik_i}\cdots a_{jk_j}\cdots a_{nk_n}) \\ &= -\Sigma(-1)^{\delta(\mathbf{k}^*)}(a_{1k_1}a_{2k_2}\cdots a_{jk_j}\cdots a_{ik_i}\cdots a_{nk_n}) \\ &= -|E_{ij}A| \end{aligned}$$

In particular,

$$|E_{ij}| = |E_{ij}I| = -|I| = -1$$

Thus for any elementary matrix E of type I,

$$|EA| = -|A| = |E||A|$$

Row Operation II

A row of A is multiplied by a nonzero constant. Let E denote the elementary matrix of type II formed from I by multiplying the ith row by the nonzero constant α. If $|EA|$ is expanded by cofactors along the ith row, then

$$\begin{aligned} |EA| &= \alpha a_{i1}A_{i1} + \alpha a_{i2}A_{i2} + \cdots + \alpha a_{in}A_{in} \\ &= \alpha(a_{i1}A_{i1} + a_{i2}A_{i2} + \cdots + a_{in}A_{in}) \\ &= \alpha|A| \end{aligned}$$

In particular,

$$|E| = |EI| = \alpha|I| = \alpha$$

and hence

$$|EA| = \alpha|A| = |E||A|$$

Row Operation III

A multiple of one row is added to another row. Let E be the elementary matrix of type III formed from I by adding c times the ith row to the jth row. Since E is triangular and its diagonal elements are all 1, it follows that

$|E| = 1$. We will show that

$$|EA| = |A| = |E||A|$$

By definition,

$$\begin{aligned} |EA| &= \sum_{\mathbf{k} \in \pi_n} (-1)^{\delta(\mathbf{k})} (a_{1k_1} a_{2k_2} \cdots (a_{jk_j} + c a_{ik_i}) \cdots a_{nk_n}) \\ &= \sum_{\mathbf{k} \in \pi_n} (-1)^{\delta(\mathbf{k})} (a_{1k_1} \cdots a_{nk_n}) \\ &\quad + c \sum_{\mathbf{k} \in \pi_n} (-1)^{\delta(\mathbf{k})} (a_{1k_1} \cdots a_{ik_i} \cdots a_{ik_i} \cdots a_{nk_n}) \\ &= |A| + c|A^*| \end{aligned}$$

where

$$A^* = \begin{bmatrix} a_{11} & a_{12} & \cdots & a_{1n} \\ \vdots & & & \\ a_{i1} & a_{i2} & \cdots & a_{in} \\ \vdots & & & \\ a_{i1} & a_{i2} & \cdots & a_{in} \\ \vdots & & & \\ a_{n1} & a_{n2} & \cdots & a_{nn} \end{bmatrix}$$

Since the ith and jth rows of A^* are the same, it follows that $|A^*| = 0$ (see Exercise 9 of Section 1). Thus

$$|EA| = |A| = |E||A|$$

In summation, if E is an elementary matrix, then $|EA| = |E||A|$, where

$$|E| = \begin{cases} -1 & \text{if } E \text{ is of type I} \\ \alpha \neq 0 & \text{if } E \text{ is of type II} \\ 1 & \text{if } E \text{ is of type III} \end{cases}$$

Similar results hold for column operations. Indeed, if E is an elementary matrix, then

$$|AE| = |(AE)^T| = |E^T A^T| = |E^T||A^T| = |E||A|$$

Note that all elementary matrices have nonzero determinants. This observation can be used to prove the following theorem.

Theorem 2.2.1. *An $n \times n$ matrix A is singular if and only if $|A| = 0$.*

PROOF. The matrix A can be reduced to row echelon form with a finite number of row operations. Thus

$$U = E_k E_{k-1} \cdots E_1 A$$

where U is in row echelon form and the E_i's are all elementary matrices.

$$\begin{aligned} |U| &= |E_k E_{k-1} \cdots E_1 A| \\ &= |E||E_{k-1}| \cdots |E_1||A| \end{aligned}$$

Since the determinants of the E_i's are all nonzero, it follows that $|A| = 0$ if and only if $|U| = 0$. If A is singular, then U has a row consisting entirely of zeros and hence $|U| = 0$. If A is nonsingular, U is triangular with 1's along the diagonal and hence $|U| = 1$.

From the foregoing proof we can obtain a method for computing $|A|$. Reduce A to row echelon form.

$$U = E_k E_{k-1} \cdots E_1 A$$

If the last row of U consists entirely of zeros, A is singular and $|A| = 0$. Otherwise, A is nonsingular and

$$|A| = (|E_k||E_{k-1}| \cdots |E_1|)^{-1}$$

Actually, if A is nonsingular, it is simpler to reduce A to triangular form. This can be done using only row operations I and III. Thus

$$T = E_m E_{m-1} \cdots E_1 A$$

and hence

$$|A| = \pm|T| = \pm t_{11} t_{22} \cdots t_{nn}$$

The sign will be positive if row operation I has been used an even number of times and negative otherwise.

EXAMPLE 1. Evaluate

$$\begin{vmatrix} 2 & 1 & 3 \\ 4 & 2 & 1 \\ 6 & -3 & 4 \end{vmatrix}$$

SOLUTION

$$\begin{aligned} \begin{vmatrix} 2 & 1 & 3 \\ 4 & 2 & 1 \\ 6 & -3 & 4 \end{vmatrix} &= \begin{vmatrix} 2 & 1 & 3 \\ 0 & 0 & -5 \\ 0 & -6 & -5 \end{vmatrix} \\ &= (-1)\begin{vmatrix} 2 & 1 & 3 \\ 0 & -6 & -5 \\ 0 & 0 & -5 \end{vmatrix} \\ &= (-1)(2)(-6)(-5) \\ &= -60 \end{aligned}$$

We now have three methods for evaluating the determinant of an $n \times n$ matrix A. If $n > 3$ and A has nonzero entries, elimination is the most efficient method in the sense that it involves less arithmetic operations than the other two methods. In Table 2.2.1 the number of arithmetic operations involved in each of the methods is given for $n = 2, 3, 4, 5$. It is not difficult to derive general formulas for the number of operations in each of the methods (see Exercises 5, 6, and 7).

We have seen that for any elementary matrix E,

$$|EA| = |E||A| = |AE|$$

This is a special case of the following theorem.

***Theorem* 2.2.2.** *If A and B are $n \times n$ matrices, then $|AB| = |A||B|$.*

PROOF. If B is singular, it follows from Theorem 1.4.3 that AB is also singular, and hence $|AB| = 0 = |A||B|$. If B is nonsingular, B can be written as a product of elementary matrices. We have already seen that the result holds for elementary matrices. Thus

$$\begin{aligned} |AB| &= |AE_kE_{k-1} \cdots E_1| \\ &= |A||E_k||E_{k-1}| \cdots |E_1| \\ &= |A||E_kE_{k-1} \cdots E_1| \\ &= |A||B| \end{aligned}$$

If A is singular, the computed value of $|A|$ using exact arithmetic must be 0. This result, however, is unlikely if the computations are done by computer. Because of the finite number system of the computer, roundoff errors are usually unavoidable. Consequently, it is more likely that the computed value of $|A|$ will only be near 0. Because of roundoff errors, it is virtually impossible to determine that a matrix is singular from computer computations. In computer applications it is often more meaningful to ask whether a matrix is "close" to being singular. In general, the value of $|A|$ is not a good indicator of nearness to singularity. In Chapter 7 we will discuss how to determine whether or not a matrix is close to being singular.

n	Additions	Multiplications	Additions	Multiplications	Additions	Multiplications and Divisions
2	1	2	1	2	1	3
3	5	12	5	9	5	10
4	23	72	23	40	14	23
5	119	480	119	205	30	44
	Definition		Cofactors		Elimination	

EXERCISES

1. Evaluate each of the following determinants by inspection.

(a) $\begin{vmatrix} 0 & 0 & 3 \\ 0 & 4 & 1 \\ 2 & 3 & 1 \end{vmatrix}$ (b) $\begin{vmatrix} 1 & 1 & 1 & 3 \\ 0 & 3 & 1 & 1 \\ 0 & 0 & 2 & 2 \\ -1 & -1 & -1 & 2 \end{vmatrix}$

(c) $\begin{vmatrix} 0 & 0 & 0 & 1 \\ 1 & 0 & 0 & 0 \\ 0 & 1 & 0 & 0 \\ 0 & 0 & 1 & 0 \end{vmatrix}$

2. Let

$$A = \begin{pmatrix} 0 & 1 & 2 & 3 \\ 1 & 1 & 1 & 1 \\ -2 & -2 & 3 & 3 \\ 1 & 2 & -2 & -3 \end{pmatrix}$$

(a) Use the elimination method to evaluate $|A|$.
(b) Use the value of $|A|$ to evaluate

$$\begin{vmatrix} 0 & 1 & 2 & 3 \\ -2 & -2 & 3 & 3 \\ 1 & 2 & -2 & -3 \\ 1 & 1 & 1 & 1 \end{vmatrix} + \begin{vmatrix} 0 & 1 & 2 & 3 \\ 1 & 1 & 1 & 1 \\ -1 & -1 & 4 & 4 \\ 2 & 3 & -1 & -2 \end{vmatrix}$$

3. For each of the following, compute the determinant and state whether the matrix is singular or nonsingular.

(a) $\begin{pmatrix} 3 & 1 \\ 6 & 2 \end{pmatrix}$ (b) $\begin{pmatrix} 3 & 1 \\ 4 & 2 \end{pmatrix}$ (c) $\begin{pmatrix} 3 & 3 & 1 \\ 0 & 1 & 2 \\ 0 & 2 & 3 \end{pmatrix}$

(d) $\begin{pmatrix} 2 & 1 & 1 \\ 4 & 3 & 5 \\ 2 & 1 & 2 \end{pmatrix}$ (e) $\begin{pmatrix} 2 & -1 & 3 \\ -1 & 2 & -2 \\ 1 & 4 & 0 \end{pmatrix}$ (f) $\begin{pmatrix} 1 & 1 & 1 & 1 \\ 2 & -1 & 3 & 2 \\ 0 & 1 & 2 & 1 \\ 0 & 0 & 7 & 3 \end{pmatrix}$

4. Let A be an $n \times n$ matrix and α a scalar. Show that

$$|\alpha A| = \alpha^n |A|$$

5. How many additions and how many multiplications are necessary to evaluate the determinant of an $n \times n$ matrix directly using the definition of the determinant?

6. Show that the elimination method of computing the value of the determinant of an $n \times n$ matrix involves $[n(n-1)(2n-1)]/6$ additions and $[(n-1)(n^2+n+3)]/3$ multiplications and divisions.

7. Show that evaluating the determinant of an $n \times n$ matrix by cofactors involves $(n!-1)$ additions and $\sum_{k=1}^{n-1} n!/k!$ multiplications.

8. Let A be a nonsingular matrix. Show that

$$|A^{-1}| = \frac{1}{|A|}$$

3. CRAMER'S RULE

In this section we will learn a method for computing the inverse of a nonsingular matrix A using determinants. We will also learn a method for solving $A\mathbf{x} = \mathbf{b}$ using determinants. Both methods depend upon the following lemma.

Lemma 2.3.1. *Let A be an $n \times n$ matrix. If A_{jk} denotes the cofactor of a_{jk} for $k = 1, \ldots, n$, then*

$$(1) \qquad a_{i1}A_{j1} + a_{i2}A_{j2} + \cdots + a_{in}A_{jn} = \begin{cases} |A| & \text{if } i = j \\ 0 & \text{if } i \neq j \end{cases}$$

PROOF. If $i = j$, (1) is just the cofactor expansion of $|A|$ along the ith row of A. To prove (1) in the case $i \neq j$, let A^* be the matrix obtained by replacing the jth row of A by the ith row of A.

$$A^* = \begin{bmatrix} a_{11} & a_{12} & \cdots & a_{1n} \\ \vdots & & & \\ a_{i1} & a_{i2} & \cdots & a_{in} \\ \vdots & & & \\ a_{i1} & a_{i2} & \cdots & a_{in} \\ \vdots & & & \\ a_{n1} & a_{n2} & \cdots & a_{nn} \end{bmatrix} \begin{matrix} \\ \\ \\ \\ j\text{th row} \\ \\ \\ \\ \end{matrix}$$

Since two rows of A^* are the same, its determinant must be zero. It follows from the cofactor expansion of $|A^*|$ along the jth row that

$$\begin{aligned} 0 = |A^*| &= a_{i1}A_{j1}^* + a_{i2}A_{j2}^* + \cdots + a_{in}A_{jn}^* \\ &= a_{i1}A_{j1} + a_{i2}A_{j2} + \cdots + a_{in}A_{jn} \end{aligned}$$

The Adjoint of a Matrix

Let A be an $n \times n$ matrix. We can define a new matrix called the adjoint of A by

$$\text{adj } A = \begin{bmatrix} A_{11} & A_{21} & \cdots & A_{n1} \\ A_{12} & A_{22} & \cdots & A_{n2} \\ \vdots & & & \\ A_{1n} & A_{2n} & \cdots & A_{nn} \end{bmatrix}$$

Thus to form the adjoint, one must replace each term by its cofactor and then transpose the resulting matrix. It follows from (1) that

$$A(\text{adj } A) = |A|I$$

If A is nonsingular, $|A|$ is a nonzero scalar and we may write

$$A\left(\frac{1}{|A|}\text{adj } A\right) = I$$

Thus

$$A^{-1} = \frac{1}{|A|}\text{adj } A$$

EXAMPLE 1. For a 2×2 matrix

$$\text{adj } A = \begin{pmatrix} a_{22} & -a_{12} \\ -a_{21} & a_{11} \end{pmatrix}$$

If A is nonsingular, then

$$A^{-1} = \frac{1}{a_{11}a_{22} - a_{12}a_{21}}\begin{pmatrix} a_{22} & -a_{12} \\ -a_{21} & a_{11} \end{pmatrix}$$

EXAMPLE 2. Let

$$A = \begin{bmatrix} 2 & 1 & 2 \\ 3 & 2 & 2 \\ 1 & 2 & 3 \end{bmatrix}$$

Compute adj A and A^{-1}.

SOLUTION

$$\operatorname{adj} A = \begin{pmatrix} \begin{vmatrix} 2 & 2 \\ 2 & 3 \end{vmatrix} & -\begin{vmatrix} 3 & 2 \\ 1 & 3 \end{vmatrix} & \begin{vmatrix} 3 & 2 \\ 1 & 2 \end{vmatrix} \\ -\begin{vmatrix} 1 & 2 \\ 2 & 3 \end{vmatrix} & \begin{vmatrix} 2 & 2 \\ 1 & 3 \end{vmatrix} & -\begin{vmatrix} 2 & 1 \\ 1 & 2 \end{vmatrix} \\ \begin{vmatrix} 1 & 2 \\ 2 & 2 \end{vmatrix} & -\begin{vmatrix} 2 & 2 \\ 3 & 2 \end{vmatrix} & \begin{vmatrix} 2 & 1 \\ 3 & 2 \end{vmatrix} \end{pmatrix}^T$$

$$= \begin{pmatrix} 2 & 1 & -2 \\ -7 & 4 & 2 \\ 4 & -3 & 1 \end{pmatrix}$$

$$A^{-1} = \frac{1}{|A|} \operatorname{adj} A$$

$$= \tfrac{1}{5} \begin{pmatrix} 2 & 1 & -2 \\ -7 & 4 & 2 \\ 4 & -3 & 1 \end{pmatrix}$$

Using the formula

$$A^{-1} = \frac{1}{|A|} \operatorname{adj} A$$

we can derive a rule for representing the solution to the system $A\mathbf{x} = \mathbf{b}$ in terms of determinants.

Theorem 2.3.2 (Cramer's Rule). *Let A be an $n \times n$ nonsingular matrix and let $\mathbf{b} \in R^n$. Let A_i be the matrix obtained by replacing the ith column of A by $\mathbf{b}$. If $\mathbf{x}$ is the unique solution to $A\mathbf{x} = \mathbf{b}$, then*

$$x_i = \frac{|A_i|}{|A|} \qquad \text{for } i = 1, 2, \ldots, n$$

PROOF. Since

$$\mathbf{x} = A^{-1}\mathbf{b} = \frac{1}{|A|}(\operatorname{adj} A)\mathbf{b}$$

it follows that

$$x_i = \frac{b_1 A_{1i} + b_2 A_{2i} + \cdots + b_n A_{ni}}{|A|}$$

$$= \frac{|A_i|}{|A|}$$

EXAMPLE 3. Use Cramer's rule to solve

$$\begin{aligned} x_1 + 2x_2 + x_3 &= 5 \\ 2x_1 + 2x_2 + x_3 &= 6 \\ x_1 + 2x_2 + 3x_3 &= 9 \end{aligned}$$

SOLUTION

$$|A| = \begin{vmatrix} 1 & 2 & 1 \\ 2 & 2 & 1 \\ 1 & 2 & 3 \end{vmatrix} = -4 \qquad |A_1| = \begin{vmatrix} 5 & 2 & 1 \\ 6 & 2 & 1 \\ 9 & 2 & 3 \end{vmatrix} = -4$$

$$|A_2| = \begin{vmatrix} 1 & 5 & 1 \\ 2 & 6 & 1 \\ 1 & 9 & 3 \end{vmatrix} = -4 \qquad |A_3| = \begin{vmatrix} 1 & 2 & 5 \\ 2 & 2 & 6 \\ 1 & 2 & 9 \end{vmatrix} = -8$$

Therefore,

$$x_1 = \frac{-4}{-4} = 1, \qquad x_2 = \frac{-4}{-4} = 1, \qquad x_3 = \frac{-8}{-4} = 2$$

Cramer's rule gives us a convenient method for writing down the solution to an $n \times n$ system of linear equations in terms of determinants. To compute the solution, however, one must evaluate $n + 1$ determinants of order n. Evaluating even two of these determinants generally involves more computation than solving the system using Gaussian elimination.

EXERCISES

1. For each of the following compute (i) $|A|$, (ii) adj A, and (iii) A^{-1}.

(a) $A = \begin{pmatrix} 1 & 2 \\ 3 & -1 \end{pmatrix}$ (b) $A = \begin{pmatrix} 3 & 1 \\ 2 & 4 \end{pmatrix}$

(c) $A = \begin{pmatrix} 1 & 3 & 1 \\ 2 & 1 & 1 \\ -2 & 2 & -1 \end{pmatrix}$ (d) $A = \begin{pmatrix} 1 & 1 & 1 \\ 0 & 1 & 1 \\ 0 & 0 & 1 \end{pmatrix}$

2. Use Cramer's rule to solve each of the following systems.

(a) $\begin{aligned} x_1 + 2x_2 &= 3 \\ 3x_1 - x_2 &= 1 \end{aligned}$

(b) $\begin{aligned} x_1 + 3x_2 + x_3 &= 1 \\ 2x_1 + x_2 + x_3 &= 5 \\ -2x_1 + 2x_2 - x_3 &= -8 \end{aligned}$

(c) $\begin{aligned} x_1 + x_2 &= 0 \\ x_2 + x_3 - 2x_4 &= 1 \\ x_1 + 2x_3 + x_4 &= 0 \\ x_1 + x_2 + x_4 &= 0 \end{aligned}$

3. Let $\mathbf{x}$ and $\mathbf{y}$ be elements of R^3 and let $\mathbf{z}$ be the vector in R^3 whose coordinates are defined by

$$z_1 = \begin{vmatrix} x_2 & x_3 \\ y_2 & y_3 \end{vmatrix}, \qquad z_2 = -\begin{vmatrix} x_1 & x_3 \\ y_1 & y_3 \end{vmatrix}, \qquad z_3 = \begin{vmatrix} x_1 & x_2 \\ y_1 & y_2 \end{vmatrix}$$

Show that

$$\mathbf{z}^T\mathbf{x} = \mathbf{z}^T\mathbf{y} = 0$$

4. Let A be a nonsingular $n \times n$ matrix. Prove that

$$|\text{adj } A| = |A|^{n-1}$$

5. Show that if A is nonsingular, then adj A is nonsingular and

$$(\text{adj } A)^{-1} = |A^{-1}|A$$

6. Suppose that Q is a matrix with the property $Q^{-1} = Q^T$. Show that

$$q_{ij} = \frac{Q_{ij}}{|Q|}$$

3

Vector Spaces

The operations of addition and scalar multiplication are used in many diverse contexts in mathematics. Regardless of the context, however, these operations usually obey the same set of arithmetic rules. Thus a general theory of mathematical systems involving addition and scalar multiplication will have application to many areas in mathematics. Mathematical systems of this form are called vector spaces or linear spaces. In this chapter the definition of a vector space will be given and some of the general theory of vector spaces will be developed.

1. VECTOR SPACES: DEFINITION AND EXAMPLES

In this section we will present the formal definition of a vector space. Before doing this, however, it is instructive to look at a number of examples. We begin with the Euclidean vector spaces R^n.

Euclidean Vector Spaces

Perhaps the most elementary vector spaces are the Euclidean vector spaces R^n, $n = 1, 2, \ldots$. For simplicity, let us consider first R^2. Nonzero vectors in R^2 can be represented geometrically by directed line segments. This geometric representation will help us to visualize how the operations of scalar multiplication and addition work in R^2. Given a nonzero vector $\mathbf{x} = \begin{pmatrix} x_1 \\ x_2 \end{pmatrix}$, we can associate it with the line segment in the plane from (0, 0) to (x_1, x_2) (see Figure 3.1.1). If we equate line segments that have the same length and direction (Figure 3.1.2), $\mathbf{x}$ can be represented by any line segment from (a, b) to $(a + x_1, b + x_2)$. For example, the vector $\mathbf{x} = \begin{pmatrix} 2 \\ 1 \end{pmatrix}$ in R^2 could just as well be represented by the directed line segment from (2, 2) to (4, 3), or from $(-1, -1)$ to (1, 0), as shown in Figure 3.1.3.

We can think of the Euclidean length of a vector $\mathbf{x} = \begin{pmatrix} x_1 \\ x_2 \end{pmatrix}$ as the length of any directed line segment representing $\mathbf{x}$. The length of the line segment from (0, 0) to (x_1, x_2) is $\sqrt{x_1^2 + x_2^2}$ (see Figure 3.1.4). For each vector

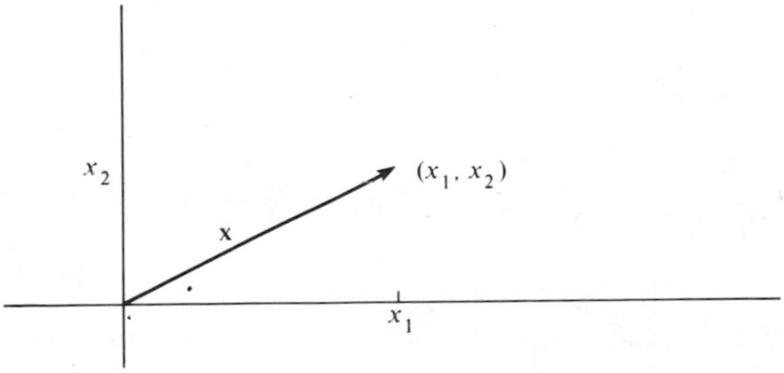

Figure 3.1.1

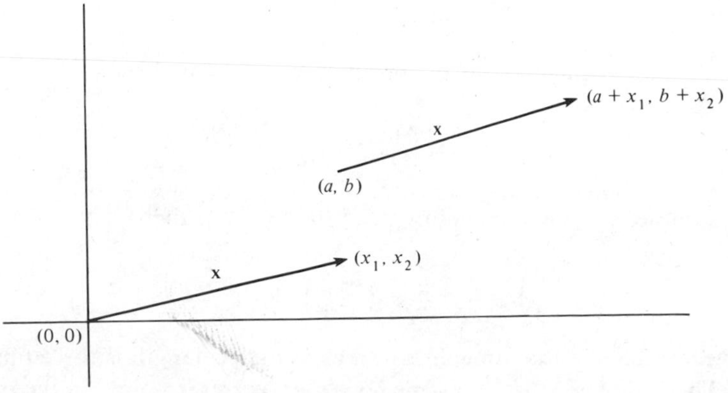

Figure 3.1.2

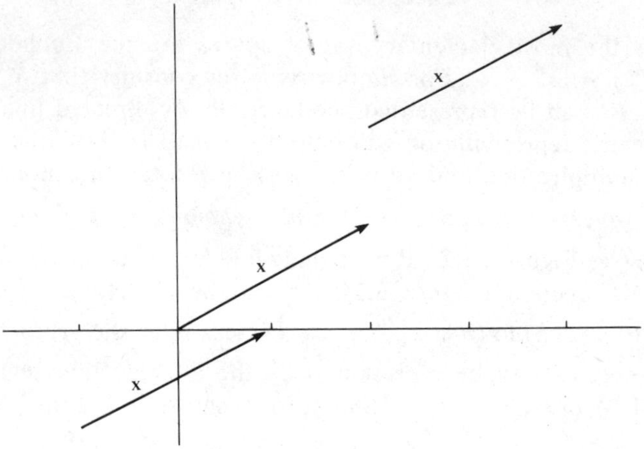

Figure 3.1.3

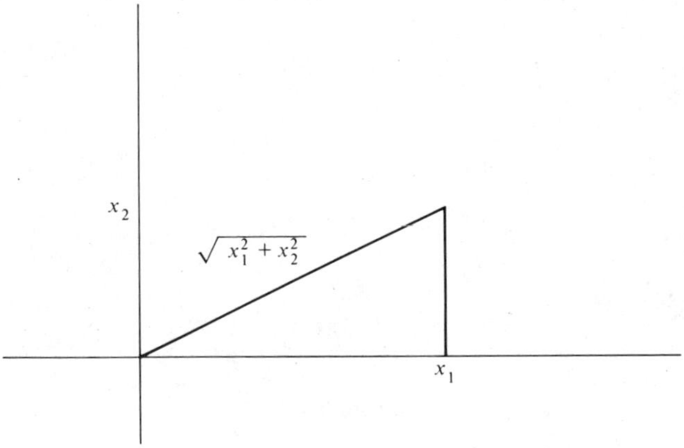

Figure 3.1.4

$\mathbf{x} = \begin{pmatrix} x_1 \\ x_2 \end{pmatrix}$ and each scalar α, the product $\alpha\mathbf{x}$ is defined by

$$\alpha \begin{pmatrix} x_1 \\ x_2 \end{pmatrix} = \begin{pmatrix} \alpha x_1 \\ \alpha x_2 \end{pmatrix}$$

For example, as shown in Figure 3.1.5, if $\mathbf{x} = \begin{pmatrix} 2 \\ 1 \end{pmatrix}$, then

$$3\mathbf{x} = \begin{pmatrix} 6 \\ 3 \end{pmatrix}, \qquad -\mathbf{x} = \begin{pmatrix} -2 \\ -1 \end{pmatrix}, \qquad -2\mathbf{x} = \begin{pmatrix} -4 \\ -2 \end{pmatrix}$$

The vector $3\mathbf{x}$ is in the same direction as $\mathbf{x}$, but its length is three times that of $\mathbf{x}$. The vector $-\mathbf{x}$ has the same length as $\mathbf{x}$, but it points in the opposite direction. The vector $-2\mathbf{x}$ is twice as long as $\mathbf{x}$ and it points in the same

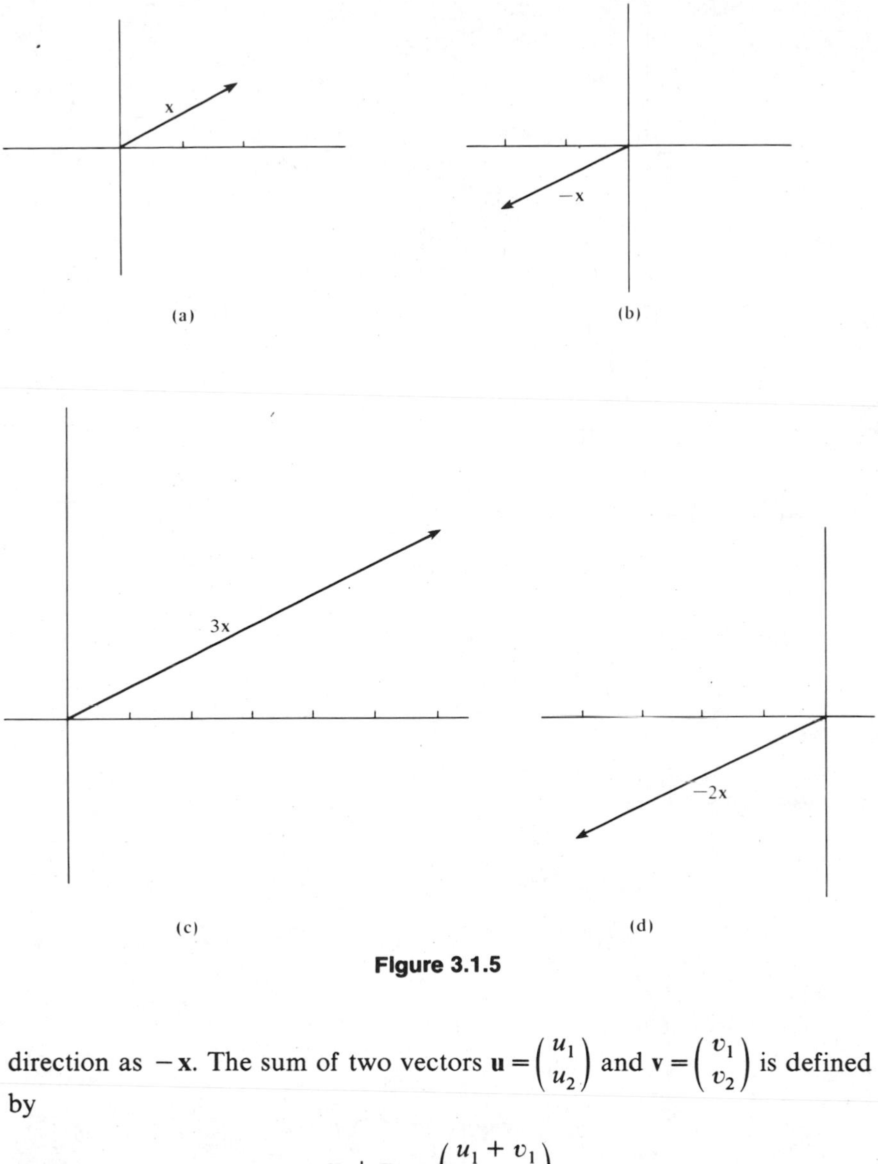

Figure 3.1.5

direction as $-\mathbf{x}$. The sum of two vectors $\mathbf{u} = \begin{pmatrix} u_1 \\ u_2 \end{pmatrix}$ and $\mathbf{v} = \begin{pmatrix} v_1 \\ v_2 \end{pmatrix}$ is defined by

$$\mathbf{u} + \mathbf{v} = \begin{pmatrix} u_1 + v_1 \\ u_2 + v_2 \end{pmatrix}$$

Note that if $\mathbf{v}$ is placed at the terminal point of $\mathbf{u}$, then $\mathbf{u} + \mathbf{v}$ is represented by the directed line segment from the initial point of $\mathbf{u}$ to the terminal point of $\mathbf{v}$ (Figure 3.1.6). If both $\mathbf{u}$ and $\mathbf{v}$ are placed at the origin and a parallelogram is formed as in Figure 3.1.7, the diagonals of the parallelogram will represent the sum $\mathbf{u} + \mathbf{v}$ and the difference $\mathbf{v} - \mathbf{u}$. In a similar manner, vectors in R^3 can be represented by directed line segments in 3-space (see Figure 3.1.8). In general, scalar multiplication and addition in

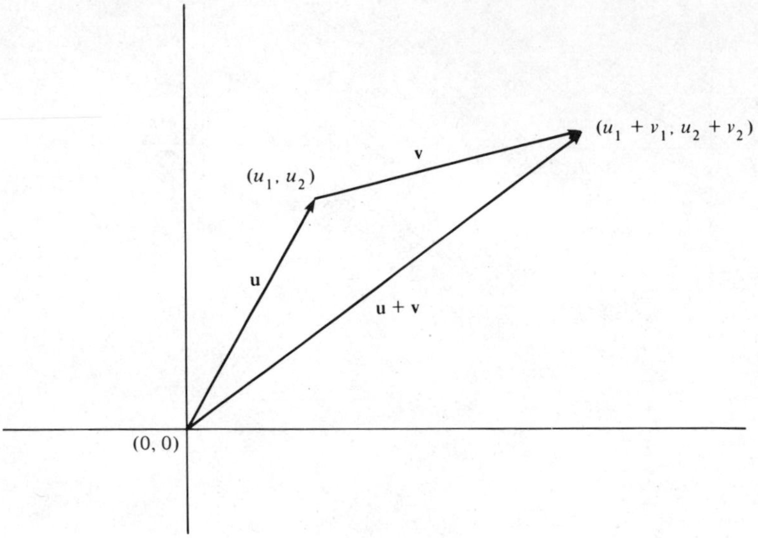

Figure 3.1.6

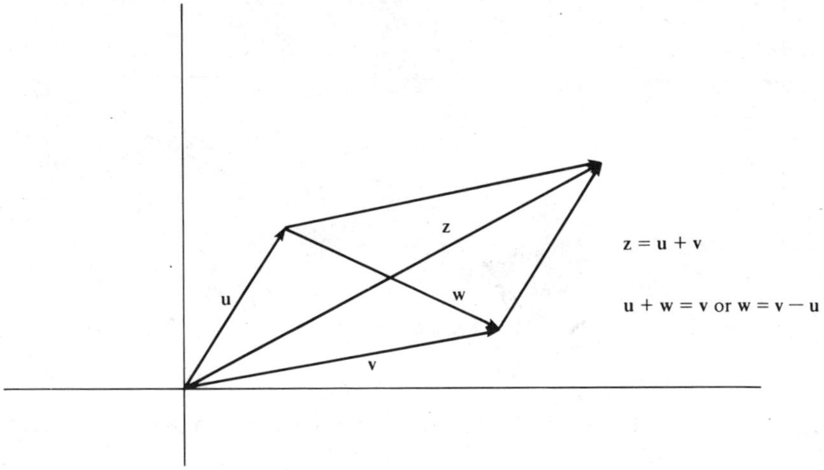

Figure 3.1.7

R^n are defined by

$$\alpha\mathbf{x} = \begin{bmatrix} \alpha x_1 \\ \alpha x_2 \\ \vdots \\ \alpha x_n \end{bmatrix} \quad \text{and} \quad \mathbf{x} + \mathbf{y} = \begin{bmatrix} x_1 + y_1 \\ x_2 + y_2 \\ \vdots \\ x_n + y_n \end{bmatrix}$$

for any $\mathbf{x}, \mathbf{y} \in R^n$ and any scalar α.

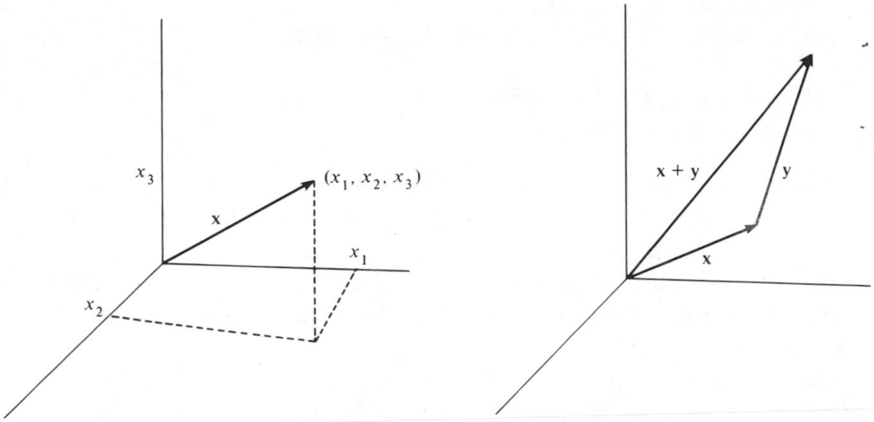

(a) (b)

Figure 3.1.8

The Vector Space $M_{m,n}$

We can also view R^n as the set of all $n \times 1$ matrices with real entries. The addition and scalar multiplication of vectors in R^n is just the usual addition and scalar multiplication of matrices. More generally, let $M_{m,n}$ denote the set of all $m \times n$ matrices with real entries. If $A = (a_{ij})$ and $B = (b_{ij})$, the sum $A + B$ is defined to be the $m \times n$ matrix $C = (c_{ij})$, where $c_{ij} = a_{ij} + b_{ij}$. Given a scalar α, one can define αA to be the $m \times n$ matrix whose ijth entry is αa_{ij}. Thus by defining operations on the set $M_{m,n}$, we have created a mathematical system. The operations of addition and scalar multiplication on $M_{m,n}$ obey certain arithmetic rules. These rules form the axioms that are used to define the concept of a vector space.

Definition. Let X be a set on which the operations of addition and scalar multiplication are defined. By this we mean that with each pair of elements $\mathbf{x}$ and $\mathbf{y}$ in X, one can associate a third element $\mathbf{x} + \mathbf{y}$ which is also in X, and with each element $\mathbf{x}$ in X and each scalar α one can associate an element $\alpha\mathbf{x}$ in X. The set X together with the operations of addition and scalar multiplication is said to form a **vector space** if the following axioms are satisfied.

A1: $\mathbf{x} + \mathbf{y} = \mathbf{y} + \mathbf{x}$ for any $\mathbf{x}$ and $\mathbf{y}$ in X.

A2: $(\mathbf{x} + \mathbf{y}) + \mathbf{z} = \mathbf{x} + (\mathbf{y} + \mathbf{z})$ for any $\mathbf{x}$, $\mathbf{y}$, $\mathbf{z}$ in X.

A3: There exists an element $\mathbf{0}$ in X such that $\mathbf{x} + \mathbf{0} = \mathbf{x}$ for each $\mathbf{x} \in X$.

A4: For each $\mathbf{x} \in X$, there exists an element $-\mathbf{x}$ in X such that $\mathbf{x} + (-\mathbf{x}) = \mathbf{0}$.

A5: $\alpha(\mathbf{x} + \mathbf{y}) = \alpha\mathbf{x} + \alpha\mathbf{y}$ for each real number α and any $\mathbf{x}$ and $\mathbf{y}$ in X.

A6: $(\alpha + \beta)\mathbf{x} = \alpha\mathbf{x} + \beta\mathbf{x}$ for any real numbers α and β and any $\mathbf{x} \in X$.

A7: $(\alpha\beta)\mathbf{x} + \alpha(\beta\mathbf{x})$ for any real numbers α and β and any $\mathbf{x} \in X$.

A8: $1 \cdot \mathbf{x} = \mathbf{x}$ for all $\mathbf{x} \in X$.

The elements of X are called *vectors* and the real numbers are called *scalars*. The symbol $\mathbf{0}$ was used in order to distinguish the zero vector from the scalar 0. In some contexts, complex numbers are used for scalars. However, in this text, scalars will usually be real numbers. Often the term *real vector space* is used to indicate that the set of scalars is the set of real numbers.

The reader may verify that R^n and $M_{m,n}$ with the usual addition and scalar multiplication of matrices are both vector spaces. There are a number of other important examples of vector spaces.

The Vector Space $C[a, b]$

Let $C[a, b]$ denote the set of all real-valued functions that are defined and continuous on the closed interval $[a, b]$. In this case our universal set is a set of functions. Thus our vectors are the functions in $C[a, b]$. The sum $f + g$ of two functions in $C[a, b]$ is defined by

$$(f + g)(x) = f(x) + g(x)$$

for all x in $[a, b]$. The new function $f + g$ is an element of $C[a, b]$, since the sum of two continuous functions is continuous. If f is a function in $C[a, b]$ and α is a real number, define αf by

$$(\alpha f)(x) = \alpha f(x)$$

for all x in $[a, b]$. Clearly, αf is in $C[a, b]$ since a constant times a continuous function is always continuous. Thus on $C[a, b]$ we have defined the operations of addition and scalar multiplication. The reader may verify that these operations satisfy the eight axioms of a vector space.

The Vector Space P_n

Let P_n denote the set of all polynomials of degree less than n. Define $p + q$ and αp by

$$(p + q)(x) = p(x) + q(x)$$

and

$$(\alpha p)(x) = \alpha p(x)$$

for all real numbers x. It is easily verified that axioms A1 through A8 hold. Thus P_n with the standard addition and scalar multiplication of functions is a vector space.

We close this section with a theorem that states three more fundamental properties of vector spaces.

Theorem 3.1.1. *If X is a vector space and $\mathbf{x}$ is any element of X, then*

(i) $0\mathbf{x} = \mathbf{0}$

(ii) $\mathbf{x} + \mathbf{y} = \mathbf{0}$ *implies that* $\mathbf{y} = -\mathbf{x}$ (*i.e., the additive inverse of* $\mathbf{x}$ *is unique*).

(iii) $(-1)\mathbf{x} = -\mathbf{x}$

PROOF. It follows from axioms A6 and A8 that

$$\mathbf{x} = 1\mathbf{x} = (1+0)\mathbf{x} = 1\mathbf{x} + 0\mathbf{x} = \mathbf{x} + 0\mathbf{x}$$

Thus

$$\text{(A2)} \qquad -\mathbf{x} + \mathbf{x} = -\mathbf{x} + (\mathbf{x} + 0\mathbf{x}) = (-\mathbf{x} + \mathbf{x}) + 0\mathbf{x}$$

$$\text{(A3 and A4)} \qquad \mathbf{0} = \mathbf{0} + 0\mathbf{x} = 0\mathbf{x}$$

To prove (ii), suppose that $\mathbf{x} + \mathbf{y} = \mathbf{0}$. Then

$$-\mathbf{x} = -\mathbf{x} + \mathbf{0} = -\mathbf{x} + (\mathbf{x} + \mathbf{y})$$

Therefore,

$$\text{(A2, A3, and A4)} \qquad -\mathbf{x} = (-\mathbf{x} + \mathbf{x}) + \mathbf{y} = \mathbf{0} + \mathbf{y} = \mathbf{y}$$

Finally, to prove (iii), note that

$$\text{((i) and A6)} \qquad \mathbf{0} = 0\mathbf{x} = (1 + (-1))\mathbf{x} = 1\mathbf{x} + (-1)\mathbf{x}$$

Thus

$$\text{(A8)} \qquad \mathbf{x} + (-1)\mathbf{x} = 0$$

and it follows from part (ii) that

$$(-1)\mathbf{x} = -\mathbf{x}$$

EXERCISES

1. Let C be the set of complex numbers. Define addition on C by

$$(a + bi) + (c + di) = (a + c) + (b + d)i$$

and define scalar multiplication by

$$\alpha(a + bi) = \alpha a + \alpha bi$$

for all real numbers α. Show that C is a vector space with these operations.

2. Let V be the set of all ordered pairs of real numbers with addition defined by

$$(x_1, x_2) + (y_1, y_2) = (x_1 + y_1, x_2 + y_2)$$

and scalar multiplication defined by

$$\alpha(x_1, x_2) = (\alpha x_1, x_2)$$

Is V a vector space with these operations? Justify your answer.

3. Show that $M_{m,n}$ with the usual addition and scalar multiplication of matrices satisfies the eight axioms of a vector space.

4. Show that $C[a, b]$ with the usual scalar multiplication and addition of functions satisfies the eight axioms of a vector space.

5. Let P be the set of all polynomials. Show that P with the usual addition and scalar multiplication of functions forms a vector space.

6. Show that the element $\mathbf{0}$ in a vector space is unique.

7. Let X be a vector space and let $\mathbf{x} \in X$.
   (a) Show that $\beta\mathbf{0} = \mathbf{0}$ for each scalar β.
   (b) Show that if $\alpha\mathbf{x} = \mathbf{0}$, then either $\alpha = 0$ or $\mathbf{x} = \mathbf{0}$.

8. Let S be the set of all ordered pairs of real numbers. Define scalar multiplication and addition on S by

$$\alpha(x_1, x_2) = (\alpha x_1, \alpha x_2)$$
$$(x_1, x_2) + (y_1, y_2) = (x_1 + y_1, 0)$$

Show that S is not a vector space. Which of the eight axioms fail to hold?

2. SUBSPACES

Given a vector space V, it is often possible to form another vector space by taking a subset S of V and using the operations of V. Since V is a vector space, the operations of addition and scalar multiplication always produce another vector in V. In order for a new system using a subset S of V as its universal set to be a vector space, the set S must be closed under the operations of addition and scalar multiplication; that is, the sum of two elements of S must always be an element of S and the product of a scalar and an element of S must always be an element of S.

EXAMPLE 1. Let $S = \left\{ \begin{pmatrix} x_1 \\ x_2 \end{pmatrix} \middle| x_2 = 2x_1 \right\}$. S is a subset of R^2. If $\begin{pmatrix} c \\ 2c \end{pmatrix}$ is any element of S and α is any scalar, then

$$\alpha \begin{pmatrix} c \\ 2c \end{pmatrix} = \begin{pmatrix} \alpha c \\ 2\alpha c \end{pmatrix}$$

which is an element of S. If $\begin{pmatrix} a \\ 2a \end{pmatrix}$ and $\begin{pmatrix} b \\ 2b \end{pmatrix}$ are any two elements of S, their sum

$$\begin{pmatrix} a + b \\ 2a + 2b \end{pmatrix} = \begin{pmatrix} a + b \\ 2(a + b) \end{pmatrix}$$

is also an element of S. It is easily seen that the mathematical system consisting of the set S (instead of R^2), together with the operations from R^2, is itself a vector space.

Definition. If S is a nonempty subset of a vector space V and S satisfies the following conditions:

(i) $\alpha \mathbf{x} \in S$ whenever $\mathbf{x} \in S$ for any scalar α

(ii) $\mathbf{x} + \mathbf{y} \in S$ whenever $\mathbf{x} \in S$ and $\mathbf{y} \in S$

then S is said to be a **subspace** of V.

Condition (i) says that S is closed under scalar multiplication; that is, whenever an element of S is multiplied by a scalar, the result is an element of S. Condition (ii) says that S is closed under addition; that is, the sum of two elements of S is always an element of S. Thus, if we do arithmetic using the operations from V and the elements of S, we will always end up with elements of S. A subspace of V, then, is a subset S that is closed under the operations of V.

Let S be a subspace of a vector space V. Using the operations of addition and scalar multiplication as defined on V, we can form a new mathematical system with S as the universal set. It is easily seen that all eight axioms will remain valid for this new system. Axioms A3 and A4 follow from Theorem 3.1.1 and condition (i) of the definition of a subspace. The remaining six axioms are valid for any elements of V, so in particular they are valid for the elements of S. Thus every subspace is a vector space in its own right.

EXAMPLE 2. Let $S = \{(x_1, x_2, x_3)^T | x_1 = x_2\}$. It follows that S is a subspace of R^3, since

(i) If $\mathbf{x} = (a, a, b)^T \in S$, then

$$\alpha \mathbf{x} = (\alpha a, \alpha a, \alpha b)^T \in S$$

(ii) If $(a, a, b)^T$ and $(c, c, d)^T$ are arbitrary elements of S, then

$$(a, a, b)^T + (c, c, d)^T = (a + c, a + c, b + d)^T \in S$$

EXAMPLE 3. Let $S = \left\{ \begin{pmatrix} x \\ 1 \end{pmatrix} \middle| \, x \text{ is a real number} \right\}$. S is not a subspace of R^2. In this case both conditions fail. S is not closed under scalar multiplication, since $\alpha \begin{pmatrix} x \\ 1 \end{pmatrix} \notin S$ unless $\alpha = 1$. S is not closed under addition, since

$$\begin{pmatrix} x \\ 1 \end{pmatrix} + \begin{pmatrix} y \\ 1 \end{pmatrix} = \begin{pmatrix} x+y \\ 2 \end{pmatrix} \notin S$$

EXAMPLE 4. Let V be any vector space. Let $S = \{\mathbf{0}\}$. We leave it to the reader to verify that S is a subspace of V. The reader may also verify that V is a subspace of itself.

EXAMPLE 5. Let S be the set of all polynomials of degree less than n with the property $p(0) = 0$. We claim S is a subspace of P_n. This follows, since:

(i) If $p(x) \in S$ and α is a scalar, then

$$\alpha p(0) = \alpha \cdot 0 = 0$$

and hence $\alpha p \in S$.

(ii) If $p(x)$ and $q(x)$ are elements of S, then

$$(p+q)(0) = p(0) + q(0) = 0 + 0 = 0$$

and hence $p + q \in S$.

EXAMPLE 6. Let $C^n[a, b]$ be the set of all functions f that have a continuous nth derivative on $[a, b]$. We leave it to the reader to verify that $C^n[a, b]$ is a subspace of $C[a, b]$.

EXAMPLE 7. Let A be the set of all f in $C^2[a, b]$ such that

$$f''(x) + f(x) = 0$$

for all x in $[a, b]$. If $f \in A$ and α is any scalar, then for any $x \in [a, b]$,

$$(\alpha f)''(x) + (\alpha f)(x) = \alpha f''(x) + \alpha f(x) = \alpha(f''(x) + f(x)) = \alpha \cdot 0 = 0$$

Thus $\alpha f \in A$. If f and g are both in A, then

$$\begin{aligned} (f+g)''(x) + (f+g)(x) &= f''(x) + g''(x) + f(x) + g(x) \\ &= [f''(x) + f(x)] + [g''(x) + g(x)] \\ &= 0 + 0 = 0 \end{aligned}$$

Thus the set of all solutions on $[a, b]$ to the differential equation $y'' + y = 0$ forms a subspace of $C^2[a, b]$. If we note that $f(x) = \sin x$ and $g(x) = \cos x$ are both in A, it follows that any function of the form $c_1 \sin x + c_2 \cos x$ must also be in A. One can easily verify that functions of this form are solutions to $y'' + y = 0$.

The Nullspace of a Matrix

Let A be an $m \times n$ matrix. Let $N(A)$ denote the set of all solutions to the homogeneous system $A\mathbf{x} = \mathbf{0}$. Thus

$$N(A) = \{\mathbf{x} \in R^n \mid A\mathbf{x} = \mathbf{0}\}$$

We claim that $N(A)$ is a subspace of R^n. If $\mathbf{x} \in N(A)$ and α is a scalar, then

$$A(\alpha\mathbf{x}) = \alpha A\mathbf{x} = \alpha\mathbf{0} = \mathbf{0}$$

and hence $\alpha\mathbf{x} \in N(A)$. If $\mathbf{x}$ and $\mathbf{y}$ are elements of $N(A)$, then

$$A(\mathbf{x}+\mathbf{y}) = A\mathbf{x} + A\mathbf{y} = \mathbf{0} + \mathbf{0} = \mathbf{0}$$

Therefore, $\mathbf{x}+\mathbf{y} \in N(A)$. It follows then that $N(A)$ is a subspace of R^n. The set of all solutions to the homogeneous system $A\mathbf{x} = \mathbf{0}$ forms a subspace of R^n. The subspace $N(A)$ is called the *nullspace* of A.

EXAMPLE 8. Determine $N(A)$ if

$$A = \begin{pmatrix} 1 & 1 & 1 & 0 \\ 2 & 1 & 0 & 1 \end{pmatrix}$$

SOLUTION. Using Gauss–Jordan reduction to solve $A\mathbf{x} = \mathbf{0}$, we obtain

$$\left(\begin{array}{cccc|c} 1 & 1 & 1 & 0 & 0 \\ 2 & 1 & 0 & 1 & 0 \end{array}\right) \rightarrow \left(\begin{array}{cccc|c} 1 & 1 & 1 & 0 & 0 \\ 0 & -1 & -2 & 1 & 0 \end{array}\right) \rightarrow \left(\begin{array}{cccc|c} 1 & 0 & -1 & 1 & 0 \\ 0 & -1 & -2 & 1 & 0 \end{array}\right) \rightarrow \left(\begin{array}{cccc|c} 1 & 0 & -1 & 1 & 0 \\ 0 & 1 & 2 & -1 & 0 \end{array}\right)$$

The reduced row echelon form involves two free variables, x_3 and x_4.

$$\begin{aligned} x_1 &= x_3 - x_4 \\ x_2 &= -2x_3 + x_4 \end{aligned}$$

Thus if we set $x_3 = \alpha$ and $x_4 = \beta$, then

$$\mathbf{x} = \begin{bmatrix} \alpha - \beta \\ -2\alpha + \beta \\ \alpha \\ \beta \end{bmatrix} = \alpha \begin{bmatrix} 1 \\ -2 \\ 1 \\ 0 \end{bmatrix} + \beta \begin{bmatrix} -1 \\ 1 \\ 0 \\ 1 \end{bmatrix}$$

is a solution to $A\mathbf{x} = \mathbf{0}$. The vector space $N(A)$ consists of all vectors of the form

$$\alpha \begin{bmatrix} 1 \\ -2 \\ 1 \\ 0 \end{bmatrix} + \beta \begin{bmatrix} -1 \\ 1 \\ 0 \\ 1 \end{bmatrix}$$

where α and β are scalars.

Definition. Let $\mathbf{v}_1, \mathbf{v}_2, \ldots, \mathbf{v}_n$ be vectors in a vector space V. A sum of the form $\alpha_1\mathbf{v}_1 + \alpha_2\mathbf{v}_2 + \cdots + \alpha_n\mathbf{v}_n$, where $\alpha_1, \ldots, \alpha_n$ are scalars, is called a **linear combination** of $\mathbf{v}_1, \mathbf{v}_2, \ldots, \mathbf{v}_n$. The set of all linear combinations of $\mathbf{v}_1, \mathbf{v}_2, \ldots, \mathbf{v}_n$ is called the **span** of $\mathbf{v}_1, \ldots, \mathbf{v}_n$. The span of $\mathbf{v}_1, \ldots, \mathbf{v}_n$ will be denoted by $\mathcal{S}(\mathbf{v}_1, \ldots, \mathbf{v}_n)$.

In Example 8, we saw that the nullspace of A was the span of the vectors $(1, -2, 1, 0)^T$ and $(-1, 1, 0, 1)^T$.

EXAMPLE 9. In R^3 the span of $\mathbf{e}_1$ and $\mathbf{e}_2$ is the set of all vectors of the form

$$\alpha\mathbf{e}_1 + \beta\mathbf{e}_2 = \begin{pmatrix} \alpha \\ \beta \\ 0 \end{pmatrix}$$

The reader may verify that $\mathcal{S}(\mathbf{e}_1, \mathbf{e}_2)$ is a subspace of R^3. The span of $\mathbf{e}_1, \mathbf{e}_2, \mathbf{e}_3$ is the set of all vectors of the form

$$\alpha_1\mathbf{e}_1 + \alpha_2\mathbf{e}_2 + \alpha_3\mathbf{e}_3 = \begin{pmatrix} \alpha_1 \\ \alpha_2 \\ \alpha_3 \end{pmatrix}$$

Thus $\mathcal{S}(\mathbf{e}_1, \mathbf{e}_2, \mathbf{e}_3) = R^3$.

Theorem 3.2.1. *If $\mathbf{v}_1, \mathbf{v}_2, \ldots, \mathbf{v}_n$ are elements of a vector space V, then $\mathcal{S}(\mathbf{v}_1, \mathbf{v}_2, \ldots, \mathbf{v}_n)$ is a subspace of V.*

PROOF. Let β be a scalar and let $\mathbf{v} = \alpha_1\mathbf{v}_1 + \alpha_2\mathbf{v}_2 + \cdots + \alpha_n\mathbf{v}_n$ be an arbitrary element of $\mathcal{S}(\mathbf{v}_1, \mathbf{v}_2, \ldots, \mathbf{v}_n)$. Since

$$\beta\mathbf{v} = (\beta\alpha_1)\mathbf{v}_1 + (\beta\alpha_2)\mathbf{v}_2 + \cdots + (\beta\alpha_n)\mathbf{v}_n$$

it follows that $\beta\mathbf{v} \in \mathcal{S}(\mathbf{v}_1, \ldots, \mathbf{v}_n)$. Next we must show that any sum of elements of $\mathcal{S}(\mathbf{v}_1, \ldots, \mathbf{v}_n)$ is in $\mathcal{S}(\mathbf{v}_1, \ldots, \mathbf{v}_n)$. Let $\mathbf{v} = \alpha_1\mathbf{v}_1 + \cdots + \alpha_n\mathbf{v}_n$ and $\mathbf{w} = \beta_1\mathbf{v}_1 + \cdots + \beta_n\mathbf{v}_n$,

$$\mathbf{v} + \mathbf{w} = (\alpha_1 + \beta_1)\mathbf{v}_1 + \cdots + (\alpha_n + \beta_n)\mathbf{v}_n \in \mathcal{S}(\mathbf{v}_1, \ldots, \mathbf{v}_n)$$

Therefore, $\mathcal{S}(\mathbf{v}_1, \ldots, \mathbf{v}_n)$ is a subspace of V.

We will refer to $\mathcal{S}(\mathbf{v}_1, \ldots, \mathbf{v}_n)$ as the subspace spanned by $\mathbf{v}_1, \mathbf{v}_2, \ldots, \mathbf{v}_n$. It many happen that $\mathcal{S}(\mathbf{v}_1, \ldots, \mathbf{v}_n) = V$, in which case we say that $\mathbf{v}_1, \ldots, \mathbf{v}_n$ span V or that $\{\mathbf{v}_1, \ldots, \mathbf{v}_n\}$ is a spanning set for V. Thus we have the following definition.

Definition. The set $\{\mathbf{v}_1, \ldots, \mathbf{v}_n\}$ is a **spanning set** for V if and only if every vector in V can be written as a linear combination of $\mathbf{v}_1, \mathbf{v}_2, \ldots, \mathbf{v}_n$.

EXAMPLE 10. Which of the following are spanning sets for R^3?
(a) $\{\mathbf{e}_1, \mathbf{e}_2, \mathbf{e}_3, (1, 2, 3)^T\}$
(b) $\{(1, 1, 1)^T, (1, 1, 0)^T, (1, 0, 0)^T\}$
(c) $\{(1, 0, 1)^T, (0, 1, 0)^T\}$
(d) $\{(1, 2, 4)^T, (2, 1, 3)^T, (4, -1, 1)^T\}$

SOLUTION. To determine whether a set spans R^3, one must determine whether an arbitrary vector $(a, b, c)^T$ in R^3 can be written as a linear combination of the vectors in the set. In part (a) it is easily seen that $(a, b, c)^T$ can be written as

$$(a, b, c)^T = a\mathbf{e}_1 + b\mathbf{e}_2 + c\mathbf{e}_3 + 0(1, 2, 3)^T$$

For part (b) we must determine whether or not it is possible to find constants $\alpha_1, \alpha_2, \alpha_3$ such that

$$\begin{pmatrix} a \\ b \\ c \end{pmatrix} = \alpha_1 \begin{bmatrix} 1 \\ 1 \\ 1 \end{bmatrix} + \alpha_2 \begin{bmatrix} 1 \\ 1 \\ 0 \end{bmatrix} + \alpha_3 \begin{bmatrix} 1 \\ 0 \\ 0 \end{bmatrix}$$

This leads to the system of equations

$$\begin{aligned} \alpha_1 + \alpha_2 + \alpha_3 &= a \\ \alpha_1 + \alpha_2 \qquad &= b \\ \alpha_1 \qquad\qquad &= c \end{aligned}$$

Since the coefficient matrix of the system is nonsingular, the system has a unique solution

$$\begin{bmatrix} \alpha_1 \\ \alpha_2 \\ \alpha_3 \end{bmatrix} = \begin{pmatrix} c \\ b - c \\ a - b \end{pmatrix}$$

Thus

$$\begin{pmatrix} a \\ b \\ c \end{pmatrix} = c \begin{bmatrix} 1 \\ 1 \\ 1 \end{bmatrix} + (b - c) \begin{bmatrix} 1 \\ 1 \\ 0 \end{bmatrix} + (a - b) \begin{bmatrix} 1 \\ 0 \\ 0 \end{bmatrix}$$

so the three vectors span R^3. For part (c), one should note that linear combinations of $(1, 0, 1)^T$ and $(0, 1, 0)^T$ produce vectors of the form $(\alpha, \beta, \alpha)^T$. Thus any vector $(a, b, c)^T$ in R^3 where $a \neq c$ would not be in the span of these two vectors. Part (d) can be done in the same manner as part (b). If

$$\begin{pmatrix} a \\ b \\ c \end{pmatrix} = \alpha_1 \begin{bmatrix} 1 \\ 2 \\ 4 \end{bmatrix} + \alpha_2 \begin{bmatrix} 2 \\ 1 \\ 3 \end{bmatrix} + \alpha_3 \begin{bmatrix} 4 \\ -1 \\ 1 \end{bmatrix}$$

then

$$\begin{aligned} \alpha_1 + 2\alpha_2 + 4\alpha_3 &= a \\ 2\alpha_1 + \alpha_2 - \alpha_3 &= b \\ 4\alpha_1 + 3\alpha_2 + \alpha_3 &= c \end{aligned}$$

In this case, however, the coefficient matrix of the system is singular. Thus it is possible to choose a, b, and c so that the system is inconsistent and hence the vectors fail to span R^3.

EXAMPLE 11. The vectors $1 - x^2$, $x + 2$, and x^2 span P_3. If $ax^2 + bx + c$ is any polynomial in P_3, then it is possible to find scalars α_1, α_2, and α_3 such that

$$ax^2 + bx + c = \alpha_1(1 - x^2) + \alpha_2(x + 2) + \alpha_3 x^2$$

Indeed,

$$\alpha_1(1 - x^2) + \alpha_2(x + 2) + \alpha_3 x^2 = (\alpha_3 - \alpha_1)x^2 + \alpha_2 x + (\alpha_1 + 2\alpha_2)$$

Setting

$$\begin{aligned} \alpha_3 - \alpha_1 &= a \\ \alpha_2 &= b \\ \alpha_1 + 2\alpha_2 &= c \end{aligned}$$

and solving yields $\alpha_1 = c - 2b$, $\alpha_2 = b$, and $\alpha_3 = a + c - 2b$.

In Example 10(a) we saw that the vectors $\mathbf{e}_1$, $\mathbf{e}_2$, $\mathbf{e}_3$, $(1, 2, 3)^T$ span R^3. Clearly, R^3 could be spanned using only the vectors $\mathbf{e}_1$, $\mathbf{e}_2$, $\mathbf{e}_3$. The vector $(1, 2, 3)^T$ is really not necessary. In the next section we will consider the problem of finding minimal spanning sets for a vector space V (i.e., spanning sets that contain the smallest possible number of vectors).

EXERCISES

1. Determine whether or not the following sets are subspaces of R^3.
(a) $\{(x_1, x_2, x_3)^T \mid x_1 + x_3 = 1\}$
(b) $\{(x_1, x_2, x_3)^T \mid x_1 = x_2 = x_3\}$
(c) $\{(x_1, x_2, x_3)^T \mid x_3 = x_1 + x_2\}$
(d) $\{(x_1, x_2, x_3)^T \mid x_3 = x_1^2 + x_2^2\}$

2. Determine whether or not the following are subspaces of $M_{2,2}$.
(a) The set of all 2×2 diagonal matrices
(b) The set of all 2×2 lower triangular matrices
(c) The set of all 2×2 matrices A such that $a_{12} = 1$

(d) The set of all 2×2 matrices B such that $b_{11} = 0$
(e) The set of all symmetric 2×2 matrices
(f) The set of all nonsingular 2×2 matrices

3. Determine the nullspace of each of the following matrices.

(a) $\begin{pmatrix} 2 & 1 \\ 3 & 2 \end{pmatrix}$ (b) $\begin{pmatrix} 1 & 2 & -3 & -1 \\ -2 & -4 & 6 & 3 \end{pmatrix}$

(c) $\begin{bmatrix} 1 & 3 & -4 \\ 2 & -1 & -1 \\ -1 & -3 & 4 \end{bmatrix}$ (d) $\begin{bmatrix} 1 & 1 & -1 & 2 \\ 2 & 2 & -3 & 1 \\ -1 & -1 & 0 & -5 \end{bmatrix}$

4. Prove that if S is a subspace of R^1, either $S = \{\mathbf{0}\}$ or $S = R^1$.

5. Determine whether or not the following are subspaces of P_4. (Be careful!)
(a) The set of polynomials in P_4 of even degree
(b) The set of all polynomials of degree 3
(c) The set of all polynomials $p(x)$ in P_4 such that $p(0) = 0$
(d) The set of all polynomials in P_4 having at least one real root

6. Determine whether or not the following are subspaces of $C[-1, 1]$.
(a) The set of functions f in $C[-1, 1]$ such that $f(-1) = f(1)$
(b) The set of odd functions in $C[-1, 1]$
(c) The set of continuous nondecreasing functions of $[-1, 1]$
(d) The set of functions f in $C[-1, 1]$ such that $f(-1) = 0$ *and* $f(1) = 0$
(e) The set of functions f in $C[-1, 1]$ such that $f(-1) = 0$ *or* $f(1) = 0$

7. Determine whether or not the following are spanning sets for R^2.

(a) $\left\{\begin{pmatrix} 2 \\ 1 \end{pmatrix}, \begin{pmatrix} 3 \\ 2 \end{pmatrix}\right\}$ (b) $\left\{\begin{pmatrix} 2 \\ 3 \end{pmatrix}, \begin{pmatrix} 4 \\ 6 \end{pmatrix}\right\}$

(c) $\left\{\begin{pmatrix} -2 \\ 1 \end{pmatrix}, \begin{pmatrix} 1 \\ 3 \end{pmatrix}, \begin{pmatrix} 2 \\ 4 \end{pmatrix}\right\}$ (d) $\left\{\begin{pmatrix} -1 \\ 2 \end{pmatrix}, \begin{pmatrix} 1 \\ -2 \end{pmatrix}, \begin{pmatrix} 2 \\ -4 \end{pmatrix}\right\}$

(e) $\left\{\begin{pmatrix} 1 \\ 2 \end{pmatrix}, \begin{pmatrix} -1 \\ 1 \end{pmatrix}\right\}$

8. Which of the following are spanning sets for R^3? Justify your answers.
(a) $\{(1, 0, 0)^T, (0, 1, 1)^T, (1, 0, 1)^T\}$
(b) $\{(1, 0, 0)^T, (0, 1, 1)^T, (1, 0, 1)^T, (1, 2, 3)^T\}$
(c) $\{(2, 1, -2)^T, (3, 2, -2)^T, (2, 2, 0)^T\}$
(d) $\{(2, 1, -2)^T, (-2, -1, 2)^T, (4, 2, -4)^T\}$
(e) $\{(1, 1, 3)^T, (0, 2, 1)^T\}$

9. Which of the following are spanning sets for P_3? Justify your answers.
(a) $\{1, x^2, x^2 - 2\}$ (b) $\{2, x^2, x, 2x + 3\}$
(c) $\{x + 2, x + 1, x^2 - 1\}$ (d) $\{x + 2, x^2 - 1\}$

10. Let U and V be subspaces of a vector space W. Prove that $U \cap V$ is also a subspace of W.

11. Let S be the subspace of R^2 spanned by $\mathbf{e}_1$ and let T be the subspace of R^2 spanned by $\mathbf{e}_2$. Is $S \cup T$ a subspace of R^2? Explain.

12. Let U and V be subspaces of a vector space W. Define

$$U + V = \{\mathbf{z} | \mathbf{z} = \mathbf{u} + \mathbf{v} \text{ where } \mathbf{u} \in U \text{ and } \mathbf{v} \in V\}$$

Show that $U + V$ is a subspace of W.

13. Let A be an $n \times n$ matrix. Prove that the following statements are equivalent.
(a) $N(A) = \{0\}$
(b) A is nonsingular.
(c) For each $\mathbf{b} \in R^n$, the system $A\mathbf{x} = \mathbf{b}$ has a unique solution.

3. LINEAR INDEPENDENCE

In this section we will look more closely at the structure of vector spaces. To begin with, we will restrict ourselves to vector spaces that can be generated from a finite set of elements. Each vector in the vector space can be built up from the elements in this generating set using only the operations of addition and scalar multiplication. The generating set is usually referred to as a spanning set. In particular, it is desirable to find a "minimal" spanning set. To do this it is necessary to consider how the vectors in the collection "depend" on each other. Consequently, we introduce the concepts of linear dependence and linear independence. These simple concepts provide the keys to understanding the structure of vector spaces.

Before presenting any formal definitions we make two observations.

(i) If $\mathbf{v}_1, \mathbf{v}_2, \ldots, \mathbf{v}_n$ span a vector space V and one of these vectors can be written as a linear combination of the other $n - 1$ vectors, then those $n - 1$ vectors span V.
(ii) Given n vectors $\mathbf{v}_1, \ldots, \mathbf{v}_n$, it is possible to write one of the vectors as a linear combination of the other $n - 1$ vectors if and only if there exist scalars $c_1, \ldots, c_n$ not all zero such that

$$c_1\mathbf{v}_1 + c_2\mathbf{v}_2 + \cdots + c_n\mathbf{v}_n = 0$$

PROOF OF (i). Suppose that $\mathbf{v}_n$ can be written as a linear combination of $\mathbf{v}_1, \mathbf{v}_2, \ldots, \mathbf{v}_{n-1}$.

$$\mathbf{v}_n = \beta_1\mathbf{v}_1 + \beta_2\mathbf{v}_2 + \cdots + \beta_n\mathbf{v}_{n-1}$$

Let $\mathbf{v}$ be any element of V. Since $\mathbf{v}_1, \ldots, \mathbf{v}_n$ span V, we can write

$$\begin{aligned}\mathbf{v} &= \alpha_1\mathbf{v}_1 + \alpha_2\mathbf{v}_2 + \cdots + \alpha_{n-1}\mathbf{v}_{n-1} + \alpha_n\mathbf{v}_n \\ &= \alpha_1\mathbf{v}_1 + \alpha_2\mathbf{v}_2 + \cdots + \alpha_{n-1}\mathbf{v}_{n-1} + \alpha_n(\beta_1\mathbf{v}_1 + \cdots + \beta_{n-1}\mathbf{v}_{n-1}) \\ &= (\alpha_1 + \alpha_n\beta_1)\mathbf{v}_1 + (\alpha_2 + \alpha_n\beta_2)\mathbf{v}_2 + \cdots + (\alpha_{n-1} + \alpha_n\beta_{n-1})\mathbf{v}_{n-1}\end{aligned}$$

Thus any vector $\mathbf{v}$ in V can be written as a linear combination of $\mathbf{v}_1, \mathbf{v}_2, \ldots, \mathbf{v}_{n-1}$, and hence these vectors span V.

Proof of (ii). Suppose that one of the vectors $\mathbf{v}_1, \mathbf{v}_2, \ldots, \mathbf{v}_n$, say $\mathbf{v}_n$, can be written as a linear combination of the others.

$$\mathbf{v}_n = \alpha_1\mathbf{v}_1 + \alpha_2\mathbf{v}_2 + \cdots + \alpha_{n-1}\mathbf{v}_{n-1}$$

If we set $c_i = \alpha_i$ for $i = 1, \ldots, n-1$ and $c_n = -1$, then it follows that

$$\sum_{i=1}^{n} c_i\mathbf{v}_i = \sum_{i=1}^{n-1} \alpha_i\mathbf{v}_i - \sum_{i=1}^{n-1} \alpha_i\mathbf{v}_i = \mathbf{0}$$

Conversely, if

$$c_1\mathbf{v}_1 + c_2\mathbf{v}_2 + \cdots + c_n\mathbf{v}_n = \mathbf{0}$$

and at least one of the c_i's, say c_n, is nonzero, then

$$\mathbf{v}_n = \frac{-c_1}{c_n}\mathbf{v}_1 + \frac{-c_2}{c_n}\mathbf{v}_2 + \cdots + \frac{-c_{n-1}}{c_n}\mathbf{v}_{n-1}$$

Definition. The vectors $\mathbf{v}_1, \mathbf{v}_2, \ldots, \mathbf{v}_n$ in a vector space V are said to be **linearly independent** if

$$c_1\mathbf{v}_1 + c_2\mathbf{v}_2 + \cdots + c_n\mathbf{v}_n = \mathbf{0}$$

implies that all of the scalars $c_1, \ldots, c_n$ must equal 0.

It follows from (i) and (ii) that if $\{\mathbf{v}_1, \mathbf{v}_1, \ldots, \mathbf{v}_n\}$ is a minimal spanning set, then $\mathbf{v}_1, \mathbf{v}_2, \ldots, \mathbf{v}_n$ are linearly independent. Conversely, if $\mathbf{v}_1, \ldots, \mathbf{v}_n$ are linearly independent and span V, then $\{\mathbf{v}_1, \ldots, \mathbf{v}_n\}$ is a minimal spanning set for V (see Exercise 10). A minimal spanning set is called a *basis*. The concept of a basis will be studied in more detail in the next section.

EXAMPLE 1. The vectors $\begin{pmatrix}1\\1\end{pmatrix}$ and $\begin{pmatrix}1\\2\end{pmatrix}$ are linearly independent, since if

$$c_1\begin{pmatrix}1\\1\end{pmatrix} + c_2\begin{pmatrix}1\\2\end{pmatrix} = \begin{pmatrix}0\\0\end{pmatrix}$$

then

$$\begin{aligned}c_1 + c_2 &= 0 \\ c_1 + 2c_2 &= 0\end{aligned}$$

and the only solution to this system is $c_1 = 0$, $c_2 = 0$.

Definition. The vectors $\mathbf{v}_1, \mathbf{v}_2, \ldots, \mathbf{v}_n$ in a vector space V are said to be **linearly dependent** if there exist scalars $c_1, c_2, \ldots, c_n$ not all zero such that

$$c_1\mathbf{v}_1 + c_2\mathbf{v}_2 + \cdots + c_n\mathbf{v}_n = \mathbf{0}$$

EXAMPLE 2. Let $\mathbf{x} = (1, 2, 3)^T$. The vectors $\mathbf{e}_1, \mathbf{e}_2, \mathbf{e}_3, \mathbf{x}$ are linearly dependent, since

$$\mathbf{e}_1 + 2\mathbf{e}_1 + 3\mathbf{e}_3 - \mathbf{x} = \mathbf{0}$$

(In this case $c_1 = 1$, $c_2 = 2$, $c_3 = 3$, $c_4 = -1$.)

If $\mathbf{x}$ and $\mathbf{y}$ are linearly dependent in R^2, then

$$c_1\mathbf{x} + c_2\mathbf{y} = \mathbf{0}$$

where c_1 and c_2 are not both 0. If, say, $c_1 \neq 0$, we can write

$$\mathbf{x} = -\frac{c_2}{c_1}\mathbf{y}$$

If two vectors in R^2 are linearly dependent, one of the vectors can be written as a scalar multiple of the other. Thus, if both vectors are placed at the origin, they will lie along the same line (see Figure 3.3.1).

If

$$\mathbf{x} = \begin{bmatrix} x_1 \\ x_2 \\ x_3 \end{bmatrix} \quad \text{and} \quad \mathbf{y} = \begin{bmatrix} y_1 \\ y_2 \\ y_3 \end{bmatrix}$$

are linearly independent in R^3, then the points (x_1, x_2, x_3) and (y_1, y_2, y_3)

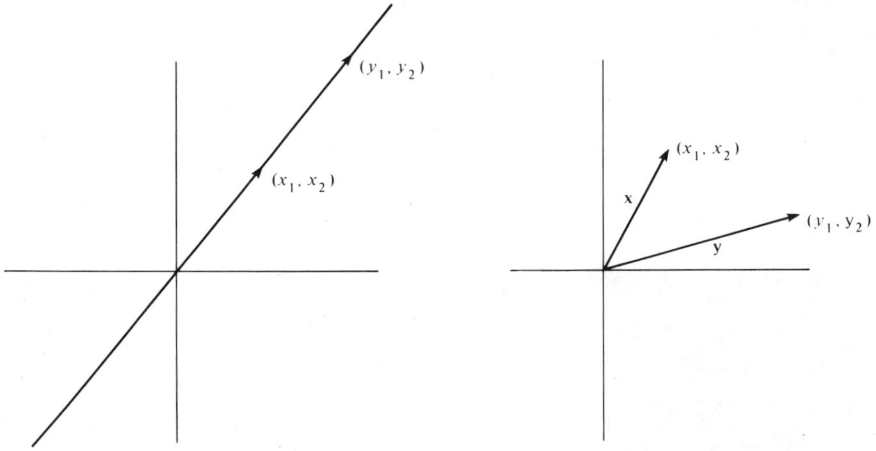

(a) x and y linearly dependent. (b) x and y linearly independent.

Figure 3.3.1

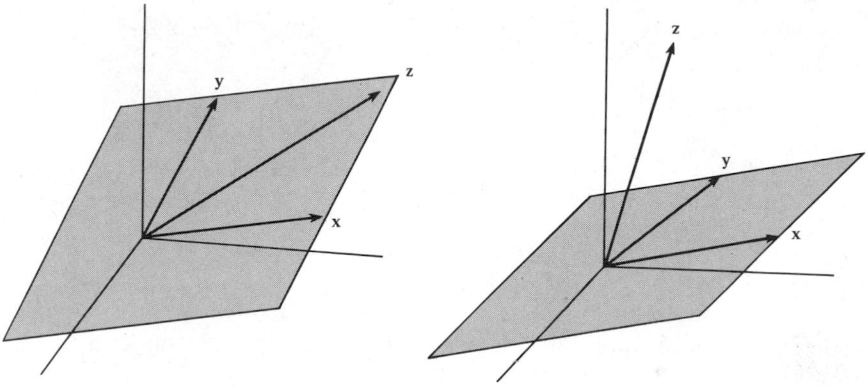

(a) x, y, z linearly dependent.

(b) x, y, z linearly independent.

Figure 3.3.2

do not lie on the same line through the origin in 3-space. Since (0, 0, 0), (x_1, x_2, x_3), and (y_1, y_2, y_3) are not collinear, they determine a plane. If (z_1, z_2, z_3) lies on this plane, the vector $\mathbf{z} = (z_1, z_2, z_3)^T$ can be written as a linear combination of **x** and **y**, and hence **x**, **y**, and **z** are linearly dependent. If (z_1, z_2, z_3) does not lie on the plane, the three vectors will be linearly independent (see Figure 3.3.2).

EXAMPLE 3. Which of the following collections of vectors are linearly independent in R^3?

(a) $(1, 1, 1)^T, (1, 1, 0)^T, (1, 0, 0)^T$.
(b) $(1, 0, 1)^T, (0, 1, 0)^T$.
(c) $(1, 2, 4)^T, (2, 1, 3)^T, (4, -1, 1)^T$.

SOLUTION

(a) These vectors are linearly independent. If

$$c_1(1, 1, 1)^T + c_2(1, 1, 0)^T + c_3(1, 0, 0)^T = (0, 0, 0)^T$$

then

$$\begin{aligned} c_1 + c_2 + c_3 &= 0 \\ c_1 + c_2 \phantom{{}+c_3} &= 0 \\ c_1 \phantom{{}+c_2+c_3} &= 0 \end{aligned}$$

and the only solution to this system is $c_1 = 0$, $c_2 = 0$, $c_3 = 0$.

(b) If

$$c_1(1, 0, 1,)^T + c_2(0, 1, 0)^T = (0, 0, 0)^T$$

then

$$(c_1, c_2, c_1)^T = (0, 0, 0)^T$$

and so $c_1 = c_2 = 0$. Therefore, the two vectors are linearly independent.

(c) If

$$c_1(1, 2, 4)^T + c_2(2, 1, 3)^T + c_3(4, -1, 1)^T = (0, 0, 0)^T$$

then

$$\begin{aligned} c_1 + 2c_2 + 4c_3 &= 0 \\ 2c_1 + c_2 - c_3 &= 0 \\ 4c_1 + 3c_2 + c_3 &= 0 \end{aligned}$$

The coefficient matrix of this system is singular. Hence the system has nontrivial solutions, and hence the vectors are linearly dependent.

Notice in Example 3, parts (a) and (c), that it was necessary to solve a 3×3 system to determine whether or not the three vectors were linearly independent. In part (a), where the coefficient matrix was nonsingular, the vectors were linearly independent, while in part (c), where the coefficient was singular, the vectors were linearly dependent. This illustrates a special case of the following theorem.

Theorem 3.3.1. *Let $\mathbf{x}_1, \mathbf{x}_2, \ldots, \mathbf{x}_n$ be n vectors in R^n with $\mathbf{x}_i = (x_{1i}, x_{2i}, \ldots, x_{ni})^T$ for $i = 1, \ldots, n$. Let $X = (x_{ij})$; that is, the $\mathbf{x}_i$'s form the columns of X. The vectors $\mathbf{x}_1, \mathbf{x}_2, \ldots, \mathbf{x}_n$ will be linearly dependent if and only if X is singular.*

PROOF. The equation

$$c_1\mathbf{x}_1 + c_2\mathbf{x}_2 + \cdots + c_n\mathbf{x}_n = \mathbf{0}$$

is equivalent to the system of equations

$$\begin{aligned} c_1x_{11} + c_2x_{12} + \cdots + c_nx_{1n} &= 0 \\ c_1x_{21} + c_2x_{22} + \cdots + c_nx_{2n} &= 0 \\ &\vdots \\ c_1x_{n1} + c_2x_{n2} + \cdots + c_nx_{nn} &= 0 \end{aligned}$$

If we let $\mathbf{c} = (c_1, c_2, \ldots, c_n)^T$, the system can be written as a matrix equation

$$X\mathbf{c} = \mathbf{0}$$

This equation will have a nontrivial solution if and only if X is singular. Thus $\mathbf{x}_1, \ldots, \mathbf{x}_n$ will be linearly dependent if and only if X is singular.

One can use Theorem 3.3.1 to test whether n vectors are linearly independent in R^n. One simply forms a matrix X whose columns are the vectors being tested. To determine whether or not X is singular, one can calculate $|X|$. If $|X| = 0$, the vectors are linearly dependent. If $|X| \neq 0$, the vectors are linearly independent.

EXAMPLE 4. Determine whether or not the vectors $(4, 2, 3)^T$, $(2, 3, 1)^T$, and $(2, -5, 3)^T$ are linearly dependent.

SOLUTION. Since

$$\begin{vmatrix} 4 & 2 & 2 \\ 2 & 3 & -5 \\ 3 & 1 & 3 \end{vmatrix} = 0$$

the vectors are linearly dependent.

Determinants can also be used to help determine if a set of vectors is linearly independent in $C^{(n-1)}[a, b]$. Indeed, let $f_1, f_2, \ldots, f_n$ be elements of $C^{(n-1)}[a, b]$. If these vectors are linearly dependent, then there exist scalars $c_1, c_2, \ldots, c_n$ not all zero such that

$$c_1 f_1(x) + c_2 f_2(x) + \cdots + c_n f_n(x) = 0 \tag{1}$$

for each x in $[a, b]$. Taking the derivative with respect to x of both sides of (1) yields

$$c_1 f_1'(x) + c_2 f_2'(x) + \cdots + c_n f_n'(x) = 0$$

If we continue taking derivatives of both sides, we end up with the system

$$\begin{aligned} c_1 f_1(x) &+ c_2 f_2(x) + \cdots + c_n f_n(x) = 0 \\ c_1 f_1'(x) &+ c_2 f_2'(x) + \cdots + c_n f_n'(x) = 0 \\ &\vdots \\ c_1 f_1^{(n-1)}(x) &+ c_2 f_2^{(n-1)}(x) + \cdots + c_n f_n^{(n-1)}(x) = 0 \end{aligned}$$

Thus for each fixed x in $[a, b]$, the matrix equation

$$\begin{bmatrix} f_1(x) & f_2(x) & \cdots & f_n(x) \\ f_1'(x) & f_2'(x) & \cdots & f_n'(x) \\ \vdots & & & \\ f_1^{(n-1)}(x) & f_2^{(n-1)}(x) & \cdots & f_n^{(n-1)}(x) \end{bmatrix} \begin{bmatrix} \alpha_1 \\ \alpha_2 \\ \vdots \\ \alpha_n \end{bmatrix} = \begin{bmatrix} 0 \\ 0 \\ \vdots \\ 0 \end{bmatrix} \tag{2}$$

has a nontrivial solution $(c_1, c_2, \ldots, c_n)^T$. Thus, if $f_1, \ldots, f_n$ are linearly dependent in $C^{(n-1)}[a, b]$, then for each fixed x in $[a, b]$, the coefficient matrix of the system (2) is singular. If the matrix is singular, its determinant is zero.

Definition. Let $f_1, f_2, \ldots, f_n$ be functions in $C^{(n-1)}[a, b]$ and define the function $W[f_1, f_2, \ldots, f_n](x)$ on $[a, b]$ by

$$W[f_1, f_2, \ldots, f_n](x) = \begin{vmatrix} f_1(x) & f_2(x) & \cdots & f_n(x) \\ f_1'(x) & f_2'(x) & \cdots & f_n'(x) \\ \vdots & & & \\ f_1^{(n-1)}(x) & f_2^{(n-1)}(x) & \cdots & f_n^{(n-1)}(x) \end{vmatrix}$$

The function $W[f_1, f_2, \ldots, f_n]$ is called the **Wronskian** of $f_1, f_2, \ldots, f_n$.

Theorem 3.3.2. *Let $f_1, f_2, \ldots, f_n$ be elements of $C^{(n-1)}[a, b]$. If there exists a point x_0 in $[a, b]$ such that $W[f_1, f_2, \ldots, f_n](x_0) \neq 0$, then $f_1, f_2, \ldots, f_n$ are linearly independent.*

PROOF. If $f_1, \ldots, f_n$ were linearly dependent, then by the preceding discussion the coefficient matrix in (2) would be singular for each x in $[a, b]$ and hence $W[f_1, f_2, \ldots, f_n](x)$ would be identically zero on $[a, b]$.

If $f_1, f_2, \ldots, f_n$ are linearly independent in $C^{(n-1)}[a, b]$, they will also be linearly independent in $C[a, b]$.

EXAMPLE 5. Show that e^x and e^{-x} are linearly independent in $C(-\infty, \infty)$.

SOLUTION

$$W[e^x, e^{-x}] = \begin{vmatrix} e^x & e^{-x} \\ e^x & -e^{-x} \end{vmatrix} = -2$$

Since $W[e^x, e^{-x}]$ is not identically zero, e^x and e^{-x} are linearly independent.

EXAMPLE 6. It is easily seen that the functions x^2 and $x|x|$ are linearly independent in $C[-1, 1]$. If

$$c_1x^2 + c_2x|x| = 0$$

for all x in $[-1, 1]$, then in particular for $x = 1$ and $x = -1$, we have

$$\begin{aligned} c_1 + c_2 &= 0 \\ c_1 - c_2 &= 0 \end{aligned}$$

and hence $c_1 = c_2 = 0$. The Wronskian, however, is identically zero.

$$W[x^2, x|x|](x) = \begin{vmatrix} x^2 & x|x| \\ 2x & 2|x| \end{vmatrix} = 0$$

This example shows that the converse to Theorem 3.3.2 is not valid.

EXAMPLE 7. Show that the vectors 1, x, x^2, x^3 are linearly independent in P_4.

SOLUTION

$$W[1, x, x^2, x^3] = \begin{vmatrix} 1 & x & x^2 & x^3 \\ 0 & 1 & 2x & 3x^2 \\ 0 & 0 & 2 & 6x \\ 0 & 0 & 0 & 6 \end{vmatrix} = 12$$

Since $W[1, x, x^2, x^3] \neq 0$, the vectors are linearly independent.

EXERCISES

1. Determine whether or not the following vectors are linearly independent in R^2.

(a) $\begin{pmatrix} 2 \\ 1 \end{pmatrix}, \begin{pmatrix} 3 \\ 2 \end{pmatrix}$

(b) $\begin{pmatrix} 2 \\ 3 \end{pmatrix}, \begin{pmatrix} 4 \\ 6 \end{pmatrix}$

(c) $\begin{pmatrix} -2 \\ 1 \end{pmatrix}, \begin{pmatrix} 1 \\ 3 \end{pmatrix}, \begin{pmatrix} 2 \\ 4 \end{pmatrix}$

(d) $\begin{pmatrix} -1 \\ 2 \end{pmatrix}, \begin{pmatrix} 1 \\ 2 \end{pmatrix}, \begin{pmatrix} 2 \\ -4 \end{pmatrix}$

(e) $\begin{pmatrix} 1 \\ 2 \end{pmatrix}, \begin{pmatrix} -1 \\ 1 \end{pmatrix}$

2. Determine whether or not the following vectors are linearly independent in R^3.

(a) $\begin{bmatrix} 1 \\ 0 \\ 0 \end{bmatrix}, \begin{bmatrix} 0 \\ 1 \\ 1 \end{bmatrix}, \begin{bmatrix} 1 \\ 0 \\ 1 \end{bmatrix}$

(b) $\begin{bmatrix} 1 \\ 0 \\ 0 \end{bmatrix}, \begin{bmatrix} 0 \\ 1 \\ 1 \end{bmatrix}, \begin{bmatrix} 1 \\ 0 \\ 1 \end{bmatrix}, \begin{bmatrix} 1 \\ 2 \\ 3 \end{bmatrix}$

(c) $\begin{bmatrix} 2 \\ 1 \\ -2 \end{bmatrix}, \begin{bmatrix} 3 \\ 2 \\ -2 \end{bmatrix}, \begin{bmatrix} 2 \\ 2 \\ 0 \end{bmatrix}$

(d) $\begin{bmatrix} 2 \\ 1 \\ -2 \end{bmatrix}, \begin{bmatrix} -2 \\ -1 \\ 2 \end{bmatrix}, \begin{bmatrix} 4 \\ 2 \\ -4 \end{bmatrix}$

(e) $\begin{bmatrix} 1 \\ 1 \\ 3 \end{bmatrix}, \begin{bmatrix} 0 \\ 2 \\ 1 \end{bmatrix}$

3. Determine whether or not the following vectors are linearly independent in P_3.

(a) $1, x^2, x^2 - 2$

(b) $2, x^2, x, 2x + 3$

(c) $x + 2, x + 1, x^2 - 1$

(d) $x + 2, x^2 - 1$

4. For each of the following show that the given vectors are linearly independent in $C[-1, 1]$.

(a) $\cos \pi x, \sin \pi x$

(b) x, e^x, e^{2x}

(c) $x^2, \ln(1 + x^2), 1 + x^2$

(d) $x^3, |x|^3$

5. Determine whether or not the vectors $\cos x$, 1, $\sin^2(x/2)$ are linearly independent in $C[-\pi, \pi]$.

6. Let $f(x) \in C^{(1)}[-1, 1]$. If x and $xf(x)$ are linearly dependent in $C^{(1)}[-1, 1]$, what can you conclude about $f(x)$?

7. Prove that any finite set of vectors that contains the zero vector must be linearly dependent.

8. Let $\mathbf{x}_1, \ldots, \mathbf{x}_k$ be linearly independent vectors in R^n and let A be a nonsingular $n \times n$ matrix. Define $\mathbf{y}_i = A\mathbf{x}_i$ for $i = 1, \ldots, k$. Show that $\mathbf{y}_1, \ldots, \mathbf{y}_k$ are linearly independent.

9. Let $\{\mathbf{v}_1, \ldots, \mathbf{v}_n\}$ be a spanning set for a vector space V and let $\mathbf{v}$ be any element of V. Show that $\mathbf{v}, \mathbf{v}_1, \ldots, \mathbf{v}_n$ are linearly dependent.

10. Let $\{\mathbf{v}_1, \ldots, \mathbf{v}_n\}$ be a spanning set for V. If $\mathbf{v}_1, \ldots, \mathbf{v}_n$ are linearly independent, show that the vectors $\mathbf{v}_2, \ldots, \mathbf{v}_n$ cannot span V.

4. BASIS AND DIMENSION

In Section 3 we showed that a spanning set for a vector space is minimal if its elements are linearly independent. The elements of a minimal spanning set form the basic building blocks for the whole vector space, and consequently we say they form a "basis" for the vector space.

Definition. The vectors $\mathbf{v}_1, \mathbf{v}_2, \ldots, \mathbf{v}_n$ form a **basis** for a vector space V if and only if

(i) $\mathbf{v}_1, \ldots, \mathbf{v}_n$ are linearly independent.
(ii) $\mathbf{v}_1, \ldots, \mathbf{v}_n$ span V.

EXAMPLE 1. The standard basis for R^3 is $\{\mathbf{e}_1, \mathbf{e}_2, \mathbf{e}_3\}$ but there are many other bases that could be used such as $\{(1, 1, 1)^T, (1, 1, 0)^T, (1, 0, 1)^T\}$ or $\{(1, 2, 3)^T, (0, 1, 1)^T, (2, 0, 1)^T\}$. We will see shortly that any basis for R^3 must have exactly three elements.

EXAMPLE 2. In $M_{2,2}$ consider the set $\{E_{11}, E_{12}, E_{21}, E_{22}\}$, where

$$E_{11} = \begin{pmatrix} 1 & 0 \\ 0 & 0 \end{pmatrix}, \quad E_{12} = \begin{pmatrix} 0 & 1 \\ 0 & 0 \end{pmatrix}, \quad E_{21} = \begin{pmatrix} 0 & 0 \\ 1 & 0 \end{pmatrix}, \quad E_{22} = \begin{pmatrix} 0 & 0 \\ 0 & 1 \end{pmatrix}$$

If

$$c_1E_{11} + c_2E_{12} + c_3E_{21} + c_4E_{22} = O$$

then

$$\begin{pmatrix} c_1 & c_2 \\ c_3 & c_4 \end{pmatrix} = \begin{pmatrix} 0 & 0 \\ 0 & 0 \end{pmatrix}$$

so $c_1 = c_2 = c_3 = c_4 = 0$. Therefore, E_{11}, E_{12}, E_{21}, E_{22} are linearly independent. If A is in $M_{2,2}$, then

$$A = a_{11}E_{11} + a_{12}E_{12} + a_{21}E_{21} + a_{22}E_{22}$$

Thus E_{11}, E_{12}, E_{21}, E_{22} span $M_{2,2}$ and hence form a basis for $M_{2,2}$.

In many applications it is necessary to find a particular subspace of a vector space V. This can be done by finding the basis elements of the subspace. For example, to find all solutions to the system

$$\begin{aligned} x_1 + x_2 + x_3 &= 0 \\ 2x_1 + x_2 + x_4 &= 0 \end{aligned}$$

one must find the nullspace of the matrix

$$A = \begin{pmatrix} 1 & 1 & 1 & 0 \\ 2 & 1 & 0 & 1 \end{pmatrix}$$

In Example 8 of Section 2, we saw that $N(A)$ is the subspace of R^4 spanned by the vectors

$$\begin{bmatrix} 1 \\ -2 \\ 1 \\ 0 \end{bmatrix} \quad \text{and} \quad \begin{bmatrix} -1 \\ 1 \\ 0 \\ 1 \end{bmatrix}$$

Since these two vectors are linearly independent, they form a basis for $N(A)$.

Theorem 3.4.1. *If $\mathbf{v}_1, \mathbf{v}_2, \ldots, \mathbf{v}_n$ form a basis for a vector space V, then any collection of m vectors in V, where $m > n$, is linearly dependent.*

PROOF. Let $\mathbf{u}_1, \mathbf{u}_2, \ldots, \mathbf{u}_m$ be m vectors in V where $m > n$. Then since $\mathbf{v}_1, \mathbf{v}_2, \ldots, \mathbf{v}_n$ span V, we have

$$\mathbf{u}_i = a_{i1}\mathbf{v}_1 + a_{i2}\mathbf{v}_2 + \cdots + a_{in}\mathbf{v}_n \qquad \text{for} \quad i = 1, 2, \ldots, m$$

A linear combination $c_1\mathbf{u}_1 + c_2\mathbf{u}_2 + \cdots + c_m\mathbf{u}_m$ can be written in the form

$$c_1 \sum_{j=1}^{n} a_{1j}\mathbf{v}_j + c_2 \sum_{j=1}^{n} a_{2j}\mathbf{v}_j + \cdots + c_m \sum_{j=1}^{n} a_{mj}\mathbf{v}_j$$

Rearranging terms, we see that

$$c_1\mathbf{u}_1 + c_2\mathbf{u}_2 + \cdots + c_m\mathbf{u}_m = \sum_{j=1}^{n} \left(\sum_{i=1}^{m} a_{ij}c_i \right) \mathbf{v}_j$$

Now consider the system of equations

$$\sum_{i=1}^{m} a_{ij}c_i = 0 \qquad j = 1, 2, \ldots, n$$

This is a homogeneous system with more unknowns than equations. Therefore, by Theorem 1.2.1, the system has a nontrivial solution $(\hat{c}_1, \hat{c}_2, \ldots, \hat{c}_m)$. But then

$$\hat{c}_1\mathbf{u}_1 + \hat{c}_2\mathbf{u}_2 + \cdots + \hat{c}_m\mathbf{u}_m = \sum_{j=1}^{n} 0 \cdot \mathbf{v}_j = \mathbf{0}$$

and hence $\mathbf{u}_1, \mathbf{u}_2, \ldots, \mathbf{u}_m$ are linearly dependent.

Corollary 3.4.2. *Any two bases for a vector space V have the same number of elements.*

PROOF. Let $\mathbf{v}_1, \mathbf{v}_2, \ldots, \mathbf{v}_n$ and $\mathbf{u}_1, \mathbf{u}_2, \ldots, \mathbf{u}_m$ both be bases for V. Since $\mathbf{u}_1, \mathbf{u}_2, \ldots, \mathbf{u}_m$ are linearly independent, it follows from Theorem 3.4.1 that $m \leqslant n$. By the same reasoning $\mathbf{u}_1, \mathbf{u}_2, \ldots, \mathbf{u}_m$ form a basis for V and $\mathbf{v}_1, \mathbf{v}_2, \ldots, \mathbf{v}_n$ are linearly independent, so $n \leqslant m$.

Definition. Let V be a vector space. If V has a basis consisting of n vectors, we say that V has **dimension** n. The subspace $\{\mathbf{0}\}$ of V is said to have dimension 0. If there is no finite set of vectors that span V, we say that V is **infinite-dimensional.**

If $\mathbf{x}$ is a nonzero vector in R^3, then $\mathbf{x}$ spans a one-dimensional subspace $\mathcal{S}(\mathbf{x}) = \{\alpha\mathbf{x} | \alpha \text{ is a scalar}\}$. A vector $(a, b, c)^T$ will be in $\mathcal{S}(\mathbf{x})$ if and only if the point (a, b, c) is on the line determined by $(0, 0, 0)$ and (x_1, x_2, x_3). Thus a one-dimensional subspace of R^3 can be represented geometrically by a line through the origin (Figure 3.4.1). If $\mathbf{x}$ and $\mathbf{y}$ are linearly independent in R^3, then $\mathcal{S}(\mathbf{x}, \mathbf{y}) = \{\alpha\mathbf{x} + \beta\mathbf{y} | \alpha \text{ and } \beta \text{ are scalars}\}$ is a two-dimensional subspace of R^3. A vector $(a, b, c)^T$ will be in $\mathcal{S}(\mathbf{x}, \mathbf{y})$ if and only if (a, b, c) lies on the plane determined by $(0, 0, 0)$, (x_1, x_2, x_3) and (y_1, y_2, y_3). Thus we can think of a two-dimensional subspace of R^3 as a plane through the origin. If $\mathbf{x}$, $\mathbf{y}$, and $\mathbf{z}$ are linearly independent in R^3, they form a basis for R^3 and $\mathcal{S}(\mathbf{x}, \mathbf{y}, \mathbf{z}) = R^3$. Thus any fourth point $(a, b, c)^T$ must lie in $\mathcal{S}(\mathbf{x}, \mathbf{y}, \mathbf{z})$.

EXAMPLE 3. Let P be the vector space of all polynomials. We claim that P is infinite-dimensional. If P were finite-dimensional, say of dimension n, any set of $n + 1$ vectors would be linearly dependent. However, $1, x, x^2, \ldots, x^n$ are linearly independent, since $W[1, x, x^2, \ldots, x^n] > 0$.

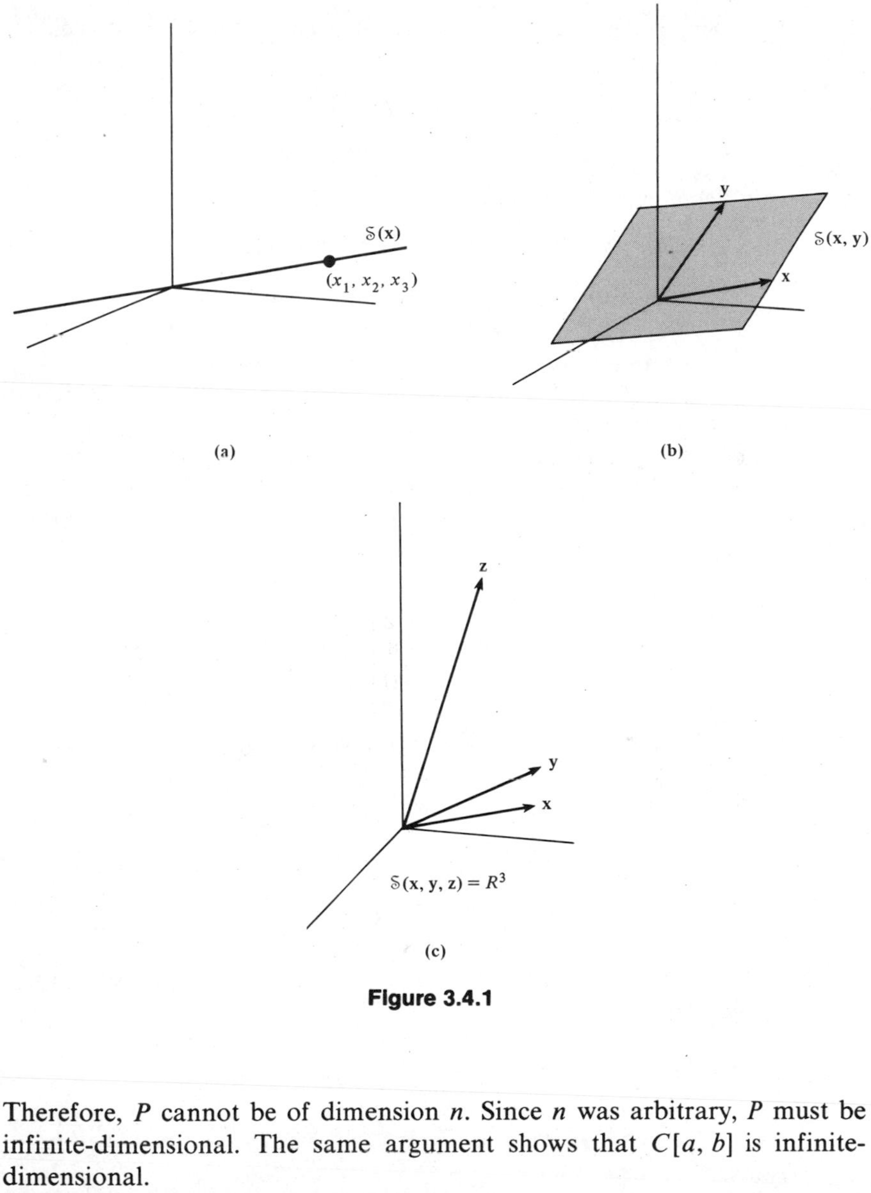

Figure 3.4.1

Therefore, P cannot be of dimension n. Since n was arbitrary, P must be infinite-dimensional. The same argument shows that $C[a, b]$ is infinite-dimensional.

Theorem 3.4.3. *If V is a vector space of dimension $n > 0$:*

(I) *Any set of n linearly independent vectors spans V*

(II) *Any n vectors that span V are linearly independent*

PROOF. To prove (I), suppose that $\mathbf{v}_1, \ldots, \mathbf{v}_n$ are linearly independent and $\mathbf{v}$ is any other vector in V. It follows from Theorem 3.4.1 that

$\mathbf{v}_1, \mathbf{v}_2, \ldots, \mathbf{v}_n, \mathbf{v}$ are linearly dependent. Thus there exist scalars $c_1, c_2, \ldots, c_n, c_{n+1}$ not all zero such that

$$(1) \qquad c_1\mathbf{v}_1 + c_2\mathbf{v}_2 + \cdots + c_n\mathbf{v}_n + c_{n+1}\mathbf{v} = \mathbf{0}$$

The scalar c_{n+1} cannot be zero, for then (1) would imply that $\mathbf{v}_1, \ldots, \mathbf{v}_n$ are linearly dependent. Thus (1) can be solved for $\mathbf{v}$.

$$\mathbf{v} = \alpha_1\mathbf{v}_1 + \alpha_2\mathbf{v}_2 + \cdots + \alpha_n\mathbf{v}_n$$

where $\alpha_i = -c_i/c_{n+1}$ for $i = 1, 2, \ldots, n$. Since $\mathbf{v}$ was an arbitrary vector in V, it follows that $\mathbf{v}_1, \mathbf{v}_2, \ldots, \mathbf{v}_n$ span V.

To prove (II), suppose that $\mathbf{v}_1, \ldots, \mathbf{v}_n$ span V. If $\mathbf{v}_1, \ldots, \mathbf{v}_n$ are linearly dependent, then one of the $\mathbf{v}_i$'s, say $\mathbf{v}_n$, can be written as a linear combination of the others. It follows that $\mathbf{v}_1, \ldots, \mathbf{v}_{n-1}$ will still span V. If $\mathbf{v}_1, \ldots, \mathbf{v}_{n-1}$ are linearly dependent, we can eliminate another vector and still have a spanning set. We can continue eliminating vectors in this way until we arrive at a linearly independent spanning set with $k < n$ elements. But this contradicts $\dim V = n$. Therefore, $\mathbf{v}_1, \ldots, \mathbf{v}_n$ must be linearly independent.

EXAMPLE 4. Show that $\left\{ \begin{pmatrix} 1 \\ 2 \\ 3 \end{pmatrix}, \begin{pmatrix} -2 \\ 1 \\ 0 \end{pmatrix}, \begin{pmatrix} 1 \\ 0 \\ 1 \end{pmatrix} \right\}$ is a basis for R^3.

SOLUTION. Since $\dim R^3 = 3$, we need only show that these three vectors are linearly independent. That follows since

$$\begin{vmatrix} 1 & -2 & 1 \\ 2 & 1 & 0 \\ 3 & 0 & 1 \end{vmatrix} = 2$$

We close this section with two final observations. These are given in the following theorem.

Theorem 3.4.4. *If V is a vector space of dimension n, then:*

(i) *No set of less than n vectors can span V.*

(ii) *Any subset of less than n linearly independent vectors can be extended to form a basis for V.*

Observation (i) follows by the same reasoning that was used to prove (II) of Theorem 3.4.3. To prove (ii), suppose that $\mathbf{v}_1, \ldots, \mathbf{v}_k$ are linearly independent and $k < n$. It follows from (i) that $\mathcal{S}(\mathbf{v}_1, \ldots, \mathbf{v}_k)$ is a proper subspace of V and hence there exists a vector $\mathbf{v}_{k+1}$ that is in V but not in $\mathcal{S}(\mathbf{v}_1, \ldots, \mathbf{v}_k)$. It follows that the vectors $\mathbf{v}_1, \mathbf{v}_2, \ldots, \mathbf{v}_k, \mathbf{v}_{k+1}$ are linearly independent. If $k + 1 < n$, then in the same manner $\{\mathbf{v}_1, \ldots, \mathbf{v}_k, \mathbf{v}_{k+1}\}$ can be extended to a set of $k + 2$ linearly independent vectors. This process may be continued until a set $\{\mathbf{v}_1, \mathbf{v}_2, \ldots, \mathbf{v}_k, \mathbf{v}_{k+1}, \ldots, \mathbf{v}_n\}$ of n linearly independent vectors is obtained.

EXERCISES

1. Determine the dimensions of the subspaces of R^2 spanned by the given vectors in Exercise 1 of Section 3. Indicate whether or not the vectors form a basis for R^2.

2. Determine the dimensions of the subspaces of R^3 spanned by the given vectors in Exercise 2 of Section 3. Indicate whether or not the vectors form a basis for R^3.

3. In Exercise 1 of Section 2, some of the sets formed subspaces of R^3. In each of these cases find a basis for the subspace and determine its dimension.

4. In Exercise 2 of Section 2, some of the sets formed subspaces $M_{2,2}$. In each of these cases find a basis for the subspace and determine its dimension.

5. In $C[-\pi, \pi]$ find the dimension of the subspace spanned by 1, $\cos 2x$, $\cos^2 x$.

6. In each of the following, find the dimension of the subspace of P_3 spanned by the given vectors.
(a) $x, x-1, x^2+1$
(b) $x, x-1, x^2+1, x^2-1$
(c) $x^2, x^2-x-1, x+1$
(d) $2x, x-2$

7. Let S be the subspace of P_3 consisting of all polynomials $p(x)$ such that $p(0)=0$ and let T be the subspace of all polynomials $q(x)$ such that $q(1)=0$. Find a basis for S, a basis for T, and a basis for $S \cap T$.

8. In R^4 let U be the subspace of all vectors of the form $(u_1, u_2, 0, 0)^T$ and let V be the subspace of all vectors of the form $(0, v_2, v_3, 0)^T$. What are the dimensions of U, V, $U \cap V$, $U+V$? Find a basis for each of these four subspaces.

9. Is it possible to find a pair of two-dimensional subspaces U and V of R^3 such that $U \cap V = \{\mathbf{0}\}$? Prove your answer. (*HINT:* Let $\{\mathbf{u}_1, \mathbf{u}_2\}$ and $\{\mathbf{v}_1, \mathbf{v}_2\}$ be bases for U and V, respectively. Show that $\mathbf{u}_1, \mathbf{u}_2, \mathbf{v}_1, \mathbf{v}_2$ are linearly dependent.) Give a geometrical interpretation of your conclusion.

5. THE ROW SPACE AND COLUMN SPACE

If A is an $m \times n$ matrix, each row of A is an n-tuple of real numbers and hence can be considered as a vector in $M_{1,n}$. The m vectors corresponding to the rows of A will be referred to as the *row vectors* of A. Similarly, each column of A can be considered as a vector in R^m and one can associate n *column vectors* with the matrix A.

Definition. If A is an $m \times n$ matrix, the subspace of $M_{1,n}$ spanned by the row vectors of A is called the **row space of** A. The subspace of R^m spanned by the column vectors of A is called the **column space of** A.

EXAMPLE 1. Let

$$A = \begin{pmatrix} 1 & 0 & 0 \\ 0 & 1 & 0 \end{pmatrix}$$

The row space of A is the set of all 3-tuples of the form

$$\alpha(1, 0, 0) + \beta(0, 1, 0) = (\alpha, \beta, 0)$$

The column space of A is the set of all vectors of the form

$$\alpha\begin{pmatrix} 1 \\ 0 \end{pmatrix} + \beta\begin{pmatrix} 0 \\ 1 \end{pmatrix} + \gamma\begin{pmatrix} 0 \\ 0 \end{pmatrix} = \begin{pmatrix} \alpha \\ \beta \end{pmatrix}$$

Thus the row space of A is a two-dimensional subspace of $M_{1,3}$ and the column space of A is R^2.

Theorem 3.5.1. *Two row equivalent matrices have the same row space.*

PROOF. If B is row equivalent to A, then B can be formed from A by a finite sequence of row operations. Thus the row vectors of B must be linear combinations of the row vectors of A. Consequently, the row space of B must be a subspace of the row space of A. Since A is row equivalent to B, by the same reasoning, the row space of A is a subspace of the row space of B.

Definition. The **rank** of a matrix A is the dimension of the row space of A.

To determine the rank of a matrix, one can reduce the matrix to row echelon form. The nonzero rows of the row echelon matrix will form a basis for the row space.

EXAMPLE 2. Let

$$A = \begin{bmatrix} 1 & -2 & 3 \\ 2 & -5 & 1 \\ 1 & -4 & -7 \end{bmatrix}$$

Reducing A to row echelon form one obtains the matrix

$$U = \begin{bmatrix} 1 & -2 & 3 \\ 0 & 1 & 5 \\ 0 & 0 & 0 \end{bmatrix}$$

Clearly, $(1, -2, 3)$ and $(0, 1, 5)$ form a basis for the row space of U. Since U and A are row equivalent, they have the same row space and hence the rank of A is 2.

The dimension of the column space of a matrix A is the same as the dimension of the row space of A^T. Thus one could determine the dimension of the column space by reducing A^T to row echelon form and counting the number of nonzero rows. More simply, one could perform column operations to reduce A to column echelon form and count the number of nonzero columns.

EXAMPLE 3. To determine the dimension of the column space of the matrix A in Example 2, we use column operations to reduce A to the form

$$\begin{bmatrix} 1 & 0 & 0 \\ 2 & 1 & 0 \\ 1 & 2 & 0 \end{bmatrix}$$

Thus $\{(1, 2, 1)^T, (0, 1, 2)^T\}$ is a basis for the column space of A.

In the examples examined so far, the dimension of the row space and the dimension of the column space have always been the same. Is this true in general? The next theorem answers this question.

Theorem 3.5.2. *If A is an $m \times n$ matrix, the dimension of the row space of A equals the dimension of the column space of A.*

PROOF. Let A be an $m \times n$ matrix of rank r, and let $\mathbf{b}_1, \mathbf{b}_2, \ldots, \mathbf{b}_r$ be a basis for the row space of A. Each $\mathbf{b}_i$ is an n-tuple of the form $(b_{i1}, b_{i2}, \ldots, b_{in})$. Each row vector of A can be written as a linear combination of the $\mathbf{b}_i$'s. Thus

$$\begin{aligned}
(a_{11}, a_{12}, \ldots, a_{1n}) &= c_{11}\mathbf{b}_1 + c_{12}\mathbf{b}_2 + \cdots + c_{1r}\mathbf{b}_r \\
(a_{21}, a_{22}, \ldots, a_{2n}) &= c_{21}\mathbf{b}_1 + c_{22}\mathbf{b}_2 + \cdots + c_{2r}\mathbf{b}_r \\
&\;\vdots \\
(a_{m1}, a_{m2}, \ldots, a_{mn}) &= c_{m1}\mathbf{b}_1 + c_{m2}\mathbf{b}_2 + \cdots + c_{mr}\mathbf{b}_r
\end{aligned}$$

For fixed j, $1 \leqslant j \leqslant n$, equate the jth components of each of these equations.

$$
\begin{aligned}
a_{1j} &= c_{11}b_{1j} + c_{12}b_{2j} + \cdots + c_{1r}b_{rj} \\
a_{2j} &= c_{21}b_{1j} + c_{22}b_{2j} + \cdots + c_{2r}b_{rj} \\
&\vdots \\
a_{mj} &= c_{m1}b_{1j} + c_{m2}b_{2j} + \cdots + c_{mr}b_{rj}
\end{aligned}
\tag{1}
$$

If we let

$$
\mathbf{a}_j = \begin{bmatrix} a_{1j} \\ a_{2j} \\ \vdots \\ a_{mj} \end{bmatrix} \quad \text{and} \quad \mathbf{c}_i = \begin{bmatrix} c_{1i} \\ c_{2i} \\ \vdots \\ c_{mi} \end{bmatrix} \quad \text{for } i = 1, \ldots, r
$$

then system (1) can be written in the form

$$
\mathbf{a}_j = b_{1j}\mathbf{c}_1 + b_{2j}\mathbf{c}_2 + \cdots + b_{rj}\mathbf{c}_r \qquad 1 \leqslant j \leqslant n
$$

Thus each column vector of A can be written as a linear combination of the vectors $\mathbf{c}_1, \ldots, \mathbf{c}_r$. It follows then that the column space of A is a subspace of the vector space spanned by $\mathbf{c}_1, \ldots, \mathbf{c}_r$, and hence the dimension of the column space must be less than or equal to r. We have proved that for any matrix the dimension of the column space is less than or equal to the dimension of the row space. Applying this result to the matrix A^T, we see that

$$
\begin{aligned}
\dim(\text{row space of } A) &= \dim(\text{column space of } A^T) \\
&\leqslant \dim(\text{row space of } A^T) \\
&= \dim(\text{column space of } A)
\end{aligned}
$$

Thus for any matrix A the dimension of the row space must equal the dimension of the column space.

The concepts of row space and column space are useful in the study of linear systems. The system $A\mathbf{x} = \mathbf{b}$ can be written in the form

$$
x_1 \begin{bmatrix} a_{11} \\ a_{21} \\ \vdots \\ a_{m1} \end{bmatrix} + x_2 \begin{bmatrix} a_{12} \\ a_{22} \\ \vdots \\ a_{m2} \end{bmatrix} + \cdots + x_n \begin{bmatrix} a_{1n} \\ a_{2n} \\ \vdots \\ a_{mn} \end{bmatrix} = \begin{bmatrix} b_1 \\ b_2 \\ \vdots \\ b_m \end{bmatrix}
\tag{2}
$$

It follows from (2) that the system $A\mathbf{x} = \mathbf{b}$ will be consistent if and only if $\mathbf{b}$ can be written as a linear combination of the column vectors of A. Thus $A\mathbf{x} = \mathbf{b}$ is consistent if and only if $\mathbf{b}$ is in the column space of A. If $\mathbf{b}$ is replaced by the zero vector, then (2) becomes

$$x_1\mathbf{a}_1 + x_2\mathbf{a}_2 + \cdots + x_n\mathbf{a}_n = \mathbf{0} \tag{3}$$

It follows from (3) that the system $A\mathbf{x} = \mathbf{0}$ will have only the trivial solution $\mathbf{x} = \mathbf{0}$ if and only if the column vectors of A are linearly independent.

Theorem 3.5.3. *Let A be an $m \times n$ matrix. The linear system $A\mathbf{x} = \mathbf{b}$ is consistent for every $\mathbf{b} \in R^m$ if and only if the column vectors of A span R^m. The system $A\mathbf{x} = \mathbf{b}$ has at most one solution for every $\mathbf{b} \in R^m$ if and only if the column vectors of A are linearly independent.*

PROOF. We have seen that the system $A\mathbf{x} = \mathbf{b}$ is consistent if and only if $\mathbf{b}$ is in the column space of A. It follows that $A\mathbf{x} = \mathbf{b}$ will be consistent for every $\mathbf{b} \in R^m$ if and only if the column vectors of A span R^m. To prove the second statement, note that if $A\mathbf{x} = \mathbf{b}$ has at most one solution for every $\mathbf{b}$, then in particular the system $A\mathbf{x} = \mathbf{0}$ can have only the trivial solution, and hence the column vectors of A must be linearly independent. Conversely, if the column vectors of A are linearly independent, $A\mathbf{x} = \mathbf{0}$ has only the trivial solution. Now if $\mathbf{x}_1$ and $\mathbf{x}_2$ were both solutions to $A\mathbf{x} = \mathbf{b}$, then $\mathbf{x}_1 - \mathbf{x}_2$ would be a solution to $A\mathbf{x} = \mathbf{0}$,

$$A(\mathbf{x}_1 - \mathbf{x}_2) = A\mathbf{x}_1 - A\mathbf{x}_2 = \mathbf{b} - \mathbf{b} = \mathbf{0}$$

It follows that $\mathbf{x}_1 - \mathbf{x}_2 = \mathbf{0}$, and hence $\mathbf{x}_1$ must equal $\mathbf{x}_2$.

Let A be an $m \times n$ matrix. If the column vectors of A span R^m, then n must be greater than or equal to m, since no set of less than m vectors could span R^m. If the columns of A are linearly independent, then n must be less than or equal to m, since every set of more than m vectors in R^m is linearly dependent. Thus, if the column vectors of A form a basis for R^m, then n must equal m.

Corollary 3.5.4. *An $n \times n$ matrix A is nonsingular if and only if the column vectors of A form a basis for R^n.*

If A is an $m \times n$ matrix of rank n, then the column vectors of A must be linearly independent and consequently $A\mathbf{x} = \mathbf{0}$ will have only the trivial solution. Thus, if A has rank n, then $N(A) = \{\mathbf{0}\}$. This is a special case of the following theorem.

Theorem 3.5.5. *If A is an $m \times n$ matrix of rank r, then* $\dim N(A) = n - r$.

PROOF. Let U be the reduced row echelon form of A. The system $A\mathbf{x} = \mathbf{0}$ is equivalent to the system $U\mathbf{x} = \mathbf{0}$. If A has rank r, then U will have r nonzero rows, and consequently the system $U\mathbf{x} = \mathbf{0}$ will involve r lead variables and $n - r$ free variables. The dimension of $N(A)$ will equal the number of free variables.

EXAMPLE 4. Let

$$A = \begin{bmatrix} 1 & 2 & -1 & 1 \\ 2 & 4 & -3 & 0 \\ 1 & 2 & 1 & 5 \end{bmatrix}$$

Find a basis for the row space of A and a basis for $N(A)$. Verify that $\dim N(A) = n - r$.

SOLUTION. The reduced row echelon form of A is given by

$$U = \begin{bmatrix} 1 & 2 & 0 & 3 \\ 0 & 0 & 1 & 2 \\ 0 & 0 & 0 & 0 \end{bmatrix}$$

Thus $\{(1, 2, 0, 3), (0, 0, 1, 2)\}$ is a basis for the row space of A and A has rank 2. Since the systems $A\mathbf{x} = \mathbf{0}$ and $U\mathbf{x} = \mathbf{0}$ are equivalent, it follows that $\mathbf{x}$ is in $N(A)$ if and only if

$$\begin{aligned} x_1 + 2x_2 + \qquad 3x_4 &= 0 \\ x_3 + 2x_4 &= 0 \end{aligned}$$

The lead variables x_1 and x_3 can be solved for in terms of the free variables x_2 and x_4.

$$\begin{aligned} x_1 &= -2x_2 - 3x_4 \\ x_3 &= -2x_4 \end{aligned}$$

Let $x_2 = \alpha$ and $x_3 = \beta$. It follows that $N(A)$ consists of all vectors of the form

$$\begin{bmatrix} x_1 \\ x_2 \\ x_3 \\ x_4 \end{bmatrix} = \begin{bmatrix} -2\alpha - 3\beta \\ \alpha \\ -2\beta \\ \beta \end{bmatrix} = \alpha \begin{bmatrix} -2 \\ 1 \\ 0 \\ 0 \end{bmatrix} + \beta \begin{bmatrix} -3 \\ 0 \\ -2 \\ 1 \end{bmatrix}$$

The vectors $(-2, 1, 0, 0)^T$ and $(-3, 0, -2, 1)^T$ form a basis for $N(A)$. Note that

$$n - r = 4 - 2 = 2 = \dim N(A)$$

EXERCISES

1. For each of the following matrices, find a basis for the row space, a basis for the column space, and a basis for the nullspace.

(a) $\begin{bmatrix} 1 & 3 & 2 \\ 2 & 1 & 4 \\ 4 & 7 & 8 \end{bmatrix}$ (b) $\begin{bmatrix} -3 & 1 & 3 & 4 \\ 1 & 2 & -1 & -2 \\ -3 & 8 & 4 & 2 \end{bmatrix}$

(c) $\begin{bmatrix} 1 & 3 & -2 & 1 \\ 2 & 1 & 3 & 2 \\ 3 & 4 & 5 & 6 \end{bmatrix}$

2. Construct 3×3 matrices of rank 0, rank 1, rank 2, and rank 3.

3. Prove Corollary 3.5.4.

4. An $m \times n$ matrix A is said to have a *right inverse* if there exists an $n \times m$ matrix C such that $AC = I_m$. A is said to have a *left inverse* if there exists an $n \times m$ matrix D such that $DA = I_n$.

(a) If A has a right inverse, show that the column vectors of A span R^m.

(b) Is it possible for an $m \times n$ matrix to have a right inverse if $n < m$? $n \geqslant m$? Explain.

5. Prove: If A is an $m \times n$ matrix and the column vectors of A span R^m, then A has a right inverse. (*HINT:* Let $\mathbf{e}_j$ denote the jth column of I_m and solve $A\mathbf{x} = \mathbf{e}_j$ for $j = 1, \ldots, m$.)

6. Show that a matrix B has a left inverse if and only if B^T has a right inverse.

7. Let B be an $n \times m$ matrix whose columns are linearly independent. Show that B has a left inverse.

8. Prove that if a matrix B has a left inverse, the columns of B are linearly independent.

9. If a matrix U is in row echelon form, show that the nonzero row vectors of U form a basis for the row space of U.

10. Let A and B be row equivalent matrices.

(a) Show that the dimension of the column space of A equals the dimension of the column space of B.

(b) Are the column spaces of the two matrices necessarily the same? Justify your answer.

4

Linear Transformations

Linear mappings from one vector space to another play an important role in mathematics. This chapter provides an introduction to the theory of such mappings. In Section 1 the definition of a linear transformation is given and a number of examples are presented. In Section 2 it is shown that each linear transformation L mapping an n-dimensional vector space V into an m-dimensional vector space W can be represented by an $m \times n$ matrix A. Thus one can work with the matrix A in place of the operator L. In Section 3 we will consider the case where L maps V into itself. The matrix A representing L will depend upon the basis chosen for V. In many applications it is desirable to choose the basis for V so that A will be either diagonal or in some other simple form.

1. DEFINITION AND EXAMPLES

In the study of vector spaces the most important types of mappings are linear transformations.

Definition. A mapping L from a vector space V into a vector space W is said to be a **linear transformation** or a **linear operator** if

$$(1) \qquad L(\alpha\mathbf{v}_1 + \beta\mathbf{v}_2) = \alpha L(\mathbf{v}_1) + \beta L(\mathbf{v}_2)$$

for all $\mathbf{v}_1, \mathbf{v}_2 \in V$ and for all scalars α and β.

If L is a linear transformation mapping a vector space V into W, it follows from (1) that

$$(2) \qquad L(\mathbf{v}_1 + \mathbf{v}_2) = L(\mathbf{v}_1) + L(\mathbf{v}_2) \qquad (\alpha = \beta = 1)$$

and

$$(3) \qquad L(\alpha\mathbf{v}) = \alpha L(\mathbf{v}) \qquad (\mathbf{v} = \mathbf{v}_1, \mathbf{v}_2 = \mathbf{0})$$

Conversely, if L satisfies (2) and (3), then

$$\begin{aligned} L(\alpha\mathbf{v}_1 + \beta\mathbf{v}_2) &= L(\alpha\mathbf{v}_1) + L(\beta\mathbf{v}_2) \\ &= \alpha L(\mathbf{v}_1) + \beta L((\mathbf{v}_2)) \end{aligned}$$

Thus L is a linear operator on V if and only if L satisfies (2) and (3).

Let us now consider some examples of linear transformations. We begin with linear operators that map R^2 into itself. In this case it is easier to see the geometric effect of the operator.

Linear Operators on R^2

EXAMPLE 1. Let L be the operator defined by

$$L(\mathbf{x}) = 3\mathbf{x}$$

for each $\mathbf{x} \in R^2$. Since

$$L(\alpha\mathbf{x}) = 3(\alpha\mathbf{x}) = \alpha(3\mathbf{x}) = \alpha L(\mathbf{x})$$

and

$$L(\mathbf{x} + \mathbf{y}) = 3(\mathbf{x} + \mathbf{y}) = (3\mathbf{x}) + (3\mathbf{y}) = L(\mathbf{x}) + L(\mathbf{y})$$

it follows that L is a linear transformation. We can think of L as a stretching by a factor of 3 (see Figure 4.1.1). In general, if α is a positive scalar, the linear transformation $F(\mathbf{x}) = \alpha\mathbf{x}$ can be thought of as a stretching or shrinking by a factor of α.

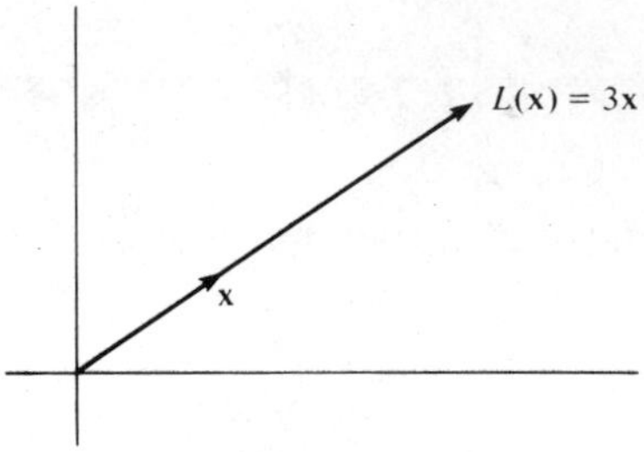

Figure 4.1.1

EXAMPLE 2. Consider the mapping L defined by

$$L(\mathbf{x}) = x_1\mathbf{e}_1$$

for each $\mathbf{x} \in R^2$. Thus if $\mathbf{x} = (x_1, x_2)^T$, then $L(\mathbf{x}) = (x_1, 0)^T$. If $\mathbf{y} = (y_1, y_2)^T$, then

$$\alpha\mathbf{x} + \beta\mathbf{y} = \begin{pmatrix} \alpha x_1 + \beta y_1 \\ \alpha x_2 + \beta y_2 \end{pmatrix}$$

and it follows that

$$\begin{aligned} L(\alpha\mathbf{x} + \beta\mathbf{y}) &= (\alpha x_1 + \beta y_1)\mathbf{e}_1 \\ &= \alpha(x_1\mathbf{e}_1) + \beta(y_1\mathbf{e}_1) \\ &= \alpha L(\mathbf{x}) + \beta L(\mathbf{y}) \end{aligned}$$

Thus L is a linear transformation. We can think of L as a projection onto the x_1 axis (see Figure 4.1.2).

EXAMPLE 3. Let L be the operator defined by

$$L(\mathbf{x}) = (x_1, -x_2)^T$$

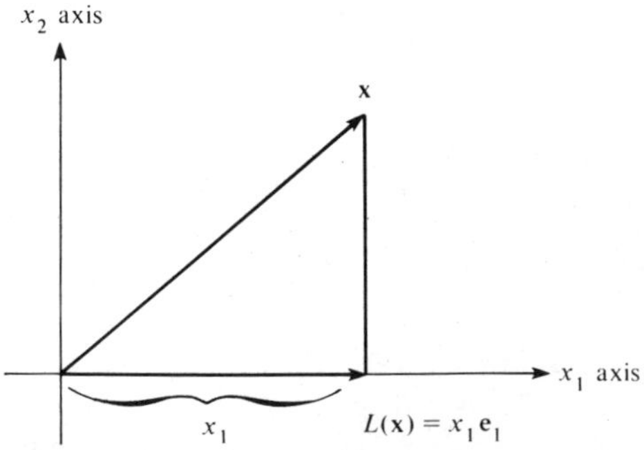

Figure 4.1.2

for each $\mathbf{x} = (x_1, x_2)^T$ in R^2. Since

$$\begin{aligned} L(\alpha\mathbf{x} + \beta\mathbf{y}) &= \begin{pmatrix} \alpha x_1 + \beta y_1 \\ -(\alpha x_2 + \beta y_2) \end{pmatrix} \\ &= \alpha\begin{pmatrix} x_1 \\ -x_2 \end{pmatrix} + \beta\begin{pmatrix} y_1 \\ -y_2 \end{pmatrix} \\ &= \alpha L(\mathbf{x}) = \beta L(\mathbf{y}) \end{aligned}$$

it follows that L is a linear transformation. The operator L has the effect of reflecting a vector $\mathbf{x}$ about the x_1 axis (see Figure 4.1.3).

EXAMPLE 4. The operator L defined by

$$L(\mathbf{x}) = (-x_2, x_1)^T$$

is linear, since

$$\begin{aligned} L(\alpha\mathbf{x} + \beta\mathbf{y}) &= \begin{pmatrix} -(\alpha x_2 + \beta y_2) \\ \alpha x_1 + \beta y_1 \end{pmatrix} \\ &= \alpha\begin{pmatrix} -x_2 \\ x_1 \end{pmatrix} + \beta\begin{pmatrix} -y_2 \\ y_1 \end{pmatrix} \\ &= \alpha L(\mathbf{x}) + \beta L(\mathbf{y}) \end{aligned}$$

The operator L has the effect of rotating each vector in R^2 by 90° in the counterclockwise direction (see Figure 4.1.4).

EXAMPLE 5. Consider the mapping M defined by

$$M(\mathbf{x}) = (x_1^2 + x_2^2)^{1/2}$$

Since

$$M(\alpha\mathbf{x}) = (\alpha^2 x_1^2 + \alpha^2 x_2^2)^{1/2} = |\alpha| M(\mathbf{x})$$

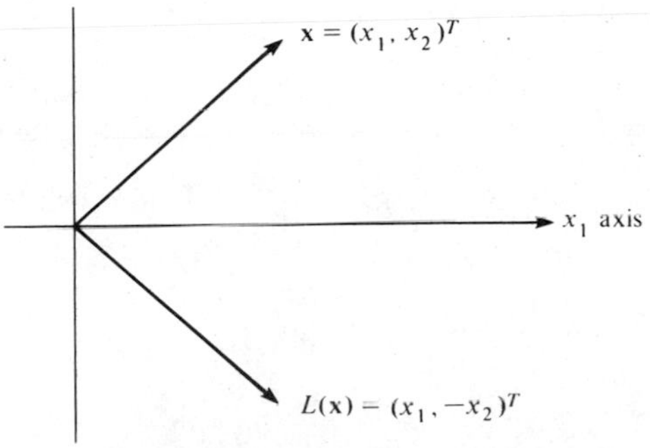

Figure 4.1.3

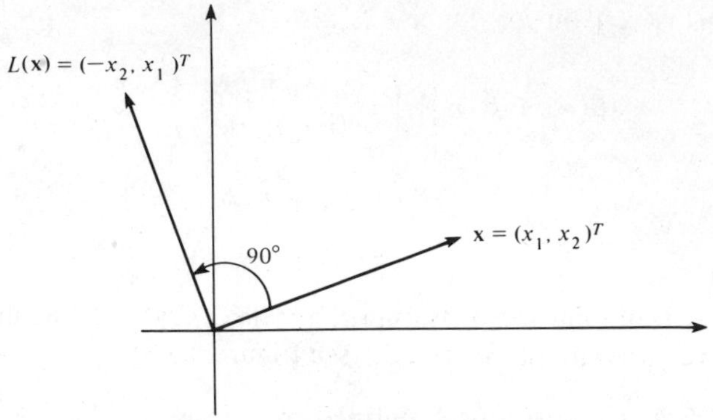

Figure 4.1.4

it follows that

$$\alpha M(\mathbf{x}) \neq M(\alpha\mathbf{x})$$

whenever $\alpha < 0$ and $\mathbf{x} \neq \mathbf{0}$. Therefore, M is not a linear transformation.

Linear Operators from R^n to R^m

EXAMPLE 6. The mapping L: $R^2 \to R^1$ defined by

$$L(\mathbf{x}) = x_1 + x_2$$

is a linear transformation, since

$$\begin{aligned} L(\alpha\mathbf{x} + \beta\mathbf{y}) &= (\alpha x_1 + \beta y_1) + (\alpha x_2 + \beta y_2) \\ &= \alpha(x_1 + x_2) + \beta(y_1 + y_2) \\ &= \alpha L(\mathbf{x}) + \beta L(\mathbf{y}) \end{aligned}$$

EXAMPLE 7. The mapping L from R^2 to R^3 defined by

$$L(\mathbf{x}) = (x_2, x_1, x_1 + x_2)^T$$

is linear, since

$$L(\alpha\mathbf{x}) = (\alpha x_2, \alpha x_1, \alpha x_1 + \alpha x_2)^T = \alpha L(\mathbf{x})$$

and

$$\begin{aligned} L(\mathbf{x} + \mathbf{y}) &= (x_2 + y_2, x_1 + y_1, x_1 + y_1 + x_2 + y_2)^T \\ &= (x_2, x_1, x_1 + x_2)^T + (y_2, y_1, y_1 + y_2)^T \\ &= L(\mathbf{x}) + L(\mathbf{y}) \end{aligned}$$

Note that if we define the matrix A by

$$A = \begin{pmatrix} 0 & 1 \\ 1 & 0 \\ 1 & 1 \end{pmatrix}$$

then

$$L(\mathbf{x}) = \begin{bmatrix} x_2 \\ x_1 \\ x_1 + x_2 \end{bmatrix} = A\mathbf{x}$$

for each $\mathbf{x} \in R^2$.

In general, if A is any $m \times n$ matrix, we can define a linear operator L_A from R^n to R^m by

$$L_A(\mathbf{x}) = A\mathbf{x}$$

for each $\mathbf{x} \in R^n$. The operator L_A is linear, since

$$\begin{aligned} L_A(\alpha\mathbf{x} + \beta\mathbf{y}) &= A(\alpha\mathbf{x} + \beta\mathbf{y}) \\ &= \alpha A\mathbf{x} + \beta A\mathbf{y} \\ &= \alpha L_A(\mathbf{x}) + \beta L_A(\mathbf{y}) \end{aligned}$$

Thus we can think of each $m \times n$ matrix A as linear operator from R^n to R^m.

In Example 6 we saw that the operator L could have been defined in terms of a matrix A. In the next section we will see that this is true for all linear operators from R^n to R^m.

Linear Transformations from V to W

If L is a linear operator mapping a vector space V into a vector space W, then

(i) $L(\mathbf{0}_V) = \mathbf{0}_W$ (where $\mathbf{0}_V$ and $\mathbf{0}_W$ are the zero vectors in V and W, respectively).

(ii) If $\mathbf{v}_1, \ldots, \mathbf{v}_n$ are elements of V and $\alpha_1, \ldots, \alpha_n$ are scalars, then

$$\begin{aligned} &L(\alpha_1\mathbf{v}_1 + \alpha_2\mathbf{v}_2 + \cdots + \alpha_n\mathbf{v}_n) \\ &\qquad = \alpha_1 L(\mathbf{v}_1) + \alpha_2 L(\mathbf{v}_2) + \cdots + \alpha_n L(\mathbf{v}_n) \end{aligned}$$

Statement (i) follows from the condition $L(\alpha\mathbf{v}) = \alpha L(\mathbf{v})$ with $\alpha = 0$. Statement (ii) can be easily proven by mathematical induction. We leave this to the reader as an exercise and conclude the section with some further examples.

EXAMPLE 8. If V is any vector space, the identity operator I is defined by

$$I(\mathbf{v}) = \mathbf{v}$$

for all $\mathbf{v} \in V$. Clearly, I is a linear transformation that maps V into itself.

$$I(\alpha\mathbf{v}_1 + \beta\mathbf{v}_2) = \alpha\mathbf{v}_1 + \beta\mathbf{v}_2 = \alpha I(\mathbf{v}_1) + \beta I(\mathbf{v}_2)$$

EXAMPLE 9. Let L be the mapping from $C[a, b]$ to R^1 defined by

$$L(f) = \int_a^b f(x)\, dx$$

If f and g are any vectors in $C[a, b]$, then

$$\begin{aligned} L(\alpha f + \beta g) &= \int_a^b (\alpha f + \beta g)(x)\, dx \\ &= \alpha \int_a^b f(x)\, dx + \beta \int_a^b g(x)\, dx \\ &= \alpha L(f) + \beta L(g) \end{aligned}$$

Therefore, L is a linear transformation.

EXAMPLE 10. Let D be the operator mapping $C^1[a, b]$ into $C[a, b]$ defined by

$$D(f) = f' \qquad \text{(the derivative of } f)$$

D is a linear transformation, since

$$D(\alpha f + \beta g) = \alpha f' + \beta g' = \alpha D(f) + \beta D(g)$$

EXERCISES

1. Show that each of the following are linear transformations from R^2 into R^2. Describe geometrically what each linear transformation accomplishes.
(a) $L(\mathbf{x}) = (-x_1, x_2)^T$
(b) $L(\mathbf{x}) = -\mathbf{x}$
(c) $L(\mathbf{x}) = (x_2, x_1)^T$
(d) $L(\mathbf{x}) = (x_1 \cos\theta - x_2 \sin\theta, x_1 \sin\theta + x_2 \cos\theta)^T$

2. Determine whether or not the following are linear transformations from R^3 into R^2.
(a) $L(\mathbf{x}) = (x_2, x_3)^T$
(b) $L(\mathbf{x}) = (0, 0)^T$
(c) $L(\mathbf{x}) = (1 + x_1, x_2)^T$
(d) $L(\mathbf{x}) = (x_3, x_1 + x_2)^T$

3. Determine whether or not the following are linear transformations from R^2 into R^3.
(a) $L(\mathbf{x}) = (x_1, x_2, 1)^T$
(b) $L(\mathbf{x}) = (x_1, x_2, x_1 + 2x_2)^T$
(c) $L(\mathbf{x}) = (x_1, 0, 0)^T$
(d) $L(\mathbf{x}) = (x_1, x_2, x_1^2 + x_2^2)^T$

4. For each $f \in C[0, 1]$ define $L(f) = F$, where

$$F(x) = \int_0^x f(t)\, dt \qquad 0 \leqslant x \leqslant 1$$

Show that L is a linear transformation from $C[0, 1]$ to $C[0, 1]$. Determine $L(e^x)$ and $L(x^2)$.

Show that L is a linear transformation from $C[0, 1]$ to $C[0, 1]$. Determine $L(e^x)$ and $L(x^2)$.

5. Determine whether or not the following are linear transformations from $C[0, 1]$ into R^1.
(a) $L(f) = f(0)$
(b) $L(f) = |f(0)|$
(c) $L(f) = [f(0) + f(1)]/2$
(d) $L(f) = \left\{ \int_0^1 [f(x)]^2 \, dx \right\}^{1/2}$

6. Determine whether or not the following are linear transformations from $M_{n,n}$ into $M_{n,n}$.
(a) $L(A) = 2A$
(b) $L(A) = A^T$
(c) $L(A) = A + I$
(d) $L(A) = A - A^T$

7. Determine whether or not the following are linear transformations from P_2 to P_3.
(a) $L(p(x)) = xp(x)$
(b) $L(p(x)) = x^2 + p(x)$
(c) $L(p(x)) = p(x) + xp(x) + x^2p'(x)$

8. If L is a linear transformation from V to W, use mathematical induction to prove that

$$L(\alpha_1 \mathbf{v}_1 + \alpha_2 \mathbf{v}_2 + \cdots + \alpha_n \mathbf{v}_n) = \alpha_1 L(\mathbf{v}_1) + \alpha_2 L(\mathbf{v}_2) + \cdots + \alpha_n L(\mathbf{v}_n)$$

9. Show that every linear transformation from R^1 to R^1 is of the form $L(\mathbf{x}) = ax$ for some scalar a.

10. Let L be a linear operator mapping a vector space V into itself. Define L^n, $n \geqslant 1$, recursively by

$$L^1 = L$$
$$L^{k+1}(\mathbf{v}) = L(L^k(\mathbf{v})) \qquad \text{for all} \quad \mathbf{v} \in V$$

Show that L^n is a linear operator on V for each $n \geqslant 1$.

2. MATRIX REPRESENTATION OF LINEAR TRANSFORMATIONS

In Section 1 it was shown that each $m \times n$ matrix A defines a linear transformation L_A from R^n to R^m where

$$L_A(\mathbf{x}) = A\mathbf{x}$$

for each $\mathbf{x} \in R^n$. In this section we will see that for every linear transformation L mapping R^n into R^m, there is an $m \times n$ matrix A such that

$$L(\mathbf{x}) = A\mathbf{x}$$

We will also see how any linear operator between finite-dimensional spaces can be represented by a matrix.

Theorem 4.2.1. *If L is a linear operator mapping R^n into R^m, then there is an $m \times n$ matrix A such that*

$$L(\mathbf{x}) = A\mathbf{x}$$

for each $\mathbf{x} \in R^n$.

PROOF. For $j = 1, \ldots, n$ define

$$\mathbf{a}_j = (a_{1j}, a_{2j}, \ldots, a_{mj})^T = L(\mathbf{e}_j)$$

Let

$$A = (a_{ij}) = (\mathbf{a}_1, \mathbf{a}_2, \ldots, \mathbf{a}_n)$$

If

$$\mathbf{x} = x_1\mathbf{e}_1 + x_2\mathbf{e}_2 + \cdots + x_n\mathbf{e}_n$$

is an arbitrary element of R^n, then

$$\begin{aligned} L(\mathbf{x}) &= x_1L(\mathbf{e}_1) + x_2L(\mathbf{e}_2) + \cdots + x_1L(\mathbf{e}_n) \\ &= x_1\mathbf{a}_1 + x_2\mathbf{a}_2 + \cdots + x_n\mathbf{a}_n \\ &= (\mathbf{a}_1, \mathbf{a}_2, \ldots, \mathbf{a}_n)\begin{pmatrix} x_1 \\ x_2 \\ \vdots \\ x_n \end{pmatrix} \\ &= A\mathbf{x} \end{aligned}$$

EXAMPLE 1. Define the operator $L\colon R^3 \to R^2$ by

$$L((x_1, x_2, x_3)^T) = (x_1 + x_2, x_2 + x_3)^T$$

for each $\mathbf{x} = (x_1, x_2, x_3)^T$ in R^3. It is easily verified that L is a linear operator. We wish to find a matrix A such that $L(\mathbf{x}) = A\mathbf{x}$ for each $\mathbf{x} \in R^3$. To do this, one must determine $L(\mathbf{e}_1)$, $L(\mathbf{e}_2)$, and $L(\mathbf{e}_3)$.

$$L(\mathbf{e}_1) = L((1, 0, 0)^T) = \begin{pmatrix} 1 \\ 0 \end{pmatrix}$$

$$L(\mathbf{e}_2) = L((0, 1, 0)^T) = \begin{pmatrix} 1 \\ 1 \end{pmatrix}$$

$$L(\mathbf{e}_3) = L((0, 0, 1)^T) = \begin{pmatrix} 0 \\ 1 \end{pmatrix}$$

We choose these vectors to be the columns of A,

$$A = \begin{pmatrix} 1 & 1 & 0 \\ 0 & 1 & 1 \end{pmatrix}$$

To check the result, compute $A\mathbf{x}$.

$$A\mathbf{x} = \begin{pmatrix} 1 & 1 & 0 \\ 0 & 1 & 1 \end{pmatrix} \begin{bmatrix} x_1 \\ x_2 \\ x_3 \end{bmatrix} = \begin{pmatrix} x_1 + x_2 \\ x_2 + x_3 \end{pmatrix}$$

Now that we have seen how matrices are used to represent linear operators from R^n to R^m, we may ask if it is possible to find a similar representation for linear operators from V into W, where V and W are vector spaces of dimension n and m, respectively. To properly phrase this question, the following definition is necessary.

Definition. Let V be a vector space with ordered basis $[\mathbf{v}_1, \mathbf{v}_2, \ldots, \mathbf{v}_n]$. If $\mathbf{v}$ is any element of V, then $\mathbf{v}$ can be written in the form

$$\mathbf{v} = x_1\mathbf{v}_1 + x_2\mathbf{v}_2 + \cdots + x_n\mathbf{v}_n$$

where $x_1, x_2, \ldots, x_n$ are scalars. Thus we can associate with each vector $\mathbf{v}$ a unique vector $\mathbf{x} = (x_1, x_2, \ldots, x_n)^T$ in R^n. The vector $\mathbf{x}$ defined in this way is called the **coordinate vector** of $\mathbf{v}$ **with respect to the ordered basis** $[\mathbf{v}_1, \mathbf{v}_2, \ldots, \mathbf{v}_n]$ and the x_i's are called the **coordinates** of $\mathbf{v}$.

Let V and W be vector spaces with ordered bases $[\mathbf{v}_1, \mathbf{v}_2, \ldots, \mathbf{v}_n]$ and $[\mathbf{w}_1, \mathbf{w}_2, \ldots, \mathbf{w}_m]$, respectively. Let L be a linear operator mapping V into W. If $\mathbf{x}$ is the coordinate vector of $\mathbf{v}$ with respect to $[\mathbf{v}_1, \ldots, \mathbf{v}_n]$ and $\mathbf{y}$ is the coordinate vector of $L(\mathbf{v})$ with respect to $[\mathbf{w}_1, \ldots, \mathbf{w}_m]$, we can think of L as an operator that associates with each vector $\mathbf{x} \in R^n$ a vector $\mathbf{y} \in R^m$. We have already seen how such an operator can be represented by an $m \times n$ matrix. Thus it should be possible to find a matrix A such that $A\mathbf{x} = \mathbf{y}$. The procedure for determining A is essentially the same as before. For $j = 1, \ldots, n$ let $\mathbf{a}_j = (a_{1j}, a_{2j}, \ldots, a_{mj})^T$ be the coordinate vector of $L(\mathbf{v}_j)$ with respect to $[\mathbf{w}_1, \mathbf{w}_2, \ldots, \mathbf{w}_m]$.

$$L(\mathbf{v}_j) = a_{1j}\mathbf{w}_1 + a_{2j}\mathbf{w}_2 + \cdots + a_{mj}\mathbf{w}_m \qquad 1 \leqslant j \leqslant n$$

Let $A = (a_{ij}) = (\mathbf{a}_1, \ldots, \mathbf{a}_n)$. If

$$\mathbf{v} = x_1\mathbf{v}_1 + x_2\mathbf{v}_2 + \cdots + x_n\mathbf{v}_n$$

then

$$\begin{aligned} L(\mathbf{v}) &= L\left(\sum_{j=1}^{n} x_j \mathbf{v}_j\right) \\ &= \sum_{j=1}^{n} x_j L(\mathbf{v}_j) \\ &= \sum_{j=1}^{n} x_j \left(\sum_{i=1}^{m} a_{ij} \mathbf{w}_i\right) \\ &= \sum_{i=1}^{m} \left(\sum_{j=1}^{n} a_{ij} x_j\right) \mathbf{w}_i \end{aligned}$$

For $i = 1, \ldots, m$, let

$$y_i = \sum_{j=1}^{n} a_{ij} x_j$$

Thus

$$\mathbf{y} = (y_1, y_2, \ldots, y_m)^T = A\mathbf{x}$$

is the coordinate vector of $L(\mathbf{v})$ with respect to $[\mathbf{w}_1, \mathbf{w}_2, \ldots, \mathbf{w}_m]$.

EXAMPLE 2. The linear operator D defined by $D(p) = p'$ maps P_3 into P_2. Given the ordered bases $[x^2, x, 1]$ and $[x, 1]$ for P_3 and P_2, respectively, we wish to determine a matrix representation for D. To do this we apply D to each of the basis elements of P_3.

$$\begin{aligned} D(x^2) &= 2x + 0 \cdot 1 \\ D(x) &= 0x + 1 \cdot 1 \\ D(1) &= 0x + 0 \cdot 1 \end{aligned}$$

In P_2, the coordinate vectors for $D(x^2)$, $D(x)$, $D(1)$ are $(2, 0)^T$, $(0, 1)^T$, $(0, 0)^T$, respectively. The matrix A is formed using these vectors as its columns.

$$A = \begin{pmatrix} 2 & 0 & 0 \\ 0 & 1 & 0 \end{pmatrix}$$

If $p(x) = ax^2 + bx + c$, then the coordinate vector of p with respect to the ordered basis of P_3 is $(a, b, c)^T$. To find the coordinate vector of $D(p)$ with respect to the ordered basis of P_2, we simply multiply

$$\begin{pmatrix} 2 & 0 & 0 \\ 0 & 1 & 0 \end{pmatrix} \begin{pmatrix} a \\ b \\ c \end{pmatrix} = \begin{pmatrix} 2a \\ b \end{pmatrix}$$

Thus $D(ax^2 + bx + c) = 2ax + b$.

EXERCISES

1. For each of the following linear transformations L mapping R^3 into R^2, find a matrix A such that $L(\mathbf{x}) = A\mathbf{x}$ for every $\mathbf{x}$ in R^3.
   (a) $L((x_1, x_2, x_3)^T) = (x_1 + x_2, 0)^T$
   (b) $L((x_1, x_2, x_3)^T) = (x_1, x_2)^T$
   (c) $L((x_1, x_2, x_3)^T) = (x_2 - x_1, x_3 - x_2)^T$

2. For each of the following linear transformations L from R^3 into R^3 find a matrix A such that $L(\mathbf{x}) = A(\mathbf{x})$ for every $\mathbf{x}$ in R^3.
   (a) $L((x_1, x_2, x_3)^T) = (x_3, x_2, x_1)^T$
   (b) $L((x_1, x_2, x_3)^T) = (x_1, x_1 + x_2, x_1 + x_2 + x_3)^T$
   (c) $L((x_1, x_2, x_3)^T) = (2x_3, x_2 + 3x_1, 2x_1 - x_3)^T$

3. Let L be the linear operator mapping P_2 into R^2 defined by

$$L(p(x)) = \begin{bmatrix} \int_0^1 p(x)\,dx \\ p(0) \end{bmatrix}$$

   Find a matrix A such that

$$L(\alpha + \beta x) = A\begin{pmatrix} \alpha \\ \beta \end{pmatrix}$$

4. The linear operator L defined by

$$L(p(x)) = p'(x) + p(0)$$

   maps P_3 into P_2. Find the matrix representation of L with respect to the ordered bases $[x^2, x, 1]$ and $[2, 1 - x]$. For each of the following vectors $p(x)$ in P_3, find the coordinates of $L(p(x))$ with respect to the ordered basis $[2, 1 - x]$.
   (a) $x^2 + 2x - 3$
   (b) $x^2 + 1$
   (c) $3x$
   (d) $4x^2 + 2x$

5. Let

$$\mathbf{y}_1 = \begin{bmatrix} 1 \\ 1 \\ 1 \end{bmatrix}, \quad \mathbf{y}_2 = \begin{bmatrix} 1 \\ 1 \\ 0 \end{bmatrix}, \quad \mathbf{y}_3 = \begin{bmatrix} 1 \\ 0 \\ 0 \end{bmatrix}$$

   and let $\mathcal{I}$ be the identity operator on R^3.
   (a) Find the coordinates of $\mathcal{I}(\mathbf{e}_1)$, $\mathcal{I}(\mathbf{e}_2)$, and $\mathcal{I}(\mathbf{e}_3)$ with respect to $[\mathbf{y}_1, \mathbf{y}_2, \mathbf{y}_3]$.
   (b) Find a matrix A such that $A\mathbf{x}$ is the coordinate vector of $\mathbf{x}$ with respect to $[\mathbf{y}_1, \mathbf{y}_2, \mathbf{y}_3]$.

6. Let S be the subspace of $C[a, b]$ spanned by e^x, xe^x, and x^2e^x. Let D be the differentiation operator on S. Find the matrix representing D with respect to $[e^x, xe^x, x^2e^x]$.

7. Let L be a linear operator mapping a vector space V into itself. Let A be the matrix representing L with respect to the order basis $[\mathbf{v}_1, \ldots, \mathbf{v}_n]$ of V [i.e., $L(\mathbf{v}_j) = \sum_{i=1}^n a_{ij}\mathbf{v}_i$, $j = 1, \ldots, n$]. Show that A^m is the matrix representing L^m with respect to $[\mathbf{v}_1, \ldots, \mathbf{v}_n]$.

3. SIMILARITY

In this section we consider operators mapping an n-dimensional vector space V into itself. The matrix representation of such operators will depend upon the ordered basis chosen for V.

Change of Basis

Let $[\mathbf{w}_1, \ldots, \mathbf{w}_n]$ and $[\mathbf{v}_1, \ldots, \mathbf{v}_n]$ be two ordered bases for a vector space V. Each vector $\mathbf{w}_j$ can then be expressed as a linear combination of the $\mathbf{v}_i$'s,

$$
\begin{aligned}
\mathbf{w}_1 &= s_{11}\mathbf{v}_1 + s_{21}\mathbf{v}_2 + \cdots + s_{n1}\mathbf{v}_n \\
\mathbf{w}_2 &= s_{12}\mathbf{v}_1 + s_{22}\mathbf{v}_2 + \cdots + s_{n2}\mathbf{v}_n \\
&\vdots \\
\mathbf{w}_n &= s_{1n}\mathbf{v}_1 + s_{2n}\mathbf{v}_2 + \cdots + s_{nn}\mathbf{v}_n
\end{aligned} \tag{1}
$$

Let $\mathbf{v} \in V$. If $\mathbf{x}$ is the coordinate vector of $\mathbf{v}$ with respect to $[\mathbf{w}_1, \ldots, \mathbf{w}_n]$, it follows from (1) that

$$
\begin{aligned}
\mathbf{v} &= x_1\mathbf{w}_1 + x_2\mathbf{w}_2 + \cdots + x_n\mathbf{w}_n \\
&= \left(\sum_{j=1}^n s_{1j}x_j\right)\mathbf{v}_1 + \left(\sum_{j=1}^n s_{2j}x_j\right)\mathbf{v}_2 + \cdots + \left(\sum_{j=1}^n s_{nj}x_j\right)\mathbf{v}_n
\end{aligned}
$$

Thus, if $\mathbf{y}$ is the coordinate vector of $\mathbf{v}$ with respect to $[\mathbf{v}_1, \ldots, \mathbf{v}_n]$, then

$$y_i = \sum_{j=1}^n s_{ij}x_j \qquad i = 1, \ldots, n$$

and hence

$$\mathbf{y} = S\mathbf{x}$$

The matrix S defined by (1) is referred to as the *transition matrix*. Once S has been determined, it is a simple matter to change coordinate systems.

To find the coordinates of $\mathbf{v} = x_1\mathbf{w}_1 + \cdots + x_n\mathbf{w}_n$ with respect to $[\mathbf{v}_1, \ldots, \mathbf{v}_n]$ one need only calculate $\mathbf{y} = S\mathbf{x}$.

The transition matrix S corresponding to the change of basis from $[\mathbf{w}_1, \ldots, \mathbf{w}_n]$ to $[\mathbf{v}_1, \ldots, \mathbf{v}_n]$ can be characterized by the condition:

$$(2) \quad S\mathbf{x} = \mathbf{y} \quad \text{if and only if} \quad x_1\mathbf{w}_1 + \cdots + x_n\mathbf{w}_n = y_1\mathbf{v}_1 + \cdots + y_n\mathbf{v}_n$$

Taking $\mathbf{y} = \mathbf{0}$ in (2), we see that $S\mathbf{x} = \mathbf{0}$ implies

$$x_1\mathbf{w}_1 + \cdots + x_n\mathbf{w}_n = \mathbf{0}$$

Since the $\mathbf{w}_i$'s are linearly independent, it follows that $\mathbf{x} = \mathbf{0}$. Thus the equation $S\mathbf{x} = \mathbf{0}$ has only the trivial solution and hence the matrix S is nonsingular. The inverse matrix is characterized by the condition

$$S^{-1}\mathbf{y} = \mathbf{x} \qquad \text{if and only if} \qquad y_1\mathbf{v}_1 + \cdots + y_n\mathbf{v}_n = x_1\mathbf{w}_1 + \cdots + x_n\mathbf{w}_n$$

Thus S^{-1} is the transition matrix used to change basis from $[\mathbf{v}_1, \ldots, \mathbf{v}_n]$ to $[\mathbf{w}_1, \ldots, \mathbf{w}_n]$.

EXAMPLE 1. Suppose that in P_3 we want to change from the ordered basis $[1, x, x^2]$ to the ordered basis $[1, 2x, 4x^2 - 2]$. To find the transition matrix, we must write the old basis elements in terms of the new ones.

$$\begin{aligned} 1 &= 1 \cdot 1 + 0 \cdot 2x + 0 \cdot (4x^2 - 2) \\ x &= 0 \cdot 1 + \tfrac{1}{2} \cdot 2x + 0 \cdot (4x^2 - 2) \\ x^2 &= \tfrac{1}{2} \cdot 1 + 0 \cdot 2x + \tfrac{1}{4} \cdot (4x^2 - 2) \end{aligned}$$

Thus

$$S = \begin{pmatrix} 1 & 0 & \frac{1}{2} \\ 0 & \frac{1}{2} & 0 \\ 0 & 0 & \frac{1}{4} \end{pmatrix}$$

Given any $p(x) = a + bx + cx^2$ in P_3 to find the coordinates of $p(x)$ with respect to $[1, 2x, 4x^2 - 2]$ one simply multiplies

$$\begin{pmatrix} 1 & 0 & \frac{1}{2} \\ 0 & \frac{1}{2} & 0 \\ 0 & 0 & \frac{1}{4} \end{pmatrix} \begin{pmatrix} a \\ b \\ c \end{pmatrix} = \begin{pmatrix} a + \frac{1}{2}c \\ \frac{1}{2}b \\ \frac{1}{4}c \end{pmatrix}$$

Thus

$$p(x) = \left(a + \tfrac{1}{2}c\right) \cdot 1 + \left(\tfrac{1}{2}b\right) \cdot 2x + \tfrac{1}{4}c \cdot (4x^2 - 2)$$

EXAMPLE 2. To change from the ordered basis $[1, 2x, 4x^2-2]$ to $[1, x, x^2]$, one computes the transition matrix as follows:

$$\begin{aligned} 1 &= 1 \cdot 1 + 0x + 0x^2 \\ 2x &= 0 \cdot 1 + 2x + 0x^2 \\ 4x^2 - 2 &= -2 \cdot 1 + 0x + 4x^2 \end{aligned}$$

The transition matrix is given by

$$T = \begin{bmatrix} 1 & 0 & -2 \\ 0 & 2 & 0 \\ 0 & 0 & 4 \end{bmatrix}$$

Note that T is the inverse of the matrix S in Example 1. Thus one could have solved Example 1 by first constructing T and then computing its inverse.

We have seen that each transition matrix is nonsingular. Actually, any nonsingular matrix can be thought of as a transition matrix. If S is an $n \times n$ nonsingular matrix and $[\mathbf{v}_1, \ldots, \mathbf{v}_n]$ is an ordered basis for V, then define $\mathbf{w}_1, \mathbf{w}_2, \ldots, \mathbf{w}_n$ by (1). To see that the $\mathbf{w}_j$'s are linearly independent, suppose

$$\sum_{j=1}^{n} x_j \mathbf{w}_j = \mathbf{0}$$

It follows from (1) that

$$\sum_{i=1}^{n} \left(\sum_{j=1}^{n} s_{ij} x_j \right) \mathbf{v}_i = \mathbf{0}$$

By the linear independence of the $\mathbf{v}_i$'s it follows that

$$\sum_{j=1}^{n} s_{ij} x_j = 0 \qquad i = 1, \ldots, n$$

or, equivalently,

$$S\mathbf{x} = \mathbf{0}$$

Since S is nonsingular, $\mathbf{x}$ must equal $\mathbf{0}$. Therefore $\mathbf{w}_1, \ldots, \mathbf{w}_n$ are linearly independent and hence they form a basis for V. The matrix S is the transition matrix corresponding to the change from the ordered basis $[\mathbf{w}_1, \ldots, \mathbf{w}_n]$ to $[\mathbf{v}_1, \ldots, \mathbf{v}_n]$.

Given a linear operator L mapping an n-dimensional vector space V into itself, the matrix representing L will depend upon the ordered basis chosen for V. If $[\mathbf{v}_1, \ldots, \mathbf{v}_n]$ is an ordered basis for V and

$$L(\mathbf{v}_j) = a_{1j}\mathbf{v}_1 + a_{2j}\mathbf{v}_2 + \cdots + a_{nj}\mathbf{v}_n \qquad j = 1, \ldots, n$$

then $A = (a_{ij})$ is the matrix representing L with respect to $[\mathbf{v}_1, \ldots, \mathbf{v}_n]$. Let

$[\mathbf{w}_1, \ldots, \mathbf{w}_n]$ be another ordered basis for V and let B be the matrix representing L with respect to $[\mathbf{w}_1, \ldots, \mathbf{w}_n]$. The following theorem shows the relation between A and B.

Theorem 4.3.1 *Let $[\mathbf{v}_1, \ldots, \mathbf{v}_n]$ and $[\mathbf{w}_1, \ldots, \mathbf{w}_n]$ be two ordered bases for a vector space V and let L be a linear operator mapping V into itself. Let S be the transition matrix representing the change from $[\mathbf{w}_1, \ldots, \mathbf{w}_n]$ to $[\mathbf{v}_1, \ldots, \mathbf{v}_n]$. If A is the matrix representing L with respect to $[\mathbf{v}_1, \ldots, \mathbf{v}_n]$ and B is the matrix representing L with respect to $[\mathbf{w}_1, \ldots, \mathbf{w}_n]$, then $B = S^{-1}AS$.*

PROOF. Let $\mathbf{x}$ be any vector in R^n and let

$$\mathbf{v} = x_1\mathbf{w}_1 + x_2\mathbf{w}_2 + \cdots + x_n\mathbf{w}_n$$

Let

$$\text{(3)} \qquad \mathbf{y} = S\mathbf{x}, \qquad \mathbf{z} = B\mathbf{x}, \qquad \mathbf{t} = A\mathbf{y}$$

It follows from the definition of S that $\mathbf{y}$ is the coordinate vector of $\mathbf{v}$ with respect to $[\mathbf{v}_1, \ldots, \mathbf{v}_n]$.

$$\mathbf{v} = y_1\mathbf{v}_1 + \cdots + y_n\mathbf{v}_n$$

Since A and B both represent L, we have

$$z_1\mathbf{w}_1 + \cdots + z_n\mathbf{w}_n = L(\mathbf{v}) = t_1\mathbf{v}_1 + \cdots + t_n\mathbf{v}_n$$

Therefore,

$$\text{(4)} \qquad S^{-1}\mathbf{t} = \mathbf{z}$$

It follows from (3) and (4) that

$$S^{-1}AS\mathbf{x} = S^{-1}A\mathbf{y} = S^{-1}\mathbf{t} = \mathbf{z} = B\mathbf{x}$$

(see Figure 4.3.1). Thus

$$S^{-1}AS\mathbf{x} = B\mathbf{x}$$

for every $\mathbf{x} \in R^n$, and hence $S^{-1}AS = B$.

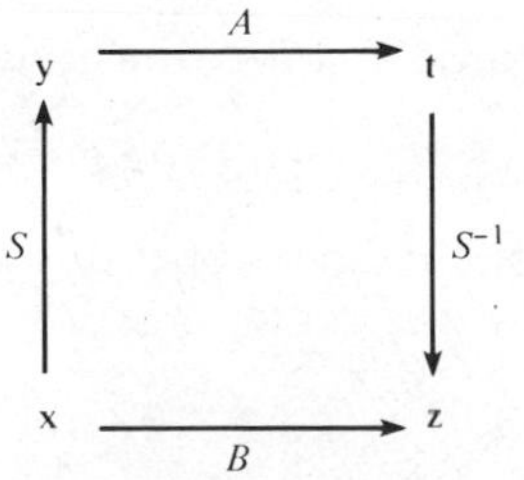

Figure 4.3.1

Definition. Let A and B be $n \times n$ matrices. B is said to be **similar** to A if there exists a nonsingular matrix S such that $B = S^{-1}AS$.

Note that if B is similar to A, then $A = SBS^{-1}$ is similar to B. Thus we may simply say that A and B are similar matrices.

It follows from Theorem 4.3.1 that if A and B are $n \times n$ matrices representing the same operator L, then A and B are similar. Conversely, suppose that A represents L with respect to the ordered basis $[\mathbf{v}_1, \ldots, \mathbf{v}_n]$ and $B = S^{-1}AS$. If $\mathbf{w}_1, \ldots, \mathbf{w}_n$ are defined by (1), then $[\mathbf{w}_1, \ldots, \mathbf{w}_n]$ is an ordered basis for V and B is the matrix representing L with respect to $[\mathbf{w}_1, \ldots, \mathbf{w}_n]$.

EXAMPLE 3. Let D be the differentiation operator on P_3. Find the matrix B representing D with respect to $[1, x, x^2]$ and the matrix A representing D with respect to $[1, 2x, 4x^2 - 2]$.

SOLUTION

$$\begin{aligned} D(1) &= 0 \cdot 1 + 0 \cdot x + 0 \cdot x^2 \\ D(x) &= 1 \cdot 1 + 0 \cdot x + 0 \cdot x^2 \\ D(x^2) &= 0 \cdot 1 + 2 \cdot x + 0 \cdot x^2 \end{aligned}$$

The matrix B is then given by

$$B = \begin{pmatrix} 0 & 1 & 0 \\ 0 & 0 & 2 \\ 0 & 0 & 0 \end{pmatrix}$$

Applying D to 1, $2x$, and $4x^2 - 2$, one obtains

$$\begin{aligned} D(1) &= 0 \cdot 1 + 0 \cdot 2x + 0 \cdot (4x^2 - 2) \\ D(2x) &= 2 \cdot 1 + 0 \cdot 2x + 0 \cdot (4x^2 - 2) \\ D(4x^2 - 2) &= 0 \cdot 1 + 4 \cdot 2x + 0 \cdot (4x^2 - 2) \end{aligned}$$

Thus

$$A = \begin{pmatrix} 0 & 2 & 0 \\ 0 & 0 & 4 \\ 0 & 0 & 0 \end{pmatrix}$$

The reader may verify that if S is the transition matrix of Example 1, then

$$B = S^{-1}AS$$

EXAMPLE 4. Let L be the linear operator mapping R^3 into R^3 defined by $L(\mathbf{x}) = A\mathbf{x}$, where

$$A = \begin{pmatrix} 2 & 2 & 0 \\ 1 & 1 & 2 \\ 1 & 1 & 2 \end{pmatrix}$$

Thus the matrix A represents L with respect to $[\mathbf{e}_1, \mathbf{e}_2, \mathbf{e}_3]$. Find the matrix representing L with respect to $[\mathbf{y}_1, \mathbf{y}_2, \mathbf{y}_3]$, where

$$\mathbf{y}_1 = \begin{pmatrix} 1 \\ -1 \\ 0 \end{pmatrix}, \qquad \mathbf{y}_2 = \begin{pmatrix} -2 \\ 1 \\ 1 \end{pmatrix}, \qquad \mathbf{y}_3 = \begin{pmatrix} 1 \\ 1 \\ 1 \end{pmatrix}$$

SOLUTION

$$\begin{aligned} L(\mathbf{y}_1) &= A\mathbf{y}_1 = \mathbf{0} = 0\mathbf{y}_1 + 0\mathbf{y}_2 + 0\mathbf{y}_3 \\ L(\mathbf{y}_2) &= A\mathbf{y}_2 = \mathbf{y}_2 = 0\mathbf{y}_1 + 1\mathbf{y}_2 + 0\mathbf{y}_3 \\ L(\mathbf{y}_3) &= A\mathbf{y}_3 = 4\mathbf{y}_3 = 0\mathbf{y}_1 + 0\mathbf{y}_2 + 4\mathbf{y}_3 \end{aligned}$$

Thus the matrix representing L with respect to $[\mathbf{y}_1, \mathbf{y}_2, \mathbf{y}_3]$ is

$$D = \begin{pmatrix} 0 & 0 & 0 \\ 0 & 1 & 0 \\ 0 & 0 & 4 \end{pmatrix}$$

In Example 4 the linear operator L is represented by a diagonal matrix D with respect to the basis $[\mathbf{y}_1, \mathbf{y}_2, \mathbf{y}_3]$. It is much simpler to work with D than with A. For example, it is easier to compute $D\mathbf{x}$ and $D^n\mathbf{x}$ than $A\mathbf{x}$ and $A^n\mathbf{x}$. Generally, it is desirable to find as simple a representation as possible for a linear operator. In particular, if the operator can be represented by a diagonal matrix, that is usually the preferred representation. The problem of finding a diagonal representation for a linear operator will be studied in Chapter 6.

EXERCISES

1. Find the transition matrix representing the change of coordinates on P_3 from the ordered basis

$$[1, x, x^2]$$

to the ordered basis

$$[1, 1 + x, 1 + x + x^2]$$

2. Let L be the operator on P_3 defined by

$$L(p(x)) = xp'(x) + p''(x)$$

(a) Find the matrix A representing L with respect to $[1, x, x^2]$.
(b) Find the matrix B representing L with respect to $[1, x, 1 + x^2]$.
(c) Find the matrix S such that $B = S^{-1}AS$.
(d) If $p(x) = a_0 + a_1x + a_2(1 + x^2)$, calculate $L^n(p(x))$.

3. Let V be the subspace of $C[a, b]$ spanned by $1, e^x, e^{-x}$ and let D be the differentiation operator on V.

(a) Find the transition matrix S representing the change of coordinates from $[1, e^x, e^{-x}]$ to $[1, \cosh x, \sinh x]$. ($\cosh x = \frac{1}{2}(e^x + e^{-x})$ and $\sinh x = \frac{1}{2}(e^x - e^{-x})$.)
(b) Find the matrix A representing D with respect to $[1, \cosh x, \sinh x]$.
(c) Find the matrix B representing D with respect to $[1, e^x, e^{-x}]$.
(d) Verify that $B = S^{-1}AS$.

4. Prove that if A is similar to B and B is similar to C, then A is similar to C.

5. Suppose that $A = S\Lambda S^{-1}$, where Λ is a diagonal matrix with diagonal elements $\lambda_1, \lambda_2, \ldots, \lambda_n$. Show that:
(a) $A\mathbf{s}_i = \lambda_i \mathbf{s}_i$, $i = 1, \ldots, n$
(b) If $\mathbf{x} = \alpha_1 \mathbf{s}_1 + \alpha_2 \mathbf{s}_2 + \cdots + \alpha_n \mathbf{s}_n$, then

$$A^k\mathbf{x} = \alpha_1\lambda_1^k\mathbf{s}_1 + \alpha_2\lambda_2^k\mathbf{s}_2 + \cdots + \alpha_n\lambda_n^k\mathbf{s}_n$$

(c) Suppose that $|\lambda_i| < 1$ for $i = 1, \ldots, n$. What happens to $A^k\mathbf{x}$ as $k \to \infty$? Explain.

6. Suppose that $A = ST$, where S is nonsingular. Let $B = TS$. Show that B is similar to A.

5

Orthogonality

One can add to the structure of a vector space by defining a scalar or inner product. Such a product is not a true vector multiplication, since to every pair of vectors it associates a scalar rather than a third vector. For example, in R^2 one can define the scalar product of two vectors $\mathbf{x}$ and $\mathbf{y}$ to be $\mathbf{x}^T\mathbf{y}$. One can think of vectors in R^2 as directed line segments initiating at the origin. It is not difficult to show that the angle between two line segments will be a right angle if and only if the scalar product of the corresponding vectors is zero. In general, if V is a vector space with a scalar product, then two vectors in V are said to be *orthogonal* if their scalar product is zero.

One can think of orthogonality as a generalization of the concept of *perpendicularity* to any vector space with an inner product. To see the significance of this, consider the following problem: Let l be a line passing through the origin and let Q be a point not on l. Find the point P on l that is closest to Q. The solution P to this problem is characterized by the condition that QP is perpendicular to OP (see Figure 5.0.1). If we think of the line l as corresponding to a subspace of R^2 and $\mathbf{v} = OQ$ as a vector in R^2, then the problem is to find a vector in the subspace that is "closest" to $\mathbf{v}$. The solution $\mathbf{p}$ will be characterized by the property that $\mathbf{p}$ is orthogonal to $\mathbf{v} - \mathbf{p}$ (see Figure 5.0.1). In the setting of a vector space with an inner product one is able to consider general "least squares" problems. In these problems one is given a vector $\mathbf{v}$ in V and a subspace W. One wishes to find a vector in W which is "closest" to $\mathbf{v}$. A solution $\mathbf{p}$ must be orthogonal to $\mathbf{v} - \mathbf{p}$. This orthogonality condition provides the key to solving the least squares problem. Least squares problems occur in many statistical applications involving data fitting.

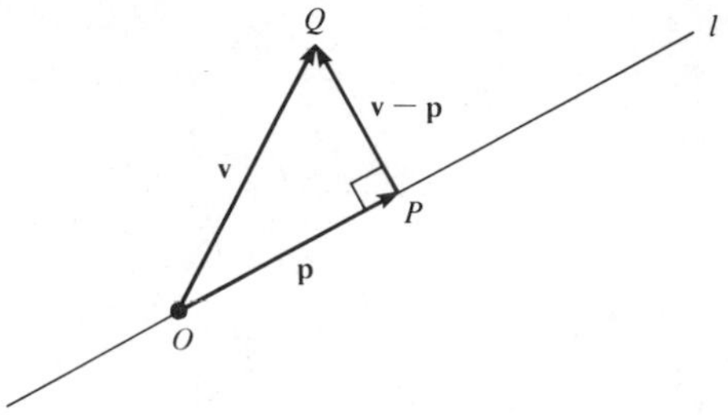

Figure 5.0.1

1. THE SCALAR PRODUCT IN R^n

Two vectors $\mathbf{x}$ and $\mathbf{y}$ in R^n may be regarded as $n \times 1$ matrices. One can then form the matrix product $\mathbf{x}^T\mathbf{y}$. This product is a 1×1 matrix which may be regarded as a vector in R^1, or more simply, as a real number. The product $\mathbf{x}^T\mathbf{y}$ is called the *scalar product* of $\mathbf{x}$ and $\mathbf{y}$. In particular, if $\mathbf{x} = (x_1, \ldots, x_n)^T$ and $\mathbf{y} = (y_1, \ldots, y_n)^T$, then

$$\mathbf{x}^T\mathbf{y} = x_1y_1 + x_2y_2 + \cdots + x_ny_n$$

EXAMPLE 1. If

$$\mathbf{x} = \begin{bmatrix} 3 \\ -2 \\ 1 \end{bmatrix} \quad \text{and} \quad \mathbf{y} = \begin{bmatrix} 4 \\ 3 \\ 2 \end{bmatrix}$$

then

$$\mathbf{x}^T\mathbf{y} = (3, -2, 1)\begin{bmatrix}4\\3\\2\end{bmatrix} = 3 \cdot 4 - 2 \cdot 3 + 1 \cdot 2 = 8$$

The Scalar Product in R^2 and R^3

In order to see the geometrical significance of the scalar product, let us restrict our attention to R^2 and R^3. Vectors in R^2 and R^3 can be represented by directed line segments. Given a vector $\mathbf{x}$ in either R^2 or R^3, its Euclidean length can be defined in terms of the scalar product.

$$\|\mathbf{x}\| = \sqrt{x_1^2 + x_2^2} = (\mathbf{x}^T\mathbf{x})^{1/2} \qquad \text{if} \quad \mathbf{x} \in R^2$$

$$\|\mathbf{x}\| = \sqrt{x_1^2 + x_2^2 + x_3^2} = (\mathbf{x}^T\mathbf{x})^{1/2} \qquad \text{if} \quad \mathbf{x} \in R^3$$

Given two nonzero vectors $\mathbf{x}$ and $\mathbf{y}$, we can think of them as line segments initiating at the same point. The angle between the two vectors is then defined as the angle θ between the line segments.

Theorem 5.1.1. *If* $\mathbf{x}$ *and* $\mathbf{y}$ *are two nonzero vectors in either* R^2 *or* R^3 *and* θ *is the angle between them, then*

(1) $$\mathbf{x}^T\mathbf{y} = \|\mathbf{x}\|\|\mathbf{y}\| \cos\theta$$

PROOF. We will prove the result for R^2. The proof for R^3 is similar. The vectors $\mathbf{x}$, $\mathbf{y}$, and $\mathbf{y} - \mathbf{x}$ may be used to form a triangle as in Figure 5.1.1. By the law of cosines,

$$\|\mathbf{y} - \mathbf{x}\|^2 = \|\mathbf{x}\|^2 + \|\mathbf{y}\|^2 - 2\|\mathbf{x}\|\|\mathbf{y}\| \cos\theta$$

or

$$\begin{aligned}\|\mathbf{x}\|\|\mathbf{y}\| \cos\theta &= \tfrac{1}{2}(\|\mathbf{x}\|^2 + \|\mathbf{y}\|^2 - \|\mathbf{x} - \mathbf{y}\|^2)\\ &= \tfrac{1}{2}\left[x_1^2 + x_2^2 + y_1^2 + y_2^2 - (x_1 - y_1)^2 - (x_2 - y_2)^2\right]\\ &= x_1y_1 + x_2y_2\\ &= \mathbf{x}^T\mathbf{y}\end{aligned}$$

Corollary 5.1.2 (Cauchy–Schwarz Inequality). *If* $\mathbf{x}$ *and* $\mathbf{y}$ *are vectors in either* R^2 *or* R^3, *then*

(2) $$|\mathbf{x}^T\mathbf{y}| \leqslant \|\mathbf{x}\|\,\|\mathbf{y}\|$$

with equality holding if and only if one of the vectors is $\mathbf{0}$ *or* $\mathbf{x} = \alpha\mathbf{y}$.

PROOF. The inequality follows from (1). If one of the vectors is $\mathbf{0}$, then both sides of (2) are 0. If $\mathbf{x} = \alpha\mathbf{y}$, then $\cos\theta = \pm 1$ in (1) and hence equality holds in (2).

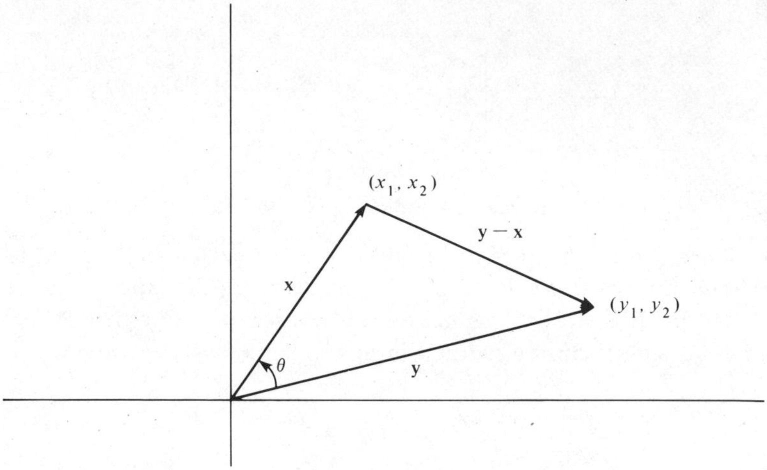

Figure 5.1.1

If $\mathbf{x}^T\mathbf{y} = 0$, it follows from Theorem 5.1.1 that either one of the vectors is the zero vector or $\cos\theta = 0$. If $\cos\theta = 0$, the angle between the vectors is a right angle.

Definition. The vectors $\mathbf{x}$ and $\mathbf{y}$ in R^2 (or R^3) are said to be **orthogonal** if $\mathbf{x}^T\mathbf{y} = 0$.

EXAMPLE 2

(a) The vector $\mathbf{0}$ is orthogonal to every vector in R^2.

(b) The vectors $\begin{pmatrix} 3 \\ 2 \end{pmatrix}$ and $\begin{pmatrix} -4 \\ 6 \end{pmatrix}$ are orthogonal in R^2.

(c) The vectors $\begin{bmatrix} 2 \\ -3 \\ 1 \end{bmatrix}$ and $\begin{bmatrix} 1 \\ 1 \\ 1 \end{bmatrix}$ are orthogonal in R^3.

Scalar and Vector Projections

The scalar product can be used to find the component of one vector in the direction of another. Let $\mathbf{x}$ and $\mathbf{y}$ be nonzero vectors in either R^2 or R^3. We would like to write $\mathbf{x}$ as a sum of the form $\mathbf{p} + \mathbf{z}$, where $\mathbf{p}$ is in the direction of $\mathbf{y}$ and $\mathbf{z}$ is orthogonal to $\mathbf{p}$ (see Figure 5.1.2). To do this, let $\mathbf{u} = (1/\|\mathbf{y}\|)\mathbf{y}$. Thus $\mathbf{u}$ is a unit vector (length 1) in the direction of $\mathbf{y}$. We wish to find α such that $\mathbf{p} = \alpha\mathbf{u}$ is orthogonal to $\mathbf{z} = \mathbf{x} - \alpha\mathbf{u}$. In order for $\mathbf{p}$ and $\mathbf{z}$ to be orthogonal, the scalar α must satisfy

$$\begin{aligned}\alpha &= \|\mathbf{x}\|\cos\theta \\ &= \frac{\|\mathbf{x}\|\|\mathbf{y}\|\cos\theta}{\|\mathbf{y}\|} \\ &= \frac{\mathbf{x}^T\mathbf{y}}{\|\mathbf{y}\|}\end{aligned}$$

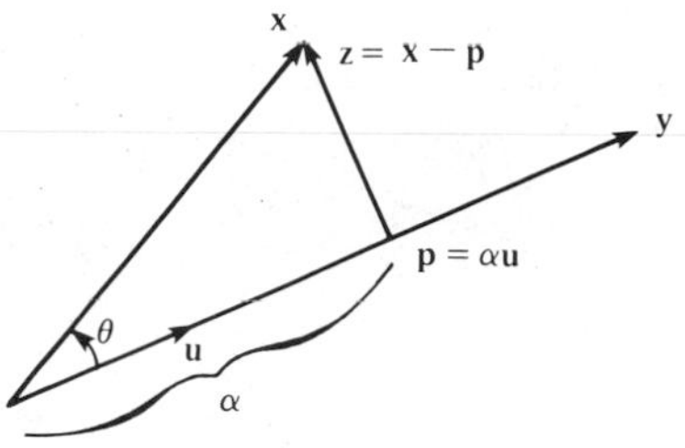

Figure 5.1.2

The scalar α is called the *scalar projection* of $\mathbf{x}$ onto $\mathbf{y}$ and the vector $\mathbf{p}$ is called the *vector projection* of $\mathbf{x}$ onto $\mathbf{y}$.

> Scalar Projection of $\mathbf{x}$ onto $\mathbf{y}$
>
> $$\alpha = \frac{\mathbf{x}^T\mathbf{y}}{\|\mathbf{y}\|}$$
>
> Vector Projection of $\mathbf{x}$ onto $\mathbf{y}$
>
> $$\mathbf{p} = \alpha\mathbf{u} = \alpha\frac{1}{\|\mathbf{y}\|}\mathbf{y} = \frac{\mathbf{x}^T\mathbf{y}}{\mathbf{y}^T\mathbf{y}}\mathbf{y}$$

EXAMPLE 3. Find the point on the line $y = \frac{1}{3}x$ that is closest to the point (1, 4) (see Figure 5.1.3).

SOLUTION. The vector $\mathbf{w} = (3, 1)^T$ is a vector in the direction of the line $y = \frac{1}{3}x$. Let $\mathbf{v} = (1, 4)^T$. If Q is the desired point, then Q^T is the vector projection of $\mathbf{v}$ onto $\mathbf{w}$.

$$Q^T = \left(\frac{\mathbf{v}^T\mathbf{w}}{\mathbf{w}^T\mathbf{w}}\right)\mathbf{w} = \frac{7}{10}\begin{pmatrix}3\\1\end{pmatrix} = \begin{pmatrix}2.1\\0.7\end{pmatrix}$$

Thus $Q = (2.1, 0.7)$ is the closest point.

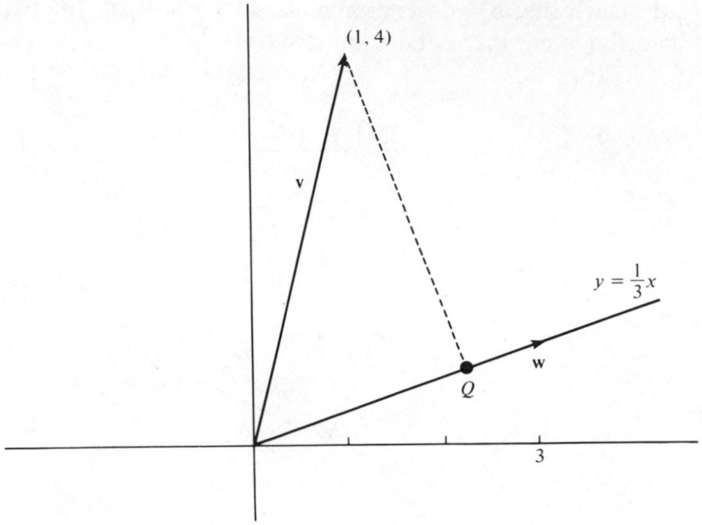

Figure 5.1.3

Notation

If P_1 and P_2 are two points in 3-space, we will denote the vector from P_1 to P_2 by $\overline{P_1P_2}$.

If $\mathbf{N}$ is a nonzero vector and P_0 is a fixed point, the set of points P such that $\overline{P_0P}$ is orthogonal to $\mathbf{N}$ forms a plane in 3-space. The vector $\mathbf{N}$ is said to be normal to the plane. Thus, if π is the plane normal to $\mathbf{N}$ which passes through P_0, then $P = (x, y, z)$ is a point on π if and only if

$$(\overline{P_0P})^T\mathbf{N} = 0$$

If $\mathbf{N} = (a, b, c)^T$ and $P_0 = (x_0, y_0, z_0)$, this equation can be written in the form

$$a(x - x_0) + b(y - y_0) + c(z - z_0) = 0$$

EXAMPLE 4. Find the equation of the plane passing through the point $(2, -1, 3)$ and normal to the vector $\mathbf{N} = (2, 3, 4)^T$.

SOLUTION. $\overline{P_0P} = (x - 2, y + 1, z - 3)^T$. The equation is $(\overline{P_0P})^T\mathbf{N} = 0$, or

$$2(x - 2) + 3(y + 1) + 4(z - 3) = 0$$

EXAMPLE 5. Find the distance from the point $(2, 0, 0)$ to the plane $x + 2y + 2z = 0$.

SOLUTION. The vector $\mathbf{N} = (1, 2, 2)^T$ is normal to the plane and the plane passes through the origin. Let $\mathbf{v} = (2, 0, 0)^T$. The distance d from $(2, 0, 0)$ to the plane is simply the absolute value of the scalar projection of $\mathbf{v}$ onto $\mathbf{N}$. Thus

$$d = \frac{|\mathbf{v}^T\mathbf{N}|}{\|\mathbf{N}\|} = \frac{2}{3}$$

Orthogonality in R^n

The definitions that have been given for R^2 and R^3 can all be generalized to R^n. Indeed, if $\mathbf{x} \in R^n$, then the Euclidean length of $\mathbf{x}$ is defined by

$$\|\mathbf{x}\| = (\mathbf{x}^T\mathbf{x})^{1/2} = \left(x_1^2 + x_2^2 + \cdots + x_n^2\right)^{1/2}$$

and the angle θ between two vectors $\mathbf{x}$ and $\mathbf{y}$ is given by

$$\cos\theta = \frac{\mathbf{x}^T\mathbf{y}}{\|\mathbf{x}\|\,\|\mathbf{y}\|}$$

The vectors $\mathbf{x}$ and $\mathbf{y}$ are said to be *orthogonal* if $\mathbf{x}^T\mathbf{y} = 0$. Often the symbol "$\perp$" is used to indicate orthogonality. Thus, if $\mathbf{x}$ and $\mathbf{y}$ are orthogonal, we will write $\mathbf{x} \perp \mathbf{y}$. Vector and scalar projections are defined in R^n in the same way that they were defined for R^2. One of the main applications of these concepts is in the solution of least squares problems. Least squares problems will be studied in Section 5.

EXERCISES

1. Find the angle between the vectors $\mathbf{v}$ and $\mathbf{w}$ in each of the following.
(a) $\mathbf{v} = (2, 1, 3)^T$, $\mathbf{w} = (6, 3, 9)^T$ (b) $\mathbf{v} = (2, -3)^T$, $\mathbf{w} = (3, 2)^T$
(c) $\mathbf{v} = (4, 1)^T$, $\mathbf{w} = (3, 2)^T$ (d) $\mathbf{v} = (-2, 3, 1)^T$, $\mathbf{w} = (1, 2, 4)^T$

2. For each of the pairs of vectors in Exercise 1, find the scalar projection of $\mathbf{v}$ onto $\mathbf{w}$. Also find the vector projection of $\mathbf{v}$ onto $\mathbf{w}$.

3. Find the point on the line $y = 2x$ that is closest to the point (5, 2).

4. Find the point on the line $y = 2x + 1$ that is closest to the point (5, 2).

5. Find the distance from the point (1, 1, 1) to the plane $2x + 2y + z = 3$.

6. If $\mathbf{x} = (x_1, x_2)^T$, $\mathbf{y} = (y_1, y_2)^T$, and $\mathbf{z} = (z_1, z_2)^T$ are arbitrary vectors in R^2, prove:
(a) $\mathbf{x}^T\mathbf{x} \geqslant 0$ (b) $\mathbf{x}^T\mathbf{y} = \mathbf{y}^T\mathbf{x}$
(c) $\mathbf{x}^T(\mathbf{y} + \mathbf{z}) = \mathbf{x}^T\mathbf{y} + \mathbf{x}^T\mathbf{z}$

7. If $\mathbf{u}$ and $\mathbf{v}$ are any vectors in R^2, show that $\|\mathbf{u}+\mathbf{v}\|^2 \leqslant (\|\mathbf{u}\| + \|\mathbf{v}\|)^2$ and hence $\|\mathbf{u}+\mathbf{v}\| \leqslant \|\mathbf{u}\| + \|\mathbf{v}\|$. When does equality hold? Give a geometric interpretation of the inequality.

8. Let l_1 be the line $y = m_1x + b_1$, $m_1 \neq 0$, and l be the line $y = mx + b$. Show that l is perpendicular to l_1 if and only if $m = -1/m_1$.

2. ORTHOGONAL SUBSPACES

Let A be an $m \times n$ matrix and let $\mathbf{x} \in N(A)$. Since $A\mathbf{x} = \mathbf{0}$, we have

$$a_{i1}x_1 + a_{i2}x_2 + \cdots + a_{in}x_n = 0 \tag{1}$$

for $i = 1, \ldots, m$. Equation (1) says that $\mathbf{x}$ is orthogonal to the ith column vector of A^T for $i = 1, \ldots, m$. Since $\mathbf{x}$ is orthogonal to each column vector of A^T, it is orthogonal to any linear combination of the column vectors of A^T. Thus, if $\mathbf{y}$ is any vector in the column space of A^T, then $\mathbf{x}^T\mathbf{y} = 0$. Thus each vector in $N(A)$ is orthogonal to every vector in the column space of A^T. When two subspaces of R^n have this property, we say that they are orthogonal.

Definition. Two subspaces X and Y of R^n are said to be **orthogonal** if $\mathbf{x}^T\mathbf{y} = 0$ for every $\mathbf{x} \in X$ and every $\mathbf{y} \in Y$. If X and Y are orthogonal, we write $X \perp Y$.

EXAMPLE 1. Let X be the subspace of R^3 spanned by $\mathbf{e}_1$ and let Y be the subspace spanned by $\mathbf{e}_2$. If $\mathbf{x} \in X$ and $\mathbf{y} \in Y$, these vectors must be of the form

$$\mathbf{x} = \begin{bmatrix} x_1 \\ 0 \\ 0 \end{bmatrix} \quad \text{and} \quad \mathbf{y} = \begin{bmatrix} 0 \\ y_2 \\ 0 \end{bmatrix}$$

Thus

$$\mathbf{x}^T\mathbf{y} = x_1 \cdot 0 + 0 \cdot y_2 + 0 \cdot 0 = 0$$

Therefore, $X \perp Y$.

EXAMPLE 2. Let X be the subspace of R^3 spanned by $\mathbf{e}_1$ and $\mathbf{e}_2$ and let Y be the subspace spanned by $\mathbf{e}_3$. If $\mathbf{x} \in X$ and $\mathbf{y} \in Y$, then

$$\mathbf{x}^T\mathbf{y} = x_1 \cdot 0 + x_2 \cdot 0 + 0 \cdot y_3 = 0$$

Thus $X \perp Y$. Furthermore, if $\mathbf{z}$ is any vector in R^3 that is orthogonal to every vector in Y, then $\mathbf{z} \perp \mathbf{e}_3$ and hence

$$z_3 = \mathbf{z}^T\mathbf{e}_3 = 0$$

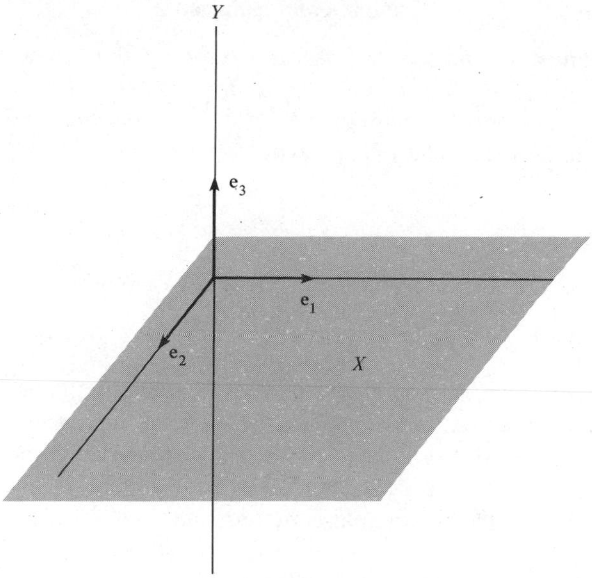

Figure 5.2.1

But if $z_3 = 0$, then $\mathbf{z} \in X$. Thus X is the set of all vectors in R^3 that are orthogonal to every vector in Y (see Figure 5.2.1).

Definition. Let Y be a subspace of R^n. The set of all vectors in R^n that are orthogonal to every vector in Y will be denoted by $Y^\perp$. Thus

$$Y^\perp = \{\mathbf{x} \in R^n \mid \mathbf{x}^T\mathbf{y} = 0 \quad \text{for every} \quad \mathbf{y} \in Y\}$$

The set $Y^\perp$ is called the **orthogonal complement** of Y.

Remarks

1. If X and Y are orthogonal subspaces of R^n, then $X \cap Y = \{\mathbf{0}\}$.
2. If Y is a subspace of R^n, then $Y^\perp$ is also a subspace of R^n.

PROOF OF (1). If $\mathbf{x} \in X \cap Y$ and $X \perp Y$, then $\|\mathbf{x}\|^2 = \mathbf{x}^T\mathbf{x} = 0$ and hence $\mathbf{x} = \mathbf{0}$.

PROOF OF (2). If $\mathbf{x} \in Y^\perp$ and α is a scalar, then for any $\mathbf{y} \in Y$,

$$(\alpha\mathbf{x})^T\mathbf{y} = \alpha(\mathbf{x}^T\mathbf{y}) = \alpha \cdot 0 = 0$$

Therefore, $\alpha\mathbf{x} \in Y^\perp$. If $\mathbf{x}_1$ and $\mathbf{x}_2$ are elements of $Y^\perp$, then

$$(\mathbf{x}_1 + \mathbf{x}_2)^T\mathbf{y} = \mathbf{x}_1^T\mathbf{y} + \mathbf{x}_2^T\mathbf{y} = 0 + 0 = 0$$

for each $\mathbf{y} \in Y$. Hence $\mathbf{x}_1 + \mathbf{x}_2 \in Y^\perp$. Therefore, $Y^\perp$ is a subspace of R^n.

Fundamental Subspaces

Let A be an $m \times n$ matrix. We saw in Chapter 3 that a vector $\mathbf{b} \in R^m$ is in the column space of A if and only if $\mathbf{b} = A\mathbf{x}$ for some $\mathbf{x} \in R^n$. If we think of A as an operator mapping R^n into R^m, then the column space of A is the same as the range of A. Let us denote the range of A by $R(A)$. Thus

$$R(A) = \{\mathbf{b} \in R^m | \mathbf{b} = A\mathbf{x} \quad \text{for some} \quad \mathbf{x} \in R^n\} = \text{the column space of } A$$

The column space of A^T, $R(A^T)$, is a subspace of R^n:

$$R(A^T) = \{\mathbf{y} \in R^n | \mathbf{y} = A^T\mathbf{x} \quad \text{for some} \quad \mathbf{x} \in R^m\}$$

The column space of $R(A^T)$ is essentially the same as the row space of A except that it consists of vectors in R^n ($n \times 1$ matrices) rather than n-tuples. Thus $\mathbf{y} \in R(A^T)$ if and only if $\mathbf{y}^T$ is in the row space of A. We have seen that $R(A^T) \perp N(A)$. The following theorem shows that $R(A^T)$ and $N(A)$ are actually orthogonal complements.

Theorem 5.2.1. *If A is an $m \times n$ matrix, then*
(i) $N(A) = R(A^T)^{\perp}$ *and* $N(A^T) = R(A)^{\perp}$
(ii) $N(A)^{\perp} = R(A^T)$ *and* $N(A^T)^{\perp} = R(A)$

PROOF. We have already seen that $N(A) \perp R(A^T)$, and this implies that $N(A) \subset R(A^T)^{\perp}$. On the other hand, if $\mathbf{x}$ is any vector in $R(A^T)^{\perp}$, then $\mathbf{x}$ is orthogonal to each of the column vectors of A^T and consequently $A\mathbf{x} = \mathbf{0}$. Thus $\mathbf{x}$ must be an element of $N(A)$ and hence $N(A) = R(A^T)^{\perp}$. This proof does not depend on the dimensions of A. In particular, the result will also hold for the matrix $B = A^T$. Thus

$$N(A^T) = N(B) = R(B^T)^{\perp} = R(A)^{\perp}$$

Statement (ii) follows, since $(S^{\perp})^{\perp} = S$ for any subspace S of either R^n or R^m (see Exercise 10).

Recall that the system $A\mathbf{x} = \mathbf{b}$ is consistent if and only if $\mathbf{b} \in R(A)$. Since $R(A) = N(A^T)^{\perp}$, we have the following corollary.

Corollary 5.2.2. *If A is an $m \times n$ matrix and $\mathbf{b} \in R^m$, then either there is a vector $\mathbf{x} \in R^n$ such that $A\mathbf{x} = \mathbf{b}$ or there is a vector $\mathbf{y} \in R^m$ such that $A^T\mathbf{y} = \mathbf{0}$ and $\mathbf{y}^T\mathbf{b} \neq 0$.*

EXAMPLE 3. Let

$$A = \begin{pmatrix} 1 & 1 & 2 \\ 0 & 1 & 1 \\ 1 & 3 & 4 \end{pmatrix}$$

Find bases for $N(A)$, $R(A^T)$, $N(A^T)$, and $R(A)$.

SOLUTION. One can find bases for $N(A)$ and $R(A^T)$ by transforming A into reduced row echelon form.

$$\begin{bmatrix} 1 & 1 & 2 \\ 0 & 1 & 1 \\ 1 & 3 & 4 \end{bmatrix} \to \begin{bmatrix} 1 & 1 & 2 \\ 0 & 1 & 1 \\ 0 & 2 & 2 \end{bmatrix} \to \begin{bmatrix} 1 & 0 & 1 \\ 0 & 1 & 1 \\ 0 & 0 & 0 \end{bmatrix}$$

Since (1, 0, 1) and (0, 1, 1) form a basis for the row space of A, it follows that $\begin{bmatrix} 1 \\ 0 \\ 1 \end{bmatrix}$ and $\begin{bmatrix} 0 \\ 1 \\ 1 \end{bmatrix}$ form a basis for $R(A^T)$. If $\mathbf{x} \in N(A)$, it follows from the reduced row echelon form of A that

$$x_1 + x_3 = 0$$
$$x_2 + x_3 = 0$$

Thus

$$x_1 = x_2 = -x_3$$

Setting $x_3 = 1$, we find that $\{(-1, -1, 1)^T\}$ is a basis for $N(A)$.

To find bases for $R(A)$ and $N(A^T)$, reduce A^T to reduced row echelon form.

$$\begin{bmatrix} 1 & 0 & 1 \\ 1 & 1 & 3 \\ 2 & 1 & 4 \end{bmatrix} \to \begin{bmatrix} 1 & 0 & 1 \\ 0 & 1 & 2 \\ 0 & 1 & 2 \end{bmatrix} \to \begin{bmatrix} 1 & 0 & 1 \\ 0 & 1 & 2 \\ 0 & 0 & 0 \end{bmatrix}$$

Thus $\begin{bmatrix} 1 \\ 0 \\ 1 \end{bmatrix}$ and $\begin{bmatrix} 0 \\ 1 \\ 2 \end{bmatrix}$ form a basis for $R(A)$. If $\mathbf{x} \in N(A^T)$, then $x_1 = -x_3$, $x_2 = -2x_3$. Thus $N(A^T)$ is the subspace of R^3 spanned by $\begin{bmatrix} -1 \\ -2 \\ 1 \end{bmatrix}$.

EXERCISES

1. For each of the following matrices, determine a basis for each of the subspaces $R(A^T)$, $N(A)$, $R(A)$, and $N(A^T)$.

(a) $A = \begin{pmatrix} 3 & 4 \\ 6 & 8 \end{pmatrix}$ (b) $A = \begin{pmatrix} 1 & 3 & 1 \\ 2 & 4 & 0 \end{pmatrix}$

(c) $A = \begin{bmatrix} 4 & -2 \\ 1 & 3 \\ 2 & 1 \\ 3 & 4 \end{bmatrix}$ (d) $A = \begin{bmatrix} 1 & 0 & 0 & 0 \\ 0 & 1 & 1 & 1 \\ 0 & 0 & 1 & 1 \\ 1 & 1 & 2 & 2 \end{bmatrix}$

2. Let S be the subspace of R^n spanned by the vectors $\mathbf{x}_1, \mathbf{x}_2, \ldots, \mathbf{x}_k$. Show that $\mathbf{y} \in S^\perp$ if and only if $\mathbf{y} \perp \mathbf{x}_i$ for $i = 1, \ldots, k$.

3. (a) Let S be the subspace of R^3 spanned by the vectors $\mathbf{x} = (x_1, x_2, x_3)^T$ and $\mathbf{y} = (y_1, y_2, y_3)^T$. Let

$$A = \begin{pmatrix} x_1 & x_2 & x_3 \\ y_1 & y_2 & y_3 \end{pmatrix}$$

Show that $S^\perp = N(A)$.

(b) Find the orthogonal complement of the subspace of R^3 spanned by $(1, 2, 1)^T$ and $(1, -1, 2)^T$.

4. Prove Corollary 5.2.2.

5. Prove: If A is an $m \times n$ matrix and $\mathbf{x} \in R^n$, then either $A\mathbf{x} = \mathbf{0}$ or there exists $\mathbf{y} \in R(A^T)$ such that $\mathbf{x}^T\mathbf{y} \neq 0$.

6. If A is an $m \times n$ matrix of rank r, what are the dimensions of $N(A)$ and $N(A^T)$? Explain.

7. Let S be a subspace of R^n. Prove $\dim S + \dim S^\perp = n$.

8. Let S be a subspace of R^n with basis $\{\mathbf{x}_1, \ldots, \mathbf{x}_r\}$. Let $\{\mathbf{x}_{r+1}, \mathbf{x}_{r+2}, \ldots, \mathbf{x}_n\}$ be a basis for $S^\perp$. Show that $\{\mathbf{x}_1, \mathbf{x}_2, \ldots, \mathbf{x}_n\}$ is a basis for R^n.

9. If U and V are subspaces of a vector space W and each $\mathbf{w} \in W$ can be written uniquely as a sum $\mathbf{u} + \mathbf{v}$, where $\mathbf{u} \in U$ and $\mathbf{v} \in V$, then we say that W is the *direct sum* of U and V and we write $W = U \oplus V$. Prove that $R^n = S \oplus S^\perp$ for any subspace S of R^n.

10. Let S be a subspace of R^n. Prove that $(S^\perp)^\perp = S$.

11. Let A and B be two $n \times n$ matrices. If $A\mathbf{x} = B\mathbf{x}$ for all $\mathbf{x} \in R^n$, show that:

(a) $N(A - B) = R^n$ (b) $A = B$

3. INNER PRODUCT SPACES

Scalar products are useful not only in R^n but in general vector spaces.

Definition. An **inner product** on a vector space V is an operation on V that assigns to each pair of vectors $\mathbf{x}$ and $\mathbf{y}$ in V a real number $\langle \mathbf{x}, \mathbf{y} \rangle$ satisfying the following conditions:

(i) $\langle \mathbf{x}, \mathbf{x} \rangle \geq 0$ with equality if and only if $\mathbf{x} = \mathbf{0}$

(ii) $\langle \mathbf{x}, \mathbf{y} \rangle = \langle \mathbf{y}, \mathbf{x} \rangle$ for all $\mathbf{x}$ and $\mathbf{y}$ in V
(iii) $\langle \alpha\mathbf{x} + \beta\mathbf{y}, \mathbf{z} \rangle = \alpha\langle \mathbf{x}, \mathbf{z} \rangle + \beta\langle \mathbf{y}, \mathbf{z} \rangle$ for all $\mathbf{x}, \mathbf{y}, \mathbf{z}$ in V and all scalars α and β

A vector space V with an inner product is called an **inner product space.**

In R^n the scalar product is an example of an inner product

$$\langle \mathbf{x}, \mathbf{y} \rangle = \mathbf{x}^T\mathbf{y}$$

In $C[a, b]$ we may define an inner product by

$$\langle f, g \rangle = \int_a^b f(x)g(x)\,dx \tag{1}$$

The reader may verify that (1) does indeed define an inner product.
If $w(x)$ is a positive continuous function on $[a, b]$, then

$$\langle f, g \rangle = \int_a^b f(x)g(x)w(x)\,dx \tag{2}$$

defines an inner product on $C[a, b]$. The function $w(x)$ is called a *weight function.* Thus it is possible to define many different inner products on $C[a, b]$.

Let $x_1, x_2, \ldots, x_n$ be distinct real numbers. For each pair of polynomials in P_n define

$$\langle p, q \rangle = \sum_{i=1}^{n} p(x_i)q(x_i) \tag{3}$$

It is easily seen that (3) satisfies conditions (ii) and (iii) of the definition of an inner product. To show that (i) holds, note that

$$\langle p, p \rangle = \sum_{i=1}^{n} (p(x_i))^2 \geqslant 0$$

If $\langle p, p \rangle = 0$, then $x_1, x_2, \ldots, x_n$ must be roots of $p(x) = 0$. Since $p(x)$ is of degree less than n, it must be the zero polynomial.
If $w(x)$ is a positive function, then

$$\langle p, q \rangle = \sum_{i=1}^{n} p(x_i)q(x_i)w(x_i) \tag{4}$$

also defines an inner product on P_n.
Two vectors $\mathbf{u}$ and $\mathbf{v}$ in an inner product space are said to be *orthogonal* if $\langle \mathbf{u}, \mathbf{v} \rangle = 0$.

EXAMPLE 1. The functions $\cos x$ and $\sin x$ are orthogonal in $C[-\pi, \pi]$ with inner product defined by (1).

$$\langle \cos x, \sin x \rangle = \int_{-\pi}^{\pi} \cos x \sin x\,dx = 0$$

EXAMPLE 2. Define an inner product on $C[-1, 1]$ by

$$\langle f, g\rangle = \int_{-1}^{1} f(x)g(x)e^{-x}\,dx$$

The functions $f(x) = x$ and $g(x) = e^x$ are orthogonal, since

$$\langle f, g\rangle = \int_{-1}^{1} xe^{x}e^{-x}\,dx = 0$$

In R^n the scalar product was used to define the length of a vector. In the same manner the length of a vector in an inner product space can be defined in terms of the inner product.

Definition. If V is an inner product space, then for each $\mathbf{v} \in V$ the **norm** of $\mathbf{v}$ is given by

$$\|\mathbf{v}\| = \sqrt{\langle \mathbf{v}, \mathbf{v}\rangle}$$

EXAMPLE 3. In $C[-\pi, \pi]$ with inner product defined by (1),

$$\|sin\ x\| = \left(\int_{-\pi}^{\pi} sin^2 x\,dx\right)^{1/2} = \sqrt{\pi}$$

EXAMPLE 4. In P_5 define an inner product by (3) with $x_i = (i-1)/4$ for $i = 1, 2, \ldots, 5$. The length of the function $p(x) = x$ is given by

$$\begin{aligned}\|x\| &= (\langle x, x\rangle)^{1/2}\\ &= \left(\sum_{i=1}^{5} x_i^2\right)^{1/2}\\ &= \left(\sum_{i=1}^{5}\left(\frac{i-1}{4}\right)^2\right)^{1/2}\\ &= \frac{\sqrt{30}}{4}\end{aligned}$$

Theorem 5.3.1 (Cauchy–Schwarz). *If* **u** *and* **v** *are any two vectors in an inner product space V, then*

$$|\langle \mathbf{u}, \mathbf{v}\rangle| \leqslant \|\mathbf{u}\|\,\|\mathbf{v}\|$$

Equality holds if and only if **u** *and* **v** *are linearly dependent.*

PROOF. Let λ be a scalar. Then

$$(5)\qquad 0 \leqslant \langle \mathbf{u} - \lambda\mathbf{v}, \mathbf{u} - \lambda\mathbf{v}\rangle \qquad [\text{from (i) of definition}]$$

$$= \langle \mathbf{u}, \mathbf{u}\rangle - 2\lambda\langle \mathbf{u}, \mathbf{v}\rangle + \lambda^2\langle \mathbf{v}, \mathbf{v}\rangle$$

$$= \|\mathbf{u}\|^2 - 2\lambda\langle \mathbf{u}, \mathbf{v}\rangle + \lambda^2\|\mathbf{v}\|$$

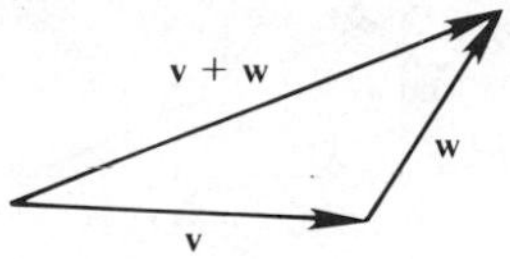

Figure 5.3.1

Let $a = \|\mathbf{v}\|^2$, $b = -2\langle \mathbf{u}, \mathbf{v}\rangle$, and $c = \|\mathbf{u}\|^2$. Equation (5) can then be viewed as a quadratic inequality,

$$a\lambda^2 + b\lambda + c \geqslant 0$$

For this inequality to be valid, the discriminant must be less than or equal to 0.

$$b^2 - 4ac \leqslant 0$$

Thus

$$4\langle \mathbf{u}, \mathbf{v}\rangle^2 - 4\|\mathbf{u}\|^2 \, \|\mathbf{v}\|^2 \leqslant 0$$

and hence

$$|\langle \mathbf{u}, \mathbf{v}\rangle| \leqslant \|\mathbf{u}\| \, \|\mathbf{v}\|$$

If equality holds in (5), then $\mathbf{u} = \lambda \mathbf{v}$.

The word *norm* in mathematics has its own meaning independent of an inner product and its use here should be justified.

Definition. A vector space V is said to be a **normed linear space** if to each vector $\mathbf{v} \in V$ there is associated a real number $\|\mathbf{v}\|$ called the **norm** of $\mathbf{v}$, satisfying:

(i) $\|\mathbf{v}\| \geqslant 0$ with equality if and only if $\mathbf{v} = \mathbf{0}$
(ii) $\|\alpha\mathbf{v}\| = |\alpha| \, \|\mathbf{v}\|$ for any scalar α
(iii) $\|\mathbf{v} + \mathbf{w}\| \leqslant \|\mathbf{v}\| + \|\mathbf{w}\|$ for all $\mathbf{v}, \mathbf{w} \in V$

The third condition is called the *triangle inequality* (see Figure 5.3.1).

Theorem 5.3.2. *If V is an inner product space, then the equation*

$$\|\mathbf{v}\| = \sqrt{\langle \mathbf{v}, \mathbf{v}\rangle} \qquad \textit{for all} \quad \mathbf{v} \in V$$

defines a norm on V.

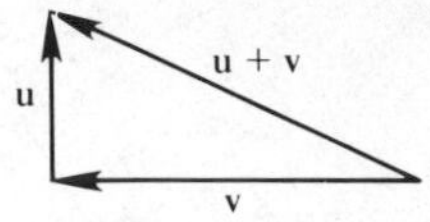

Figure 5.3.2

PROOF. It is easily seen that conditions (i) and (ii) of the definition are satisfied. We leave this for the reader to verify and proceed to show that condition (iii) is satisfied.

$$\begin{aligned}\|\mathbf{u}+\mathbf{v}\|^2 &= \langle \mathbf{u}+\mathbf{v}, \mathbf{u}+\mathbf{v}\rangle \\ &= \langle \mathbf{u}, \mathbf{u}\rangle + 2\langle \mathbf{u}, \mathbf{v}\rangle + \langle \mathbf{v}, \mathbf{v}\rangle \\ &\leqslant \|\mathbf{u}\|^2 + 2\|\mathbf{u}\|\,\|\mathbf{v}\| + \|\mathbf{v}\|^2 \qquad \text{(Cauchy–Schwarz)} \\ &= (\|\mathbf{u}\| + \|\mathbf{v}\|)^2\end{aligned}$$

Thus

$$\|\mathbf{u}+\mathbf{v}\| \leqslant \|\mathbf{u}\| + \|\mathbf{v}\|$$

As a consequence of the foregoing proof we have the Pythagorean law. If **u** and **v** are orthogonal, then

$$\|\mathbf{u}+\mathbf{v}\|^2 = \|\mathbf{u}\|^2 + \|\mathbf{v}\|^2$$

Interpreted in R^2, this is just the familiar Pythagorean theorem as shown in Figure 5.3.2.

It is possible to define many different norms on a given vector space. For example, in R^n one could define

$$\|\mathbf{x}\|_1 = \sum_{i=1}^{n} |x_i|$$

for every $\mathbf{x} = (x_1, x_2, \ldots, x_n)^T$. It is easily verified that $\|\cdot\|_1$ defines a norm on R^n. Another important norm on R^n is the *uniform norm*, which is defined by

$$\|\mathbf{x}\|_\infty = \max_{1 \leqslant i \leqslant n} |x_i|$$

More generally, one could define a norm on R^n by

$$\|\mathbf{x}\|_p = \sum_{i=1}^{n} (|x_i|^p)^{1/p}$$

for any real number $p \geqslant 1$. In particular, if $p = 2$, then

$$\|\mathbf{x}\|_2 = \left(\sum_{i-1}^{n} |x_i|^2\right)^{1/2} = \sqrt{\langle \mathbf{x}, \mathbf{x}\rangle}$$

The norm $\|\cdot\|_2$ is the norm on R^n derived from the inner product.

EXAMPLE 5. Let $\mathbf{x}$ be the vector $(4, -5, 3)^T$ in R^3. Compute $\|\mathbf{x}\|_1$, $\|\mathbf{x}\|_2$, and $\|\mathbf{x}\|_\infty$.

$$\|\mathbf{x}\|_1 = |4| + |-5| + |3| = 12$$
$$\|\mathbf{x}\|_2 = \sqrt{16 + 25 + 9} = 5\sqrt{2}$$
$$\|\mathbf{x}\|_\infty = \max(|4|, |-5|, |3|) = 5$$

In general, a norm provides a way of measuring the distance between vectors.

Definition. Let $\mathbf{x}$ and $\mathbf{y}$ be vectors in a normed linear space. The distance between $\mathbf{x}$ and $\mathbf{y}$ is defined to be the number $\|\mathbf{x} - \mathbf{y}\|$.

EXAMPLE 6. In the vector space R^2 with norm $\|\cdot\|_2$, let $\mathbf{x} = (x_1, x_2)^T$ and $\mathbf{y} = (y_1, y_2)^T$. The distance between $\mathbf{x}$ and $\mathbf{y}$ is the length of $\mathbf{x} - \mathbf{y}$ (see Figure 5.3.3).

$$\|\mathbf{x} - \mathbf{y}\|_2 = \sqrt{(x_1 - y_1)^2 + (x_2 - y_2)^2}$$

This is the standard formula used in analytic geometry for the distance between two points in the plane.

In the next section we will define norms on the vector space $M_{m,n}$ of all $m \times n$ matrices. The norm of an $m \times n$ matrix A is useful in determining the accuracy of calculated solutions to linear systems of the form $A\mathbf{x} = \mathbf{b}$.

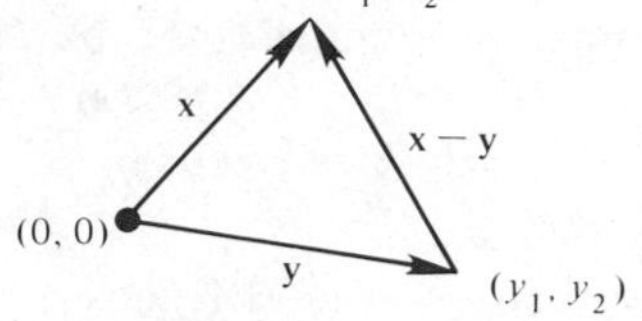

Figure 5.3.3

EXERCISES

1. Show that

(1) $$\langle f, g\rangle = \int_a^b f(x)g(x)\,dx$$

defines an inner product on $C[a, b]$.

2. In $C[0, 1]$ with inner product defined by (1), compute:
(a) $\langle e^x, e^{-x}\rangle$ (b) $\langle x, \sin \pi x\rangle$ (c) $\langle x^2, x^3\rangle$

3. In $C[-\pi, \pi]$ with inner product defined by (1), show that $\cos mx$ and $\sin nx$ are orthogonal for any integers m and n.

4. Show that the functions x and x^2 are orthogonal in P_5 with inner product defined by (3), where $x_i = (i-3)/2$ for $i = 1, \ldots, 5$.

5. If V is an inner product space, show that

$$\|\mathbf{v}\| = \sqrt{\langle \mathbf{v}, \mathbf{v}\rangle}$$

satisfies the first two conditions in the definition of a norm.

6. Show that

$$\|\mathbf{x}\|_1 = \sum_{i=1}^{n} |x_i|$$

defines a norm on R^n.

7. Show that

$$\|\mathbf{x}\|_\infty = \max_{1\leqslant i\leqslant n} |x_i|$$

defines a norm on R^n.

8. Compute $\|\mathbf{x}\|_1$, $\|\mathbf{x}\|_2$, and $\|\mathbf{x}\|_\infty$ for each of the following vectors in R^3.
(a) $x = (-3, 4, 0)^T$ (b) $x = (-1, -1, 2)^T$
(c) $x = (1, 1, 1)^T$

9. In $C[-\pi, \pi]$ with inner product defined by (1) and norm defined by

$$\|f\| = \sqrt{\langle f, f\rangle}$$

compute:
(a) $\|\sin nx\|$ (b) $\|\cos mx\|$
(c) The distance between $\sin nx$ and $\cos mx$.

10. In P_5 with inner product as in Exercise 4 and norm defined by

$$\|p\| = \sqrt{\langle p, p\rangle} = \left\{\sum_{i=1}^{5} [p(x_i)]^2\right\}^{1/2}$$

compute:
(a) $\|x\|$ (b) $\|x^2\|$ (c) The distance between x and x^2

11. In R^n with inner product

$$\langle \mathbf{x}, \mathbf{y}\rangle = \mathbf{x}^T\mathbf{y}$$

derive a formula for the distance between two vectors $\mathbf{x} = (x_1, \ldots, x_n)^T$ and $\mathbf{y} = (y_1, \ldots, y_n)^T$.

12. Show that in any vector space with a norm,

$$\|\mathbf{0}\| = 0$$

13. Show that for any **u** and **v** in a normed vector space,

$$\|\mathbf{u} + \mathbf{v}\| \geqslant |\|\mathbf{u}\| - \|\mathbf{v}\||$$

14. Prove that for any **u** and **v** in an inner product space V,

$$\|\mathbf{u} + \mathbf{v}\|^2 + \|\mathbf{u} - \mathbf{v}\|^2 = 2\|\mathbf{u}\|^2 + 2\|\mathbf{v}\|^2$$

Give a geometric interpretation of this result for the vector space R^2.

15. Determine whether or not the following define norms on $C[a, b]$.
(a) $\|f\| = |f(a)| + |f(b)|$
(b) $\|f\| = \int_a^b |f(x)|\,dx$
(c) $\|f\| = \max_{a \leqslant x \leqslant b} |f(x)|$

16. Let $\mathbf{x} \in R^n$ and show that
(a) $\|\mathbf{x}\|_1 \leqslant n\|\mathbf{x}\|_\infty$ (b) $\|\mathbf{x}\|_2 \leqslant \sqrt{n}\,\|\mathbf{x}\|_\infty$
Give examples of vectors in R^n for which equality holds in (a) and (b).

17. Sketch the set of points $(x_1, x_2) = \mathbf{x}^T$ in R^2 such that:
(a) $\|\mathbf{x}\|_2 = 1$ (b) $\|\mathbf{x}\|_1 = 1$ (c) $\|\mathbf{x}\|_\infty = 1$

18. Let $M_{2,2}$ denote the vector space of all 2×2 matrices and define

$$A * B = a_{11}b_{11} + a_{12}b_{12} + a_{21}b_{21} + a_{22}b_{22}$$

Determine whether or not the operation $*$ is an inner product on $M_{2,2}$.

19. Consider the vector space R^n with inner product $\langle \mathbf{x}, \mathbf{y} \rangle = \mathbf{x}^T\mathbf{y}$. For any $n \times n$ matrix A, show that:
(a) $\langle A\mathbf{x}, \mathbf{y} \rangle = \langle \mathbf{x}, A^T\mathbf{y} \rangle$ (b) $\langle A^TA\mathbf{x}, \mathbf{x} \rangle = \|A\mathbf{x}\|^2$

20. Let A be an $m \times n$ matrix. Prove that $N(A^TA) = N(A)$.

4. MATRIX NORMS

In the last section we defined a number of different norms on R^n. The most natural norm to use is the Euclidean norm.

(1) $$\|\mathbf{x}\|_2 = \left(\sum_{i=1}^{n} x_i^2 \right)^{1/2}$$

Actually, (1) defines a whole family of norms, since it defines a norm on R^n for $n = 1, 2, \ldots$. In a similar manner we can define a family of matrix norms by

(2) $$\|A\|_F = \left(\sum_{j=1}^{n} \sum_{i=1}^{m} a_{ij}^2 \right)^{1/2}$$

for each $m \times n$ matrix. Equation (2) defines a family of norms on the vector spaces $M_{m,n}$. Clearly, $\|A\|_F \geqslant 0$ and $\|A\|_F = 0$ if and only if $A = O$. It is easily seen that

$$\|\alpha A\|_F = \left(\sum_{j=1}^{n} \sum_{i=1}^{m} (\alpha a_{ij})^2 \right)^{1/2} = |\alpha| \|A\|_F$$

Thus to show that $\|\cdot\|_F$ actually defines a norm on $M_{m,n}$ we must show that

$$\|A + B\|_F \leqslant \|A\|_F + \|B\|_F$$

If we think of the matrices in $M_{m,n}$ as being vectors in R^{mn}, this result follows immediately. If we set

$$\mathbf{a} = (a_{11}, a_{12}, \ldots, a_{1n}, a_{21}, a_{22}, \ldots, a_{2n}, \ldots, a_{m1}, a_{m2}, \ldots, a_{mn})^T$$
$$\mathbf{b} = (b_{11}, b_{12}, \ldots, b_{1n}, b_{21}, b_{22}, \ldots, b_{2n}, \ldots, b_{m1}, b_{m2}, \ldots, b_{mn})^T$$

then $\mathbf{a}$ and $\mathbf{b}$ are vectors in R^{mn} and it follows that

$$\|A + B\|_F = \|\mathbf{a} + \mathbf{b}\|_2 \leqslant \|\mathbf{a}\|_2 + \|\mathbf{b}\|_2 = \|A\|_F + \|B\|_F$$

Thus $\|\cdot\|_F$ defines a norm on $M_{m,n}$ for each ordered pair of positive integers (m, n). We will refer to $\|A\|_F$ as the Frobenius norm of A. The $\|\cdot\|_F$ has a number of important properties, which we list below.

(I) If $\mathbf{a}_j$ represents the jth column vector of A, then

$$\|A\|_F = \left(\sum_{j=1}^{n}\sum_{i=1}^{m} a_{ij}^2\right)^{1/2} = \left(\sum_{j=1}^{n} \|\mathbf{a}_j\|_2^2\right)^{1/2}$$

(II) If $(\mathbf{a}_i)_r$ represents the ith column vector of A^T, then

$$\|A\|_F = \left(\sum_{i=1}^{m}\sum_{j=1}^{n} a_{ij}^2\right)^{1/2} = \left(\sum_{i=1}^{m} \|(\mathbf{a}_i)_r\|_2^2\right)^{1/2}$$

(III) If $\mathbf{x} \in R^n$, then

$$\begin{aligned}\|A\mathbf{x}\|_2 &= \left[\sum_{i=1}^{m}\left(\sum_{j=1}^{n} a_{ij}x_j\right)^2\right]^{1/2} = \left[\sum_{i=1}^{m} \left((\mathbf{a}_i)_r^T\mathbf{x}\right)^2\right]^{1/2} \\ &\leqslant \left[\sum_{i=1}^{m} \|\mathbf{x}\|_2^2\|(\mathbf{a}_i)_r\|^2\right]^{1/2} \qquad \text{(Cauchy–Schwarz)} \\ &= \|A\|_F\|\mathbf{x}\|_2\end{aligned}$$

(IV) If $B = (\mathbf{b}_1, \ldots, \mathbf{b}_r)$ is an $n \times r$ matrix, it follows from properties (I) and (III) that

$$\begin{aligned}\|AB\|_F &= \|(A\mathbf{b}_1, A\mathbf{b}_2, \ldots, A\mathbf{b}_r)\|_F \\ &= \left(\sum_{i=1}^{r} \|A\mathbf{b}_i\|_2^2\right)^{1/2} \\ &\leqslant \|A\|_F\left(\sum_{i=1}^{r} \|\mathbf{b}_i\|_2^2\right)^{1/2} \\ &= \|A\|_F\|B\|_F\end{aligned}$$

In view of property (III), we say that the matrix norm $\|\cdot\|_F$ and the vector norm $\|\cdot\|_2$ are compatible. In general, a matrix norm $\|\cdot\|_m$ on $M_{m,n}$ and a vector norm $\|\cdot\|_v$ on R^n are said to be *compatible* if

$$\|A\mathbf{x}\|_v \leqslant \|A\|_m\|\mathbf{x}\|_v$$

for every $\mathbf{x} \in R^n$.

There are many other norms that one could put on $M_{m,n}$. The families of norms that turn out to be most useful generally satisfy the property

(iv) $$\|AB\| \leqslant \|A\|\,\|B\|$$

Consequently, we will only consider families of norms that have this property. One important consequence of property (iv) is that

$$\|A^n\| \leqslant \|A\|^n$$

In particular, if $\|A\| < 1$, then $\|A^n\| \to 0$ as $n \to \infty$.

Subordinate Matrix Norms

We can think of each $m \times n$ matrix as a linear operator from R^n to R^m. For any family of vector norms we can define an *operator norm* by comparing $\|A\mathbf{x}\|$ and $\|\mathbf{x}\|$ for each nonzero $\mathbf{x}$ and taking

$$(3) \qquad \|A\| = \max_{\mathbf{x} \neq \mathbf{0}} \frac{\|A\mathbf{x}\|}{\|\mathbf{x}\|}$$

It can be shown that there is a particular $\mathbf{x}_0$ in R^n which maximizes $\|A\mathbf{x}\|/\|\mathbf{x}\|$, but the proof is beyond the scope of this text. Assuming that $\|A\mathbf{x}\|/\|\mathbf{x}\|$ can always be maximized, we will show that (3) actually does define a norm on $M_{m,n}$.

(i) For each $\mathbf{x} \neq \mathbf{0}$,

$$\frac{\|A\mathbf{x}\|}{\|\mathbf{x}\|} \geqslant 0$$

and, consequently,

$$\|A\| = \max_{\mathbf{x} \neq \mathbf{0}} \frac{\|A\mathbf{x}\|}{\|\mathbf{x}\|} \geqslant 0$$

If $\|A\| = 0$, then $A\mathbf{x} = \mathbf{0}$ for every $\mathbf{x} \in R^n$. This implies that $\mathbf{a}_j = A\mathbf{e}_j = \mathbf{0}$ for $j = 1, \ldots, n$, and hence A must be the zero matrix.

(ii) $\|\alpha A\| = \mathit{max}_{\mathbf{x} \neq \mathbf{0}} \dfrac{\|\alpha A\mathbf{x}\|}{\mathbf{x}} = |\alpha| \mathit{max}_{\mathbf{x} \neq \mathbf{0}} \dfrac{\|A\mathbf{x}\|}{\|\mathbf{x}\|} = |\alpha| \|A\|$

(iii)

$$\begin{aligned} \|A + B\| &= \max_{\mathbf{x} \neq \mathbf{0}} \frac{\|A\mathbf{x} + B\mathbf{x}\|}{\|\mathbf{x}\|} \\ &\leqslant \max_{\mathbf{x} \neq \mathbf{0}} \left(\frac{\|A\mathbf{x}\|}{\|\mathbf{x}\|} + \frac{\|B\mathbf{x}\|}{\|\mathbf{x}\|} \right) \\ &\leqslant \max_{\mathbf{x} \neq \mathbf{0}} \frac{\|A\mathbf{x}\|}{\|\mathbf{x}\|} + \max_{\mathbf{x} \neq \mathbf{0}} \frac{\|B\mathbf{x}\|}{\|\mathbf{x}\|} \\ &= \|A\| + \|B\| \end{aligned}$$

Thus (3) defines a norm on $M_{m,n}$. For each family of vector norms $\|\cdot\|$, one can then define a family of matrix norms by (3). The matrix norms defined by (3) are said to be *subordinate* to the vector norms $\|\cdot\|$.

Theorem 5.4.1. *If the family of matrix norms* $\|\cdot\|_m$ *is subordinate to the family of vector norms* $\|\cdot\|$, *then* $\|\cdot\|_m$ *and* $\|\cdot\|$ *are compatible and the matrix norms* $\|\cdot\|_m$ *satisfy property (iv).*

PROOF. If $\mathbf{x}$ is any nonzero vector in R^n, then

$$\frac{\|A\mathbf{x}\|}{\|\mathbf{x}\|} \leqslant \max_{\mathbf{y}\neq\mathbf{0}} \frac{\|A\mathbf{y}\|}{\|\mathbf{y}\|} = \|A\|_m$$

and hence

$$\|A\mathbf{x}\| \leqslant \|A\|_m\|\mathbf{x}\|$$

Since this last inequality is also valid if $\mathbf{x} = \mathbf{0}$, it follows that $\|\cdot\|_m$ and $\|\cdot\|$ are compatible. If B is an $n \times r$ matrix, then since $\|\cdot\|_m$ and $\|\cdot\|$ are compatible, we have

$$\|AB\mathbf{x}\| \leqslant \|A\|_m\|B\mathbf{x}\| \leqslant \|A\|_m\|B\|_m\|\mathbf{x}\|$$

Thus for all $\mathbf{x} \neq \mathbf{0}$,

$$\frac{\|AB\mathbf{x}\|}{\|\mathbf{x}\|} \leqslant \|A\|_m\|B\|_m$$

and hence

$$\|AB\|_m = \max_{\mathbf{x}\neq\mathbf{0}} \frac{\|AB\mathbf{x}\|}{\|\mathbf{x}\|} \leqslant \|A\|_m\|B\|_m$$

It is a simple matter to compute the Frobenius norm of a matrix. For example, if

$$A = \begin{pmatrix} 4 & 2 \\ 0 & 4 \end{pmatrix}$$

then

$$\|A\|_F = (4^2 + 0^2 + 2^2 + 4^2)^{1/2} = 6$$

On the other hand, it is not so obvious how to compute $\|A\|$ if $\|\cdot\|$ is a subordinate matrix norm. It turns out that the matrix norm

$$\|A\|_2 = \max_{\mathbf{x}\neq\mathbf{0}} \frac{\|A\mathbf{x}\|_2}{\|\mathbf{x}\|_2}$$

is difficult to compute; however,

$$\|A\|_1 = \max_{\mathbf{x}\neq\mathbf{0}} \frac{\|A\mathbf{x}\|_1}{\|\mathbf{x}\|_1}$$

and

$$\|A\|_\infty = \max_{\mathbf{x}\neq\mathbf{0}} \frac{\|A\mathbf{x}\|_\infty}{\|\mathbf{x}\|_\infty}$$

can be easily calculated.

Theorem 5.4.2. *If A is an $m \times n$ matrix, then*

$$\|A\|_1 = \max_{1 \leqslant j \leqslant n} \left(\sum_{i=1}^{m} |a_{ij}| \right)$$

and

$$\|A\|_\infty = \max_{1 \leqslant i \leqslant m} \left(\sum_{j=1}^{n} |a_{ij}| \right)$$

PROOF. We will prove that

$$\|A\|_1 = \max_{1 \leqslant j \leqslant n} \left(\sum_{i=1}^{m} |a_{ij}| \right)$$

and leave the proof of the second statement as an exercise. Let

$$\alpha = \max_{1 \leqslant j \leqslant n} \sum_{i=1}^{m} |a_{ij}| = \sum_{i=1}^{m} |a_{ik}|$$

That is, k is the index of the column where the maximum occurs. Let $\mathbf{x}$ be an arbitrary vector in R^n; then

$$A\mathbf{x} = \left(\sum_{j=1}^{n} a_{1j}x_j, \sum_{j=1}^{n} a_{2j}x_j, \ldots, \sum_{j=1}^{n} a_{mj}x_j \right)^T$$

and it follows that

$$\begin{aligned} \|A\mathbf{x}\|_1 &= \sum_{i=1}^{m} \left| \sum_{j=1}^{n} a_{ij}x_j \right| \\ &\leqslant \sum_{i=1}^{m} \sum_{j=1}^{n} |a_{ij}x_j| \\ &= \sum_{j=1}^{n} \left(|x_j| \sum_{i=1}^{m} |a_{ij}| \right) \\ &\leqslant \alpha \sum_{j=1}^{n} |x_j| \\ &= \alpha \|\mathbf{x}\|_1 \end{aligned}$$

Thus for any nonzero $\mathbf{x}$ in R^n,

$$\frac{\|A\mathbf{x}\|_1}{\|\mathbf{x}\|_1} \leqslant \alpha$$

and hence

$$\|A\| = \max_{\mathbf{x} \neq \mathbf{0}} \frac{\|A\mathbf{x}\|_1}{\|\mathbf{x}\|_1} \leqslant \alpha \tag{4}$$

On the other hand,

$$\|A\mathbf{e}_k\|_1 = \|\mathbf{a}_k\|_1 = \alpha$$

Since $\|\mathbf{e}_k\|_1 = 1$, it follows that

$$\|A\|_1 = \max_{\mathbf{x}\neq\mathbf{0}} \frac{\|A\mathbf{x}\|_1}{\|\mathbf{x}\|_1} \geqslant \frac{\|A\mathbf{e}_k\|_1}{\|\mathbf{e}_k\|_1} = \alpha \tag{5}$$

Together (4) and (5) imply $\|A\|_1 = \alpha$.

EXAMPLE 1. Let

$$A = \begin{bmatrix} -3 & 2 & 4 & -3 \\ 5 & -2 & -3 & 5 \\ 2 & 1 & -6 & 4 \\ 1 & 1 & 1 & 1 \end{bmatrix}$$

Then

$$\|A\|_1 = |4| + |-3| + |-6| + |1| = 14$$

and

$$\|A\|_\infty = |5| + |-2| + |-3| + |5| = 15$$

EXERCISES

1. Determine $\|\cdot\|_F$, $\|\cdot\|_\infty$, and $\|\cdot\|_1$ for each of the following matrices.

(a) $\begin{pmatrix} 1 & 0 \\ 0 & 1 \end{pmatrix}$ (b) $\begin{pmatrix} 1 & 4 \\ -2 & 2 \end{pmatrix}$ (c) $\begin{bmatrix} \frac{1}{2} & \frac{1}{2} \\ \frac{1}{2} & \frac{1}{2} \end{bmatrix}$

(d) $\begin{bmatrix} 0 & 5 & 1 \\ 2 & 3 & 1 \\ 1 & 2 & 2 \end{bmatrix}$ (e) $\begin{bmatrix} 5 & 0 & 5 \\ 4 & 1 & 0 \\ 3 & 2 & 1 \end{bmatrix}$

2 Let

$$A = \begin{pmatrix} 1 & 0 \\ 0 & 0 \end{pmatrix}$$

Show that $\|A\|_2 = 1$.

3. Show that $\|A\|_F = \|A^T\|_F$.

4. Let $\|\cdot\|$ denote a family of vector norms and let $\|\cdot\|_m$ be a subordinate matrix norm. Show that

$$\|A\|_m = \max_{\|\mathbf{x}\|=1} \|A\mathbf{x}\|$$

5. Let A be an $m \times n$ matrix. Prove that

$$\|A\|_\infty = \max_{1 \leqslant i \leqslant m} \left(\sum_{j=1}^{n} |a_{ij}| \right)$$

6. Let A be a symmetric $n \times n$ matrix. Show that $\|A\|_\infty = \|A\|_1$.

7. Let A be an $m \times n$ matrix and $\mathbf{x} \in R^n$. Prove:
(a) $\|A\mathbf{x}\|_\infty \leqslant n^{1/2}\|A\|_2\|\mathbf{x}\|_\infty$
(b) $\|A\mathbf{x}\|_2 \leqslant n^{1/2}\|A\|_\infty\|\mathbf{x}\|_2$
(c) $n^{-1/2}\|A\|_2 \leqslant \|A\|_\infty \leqslant n^{1/2}\|A\|_2$

5. LEAST SQUARES PROBLEMS

So far we have been concerned mainly with consistent systems of linear equations. In this section we turn our attention to overdetermined systems, that is, systems involving more equations than unknowns. Such systems are usually inconsistent. Thus, given an $m \times n$ system $A\mathbf{x} = \mathbf{b}$ with $m > n$, we cannot expect in general to find a vector $\mathbf{x} \in R^n$ for which $A\mathbf{x}$ equals $\mathbf{b}$. Instead, one can look for a vector $\mathbf{x}$ for which $A\mathbf{x}$ is "closest" to $\mathbf{b}$. As you might expect, orthogonality plays an important role in finding such an $\mathbf{x}$.

Let A be an $m \times n$ matrix with $m > n$. For each $\mathbf{b} \in R^m$, define

$$\|\mathbf{b}\| = \sqrt{\langle \mathbf{b}, \mathbf{b} \rangle} = \sqrt{\mathbf{b}^T\mathbf{b}}$$

Consider the system of equations $A\mathbf{x} = \mathbf{b}$. For each $\mathbf{x} \in R^n$, we can form a *residual*

$$r(\mathbf{x}) = \mathbf{b} - A\mathbf{x}$$

The distance between $\mathbf{b}$ and $A\mathbf{x}$ is given by

$$\|\mathbf{b} - A\mathbf{x}\| = \|r(\mathbf{x})\|$$

We wish to find a vector $\mathbf{x} \in R^n$ for which $\|r(\mathbf{x})\|$ will be a minimum. Minimizing $\|r(\mathbf{x})\|$ is equivalent to minimizing $\|r(\mathbf{x})\|^2$. A vector $\mathbf{x}$ that accomplishes this is said to be a *least squares solution* to the system $A\mathbf{x} = \mathbf{b}$.

Let $\mathbf{x}$ and $\mathbf{y}$ be elements of R^n.

$$\text{(1)} \quad \begin{aligned} \|r(\mathbf{y})\|^2 &= \|(\mathbf{b} - A\mathbf{x}) + (A\mathbf{x} - A\mathbf{y})\|^2 \\ &= \|\mathbf{b} - A\mathbf{x}\|^2 + 2\langle \mathbf{b} - A\mathbf{x}, A\mathbf{x} - A\mathbf{y} \rangle + \|A\mathbf{x} - A\mathbf{y}\|^2 \end{aligned}$$

It follows that

$$\|r(\mathbf{y})\|^2 \geqslant \|r(\mathbf{x})\|^2 + 2\langle r(\mathbf{x}), A(\mathbf{x} - \mathbf{y}) \rangle$$

If for some $\mathbf{x} = \hat{\mathbf{x}}$, $r(\hat{\mathbf{x}}) \in N(A^T) = R(A)^\perp$, then $r(\hat{\mathbf{x}})$ will be orthogonal to $A(\hat{\mathbf{x}} - \mathbf{y})$ and the inequality becomes

$$\|r(\mathbf{y})\|^2 \geq \|r(\hat{\mathbf{x}})\|^2 \tag{2}$$

Thus, if $r(\hat{\mathbf{x}}) \in N(A^T)$, (2) will be valid for any $\mathbf{y} \in R^n$ and hence $\hat{\mathbf{x}}$ will be a solution to the least squares problem. The residual $r(\mathbf{x})$ will be in $N(A^T)$ if and only if

$$\mathbf{0} = A^T r(\mathbf{x}) = A^T(\mathbf{b} - A\mathbf{x})$$

or

$$A^T A\mathbf{x} = A^T\mathbf{b} \tag{3}$$

Equation (3) represents an $n \times n$ system of linear equations. These equations are called the *normal equations*. A solution $\hat{\mathbf{x}}$ to the normal equations will be a least squares solution to the system $A\mathbf{x} = \mathbf{b}$.

Theorem 5.5.1. *If A is an $m \times n$ matrix of rank n, the normal equations*

$$A^T A\mathbf{x} = A^T\mathbf{b}$$

have a unique solution

$$\hat{\mathbf{x}} = (A^T A)^{-1} A^T\mathbf{b}$$

and $\hat{\mathbf{x}}$ is the unique least squares solution to the system $A\mathbf{x} = \mathbf{b}$.

PROOF. We will first show that $A^T A$ is nonsingular. To prove this, let $\mathbf{z}$ be a solution to

$$A^T A\mathbf{x} = \mathbf{0} \tag{4}$$

Then $A\mathbf{z} \in N(A^T)$. Clearly, $A\mathbf{z} \in R(A) = N(A^T)^\perp$. Since $N(A^T) \cap N(A^T)^\perp = \{\mathbf{0}\}$, it follows that $A\mathbf{z} = \mathbf{0}$. If A has rank n, the column vectors of A are linearly independent and consequently $A\mathbf{x} = \mathbf{0}$ has only the trivial solution. Thus $\mathbf{z} = \mathbf{0}$ and (4) has only the trivial solution. Therefore, by Theorem 1.4.3, $A^T A$ is nonsingular. It follows that $\hat{\mathbf{x}} = (A^T A)^{-1} A^T\mathbf{b}$ is the solution to the normal equations and hence is a solution to the least squares problem.

To show that the solution $\hat{\mathbf{x}}$ is unique, let $\mathbf{y}$ be any other vector in R^n. Since $\hat{\mathbf{x}}$ satisfies the normal equations, it follows that $r(\hat{\mathbf{x}}) \in N(A^T) = R(A)^\perp$ and hence

$$\langle \mathbf{b} - A\hat{\mathbf{x}}, A(\hat{\mathbf{x}} - \mathbf{y}) \rangle = 0$$

It follows from (1) that

$$\|r(\mathbf{y})\|^2 = \|r(\hat{\mathbf{x}})\|^2 + \|A(\hat{\mathbf{x}} - \mathbf{y})\|^2$$

(a) $b \in R^2$ and A is a 2×1 matrix of rank 1. (b) $b \in R^3$ and A is a 3×2 matrix of rank 2.

Figure 5.5.1

Since $\hat{\mathbf{x}} - \mathbf{y} \neq \mathbf{0}$ and the columns of A are linearly independent, it follows that $\|A(\hat{\mathbf{x}} - \mathbf{y})\|^2 > 0$ and hence

$$\|r(\mathbf{y})\|^2 > \|r(\hat{\mathbf{x}})\|^2$$

Therefore, $\hat{\mathbf{x}}$ is the unique least squares solution.

The vector

$$\mathbf{p} = A\hat{\mathbf{x}} = A(A^TA)^{-1}A^T\mathbf{b}$$

is the element of $R(A)$ that is closest to $\mathbf{b}$ in the least squares sense. Since $\mathbf{p} \in R(A)$ and the residual $r(\hat{\mathbf{x}}) \in N(A^T)$, it follows that $\mathbf{p} \perp r(\hat{\mathbf{x}})$ (see Figure 5.5.1). The vector $\mathbf{p}$ is called the *projection of* $\mathbf{b}$ *onto* $R(A)$. The matrix $P = A(A^TA)^{-1}A^T$ is called a *projection matrix*.

Application

Hooke's law states that the force applied to a spring is proportional to the distance the spring is stretched. Thus, if F is the force applied and x is the distance the spring has been stretched, then $F = kx$. The proportionality constant k is called the *spring constant*.

Some physics students want to determine the spring constant for a given spring. They apply forces of 3, 5, and 8 pounds, which have the effect of stretching the spring 4, 7, and 11 inches, respectively. Using Hooke's law, they derive the following system of equations:

$$\begin{aligned} 4k &= 3 \\ 7k &= 5 \\ 11k &= 8 \end{aligned}$$

The system is clearly inconsistent, since each of the equations yields a different value of k. Rather than use any one of these values, the students decide to compute the least squares solution to the system.

$$(4, 7, 11)\begin{bmatrix} 4 \\ 7 \\ 11 \end{bmatrix}(k) = (4, 7, 11)\begin{bmatrix} 3 \\ 5 \\ 8 \end{bmatrix}$$

$$186k = 135$$

$$k \approx 0.726$$

EXAMPLE 1. Find the least squares solution to the system

$$\begin{aligned} x_1 + x_2 &= 3 \\ -2x_1 + 3x_2 &= 1 \\ 2x_1 - x_2 &= 2 \end{aligned}$$

SOLUTION. The normal equations for this system are

$$\begin{pmatrix} 1 & -2 & 2 \\ 1 & 3 & -1 \end{pmatrix}\begin{bmatrix} 1 & 1 \\ -2 & 3 \\ 2 & -1 \end{bmatrix}\begin{pmatrix} x_1 \\ x_2 \end{pmatrix} = \begin{pmatrix} 1 & -2 & 2 \\ 1 & 3 & -1 \end{pmatrix}\begin{bmatrix} 3 \\ 1 \\ 2 \end{bmatrix}$$

This simplifies to the 2×2 system

$$\begin{pmatrix} 9 & -7 \\ -7 & 11 \end{pmatrix}\begin{pmatrix} x_1 \\ x_2 \end{pmatrix} = \begin{pmatrix} 5 \\ 4 \end{pmatrix}$$

The solution to the 2×2 system is $\left(\frac{83}{50}, \frac{71}{50}\right)^T$.

Scientists often collect data and try to find a functional relationship among the variables. For example, the data may involve temperatures $T_0, T_1, \ldots, T_n$ of a liquid measured at times $t_0, t_1, \ldots, t_n$, respectively. If the temperature T can be represented as a function of the time t, this function can be used to predict the temperatures at future times. If the data consist of $n + 1$ points in the plane, it is possible to find a polynomial of degree n or less passing through all the points. Such a polynomial is called an *interpolating polynomial*. Actually, since the data usually involve experimental error, there is no reason to require that the function pass through all the points. Indeed, lower-degree polynomials that do not pass through the points exactly usually give a truer description of the relationship between the variables. If, for example, the relationship between the variables is actually linear and the data involve slight errors, it would be disastrous to use an interpolating polynomial (see Figure 5.5.2).

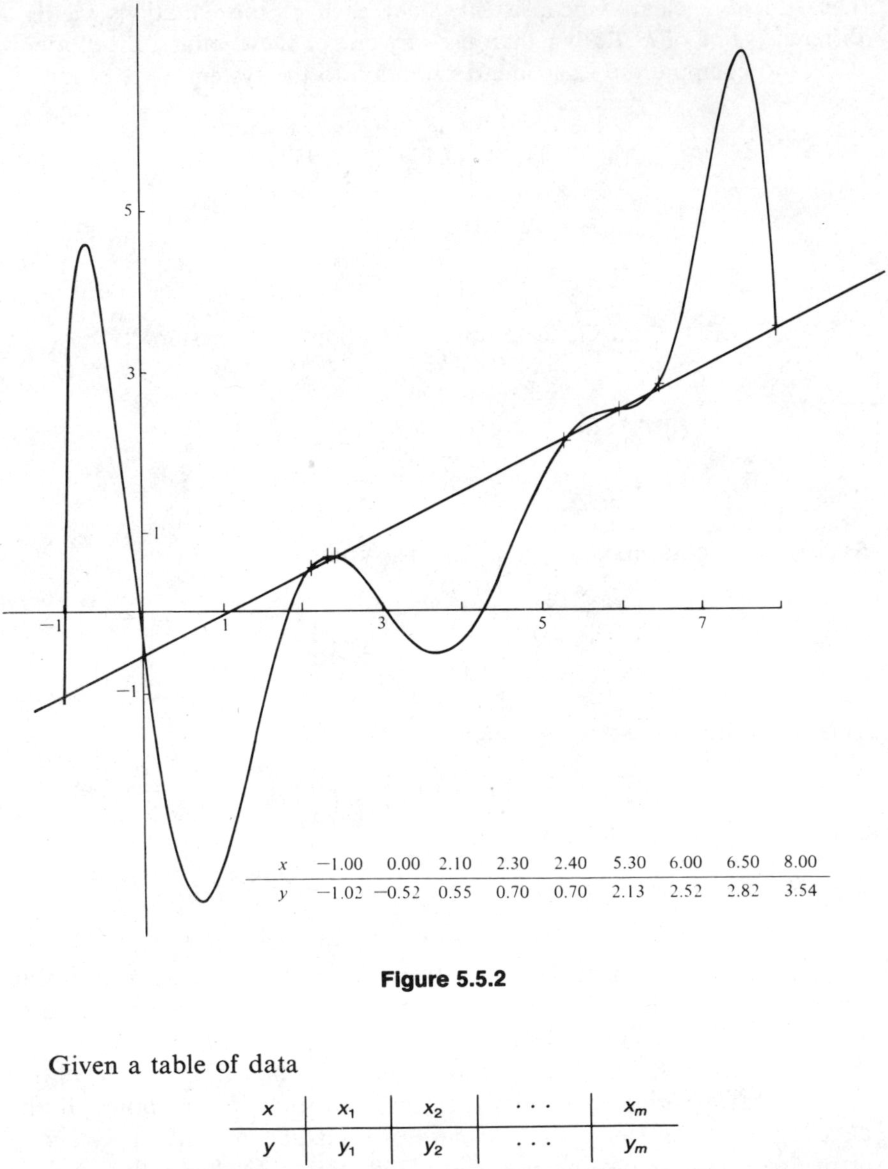

x	−1.00	0.00	2.10	2.30	2.40	5.30	6.00	6.50	8.00
y	−1.02	−0.52	0.55	0.70	0.70	2.13	2.52	2.82	3.54

Figure 5.5.2

Given a table of data

x	x_1	x_2	$\cdots$	x_m
y	y_1	y_2	$\cdots$	y_m

we wish to find a linear function

$$y = c_0 + c_1 x$$

which best fits the data in the least squares sense. If we require that

$$y_i = c_0 + c_1 x_i \qquad \text{for} \quad i = 1, \ldots, m$$

we get a system of m equations in two unknowns.

$$\begin{bmatrix} 1 & x_1 \\ 1 & x_2 \\ \vdots & \vdots \\ 1 & x_m \end{bmatrix} \begin{pmatrix} c_0 \\ c_1 \end{pmatrix} = \begin{bmatrix} y_1 \\ y_2 \\ \vdots \\ y_m \end{bmatrix} \tag{5}$$

The linear function whose coefficients are the least squares solution to (5) is said to be the best least squares fit to the data by a linear function.

EXAMPLE 2. Given the data

x	0	3	6
y	1	4	5

Find the best least squares fit by a linear function.

SOLUTION. For this example the system (5) becomes

$$A\mathbf{c} = \mathbf{y}$$

where

$$A = \begin{bmatrix} 1 & 0 \\ 1 & 3 \\ 1 & 6 \end{bmatrix}, \qquad \mathbf{c} = \begin{pmatrix} c_0 \\ c_1 \end{pmatrix}, \qquad \text{and} \qquad \mathbf{y} = \begin{bmatrix} 1 \\ 4 \\ 5 \end{bmatrix}$$

The normal equations

$$A^T A\mathbf{c} = A^T\mathbf{y}$$

simplify to

$$\begin{pmatrix} 3 & 9 \\ 9 & 45 \end{pmatrix} \begin{pmatrix} c_0 \\ c_1 \end{pmatrix} = \begin{pmatrix} 10 \\ 42 \end{pmatrix} \tag{6}$$

The solution of this system is $(\frac{4}{3}, \frac{2}{3})$. Thus the best linear least squares fit is given by

$$y = \tfrac{4}{3} + \tfrac{2}{3}x$$

Example 2 could also have been solved using calculus. The residual $r(\mathbf{c})$ is given by

$$r(\mathbf{c}) = \mathbf{y} - A\mathbf{c}$$

and

$$\begin{aligned}\|r(\mathbf{c})\|^2 &= \|\mathbf{y} - A\mathbf{c}\|^2 \\ &= [1 - (c_0 + 0c_1)]^2 + [4 - (c_0 + 3c_1)]^2 \\ &\quad + [5 - (c_0 + 6c_1)]^2 \\ &= f(c_0, c_1)\end{aligned}$$

Thus $\|r(\mathbf{c})\|^2$ can be thought of as a function of two variables, $f(c_0, c_1)$. The minimum of this function will occur when its partials are zero.

$$\frac{\partial f}{\partial c_0} = -2(10 - 3c_0 - 9c_1) = 0$$

$$\frac{\partial f}{\partial c_1} = -6(14 - 3c_0 - 15c_1) = 0$$

Dividing both equations through by 2 gives the same system as in (6) (see Figure 5.5.3).

If the data do not resemble a linear function, one could use a higher degree polynomial. To find the coefficients $c_0, c_1, \ldots, c_n$ of the best least squares fit to the data

x	x_1	x_2	$\cdots$	x_m
y	y_1	y_2	$\cdots$	y_m

by a polynomial of degree n, we must find the least squares solution to the system

(7)
$$\begin{pmatrix} 1 & x_1 & x_1^2 & \cdots & x_1^n \\ 1 & x_2 & x_2^2 & \cdots & x_2^n \\ \vdots & & & & \\ 1 & x_m & x_m^2 & \cdots & x_m^n \end{pmatrix} \begin{pmatrix} c_0 \\ c_1 \\ \vdots \\ c_n \end{pmatrix} = \begin{pmatrix} y_1 \\ y_2 \\ \vdots \\ y_m \end{pmatrix}$$

EXAMPLE 3. Find the best quadratic least squares fit to the data

x	0	1	2	3
y	3	2	4	4

SOLUTION. For this example the system (7) becomes

$$\begin{pmatrix} 1 & 0 & 0 \\ 1 & 1 & 1 \\ 1 & 2 & 4 \\ 1 & 3 & 9 \end{pmatrix} \begin{pmatrix} c_0 \\ c_1 \\ c_2 \end{pmatrix} = \begin{pmatrix} 3 \\ 2 \\ 4 \\ 4 \end{pmatrix}$$

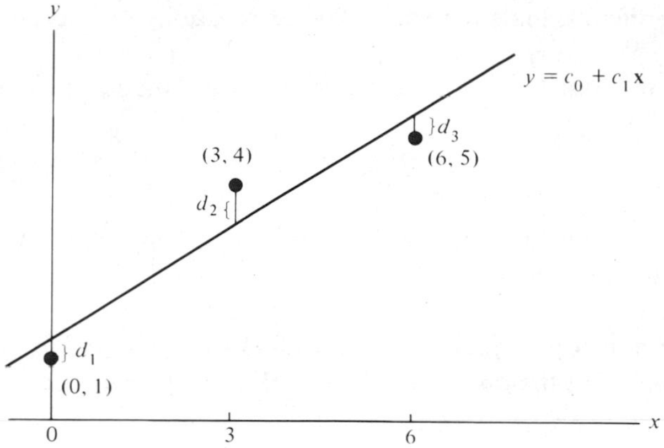

$\|r(\mathbf{c})\|^2 = d_1^2 + d_2^2 + d_3^2$

Figure 5.5.3

Thus the normal equations are

$$\begin{pmatrix} 1 & 1 & 1 & 1 \\ 0 & 1 & 2 & 3 \\ 0 & 1 & 4 & 9 \end{pmatrix} \begin{pmatrix} 1 & 0 & 0 \\ 1 & 1 & 1 \\ 1 & 2 & 4 \\ 1 & 3 & 9 \end{pmatrix} \begin{pmatrix} c_0 \\ c_1 \\ c_2 \end{pmatrix} = \begin{pmatrix} 1 & 1 & 1 & 1 \\ 0 & 1 & 2 & 3 \\ 0 & 1 & 4 & 9 \end{pmatrix} \begin{pmatrix} 3 \\ 2 \\ 4 \\ 4 \end{pmatrix}$$

These simplify to

$$\begin{pmatrix} 4 & 6 & 14 \\ 6 & 14 & 36 \\ 14 & 36 & 98 \end{pmatrix} \begin{pmatrix} c_0 \\ c_1 \\ c_2 \end{pmatrix} = \begin{pmatrix} 13 \\ 22 \\ 54 \end{pmatrix}$$

The solution to this system is (2.75, −0.25, 0.25). The quadratic polynomial that gives the best least square fit to the data is

$$p(x) = 2.75 - 0.25x + 0.25x^2$$

EXERCISES

1. Show that A^TA and AA^T are symmetric.

2. Find the least squares solution to each of the following systems.

(a) $\begin{aligned} x_1 + x_2 &= 3 \\ 2x_1 - 3x_2 &= 1 \\ 0x_1 + 0x_2 &= 2 \end{aligned}$

(b) $\begin{aligned} -x_1 + x_2 &= 10 \\ 2x_1 + x_2 &= 5 \\ x_1 - 2x_2 &= 20 \end{aligned}$

(c) $\begin{aligned} x_1 + x_2 + x_3 &= 4 \\ -x_1 + x_2 + x_3 &= 0 \\ -x_2 + x_3 &= 1 \\ x_1 \qquad + x_3 &= 2 \end{aligned}$

3. Calculate the residuals for each of your solutions in Exercise 2.

4. (a) Find the best least squares fit by a linear function to the data

x	-1	0	1	2
y	0	1	3	9

(b) Plot your linear function from part (a) along with the data on a coordinate system.

5. Find the best least square fit to the data in Exercise 4 by a quadratic polynomial. Plot the points $x = -1, 0, 1, 2$ for your function and sketch the graph.

6. Let A be an $m \times n$ matrix of rank n and let $P = A(A^TA)^{-1}A^T$.
   (a) Show that $P\mathbf{b} = \mathbf{b}$ for every $\mathbf{b} \in R(A)$. Explain this in terms of projections.
   (b) If $\mathbf{b} \in R(A)^\perp$, show that $P\mathbf{b} = \mathbf{0}$.
   (c) Give a geometric illustration of parts (a) and (b) if $R(A)$ is a plane through the origin in R^3.

7. Let $P = A(A^TA)^{-1}A^T$, where A is an $m \times n$ matrix of rank n.
   (a) Show that $P^2 = P$.
   (b) Prove $P^k = P$ for $k = 1, 2, \ldots$.
   (c) Show that P is symmetric.
   [*RECALL:* If B is nonsingular, then $(B^{-1})^T = (B^T)^{-1}$.]

6. ORTHONORMAL SETS

In R^2 it is generally more convenient to use the standard basis $\{\mathbf{e}_1, \mathbf{e}_2\}$ than to use some other basis such as $\{(2, 1)^T, (3, 5)^T\}$. For example, it would be easier to find the coordinates of $(x_1, x_2)^T$ with respect to the standard basis. The elements of the standard basis are orthogonal unit vectors. In working with an inner product space V, it is generally desirable to have a basis of mutually orthogonal unit vectors. This is convenient not only in finding coordinates of vectors but in solving least squares problems.

Definition. Let $\mathbf{v}_1, \mathbf{v}_2, \ldots, \mathbf{v}_n$ be vectors in an inner products space V. If $\langle \mathbf{v}_i, \mathbf{v}_j \rangle = 0$ whenever $i \neq j$, then $\{\mathbf{v}_1, \mathbf{v}_2, \ldots, \mathbf{v}_n\}$ is said to be an orthogonal set of vectors.

EXAMPLE 1. The set $\{(1, 1, 1)^T, (2, 1, -3)^T, (4, -5, 1)^T\}$ is an orthogonal set in R^3, since

$$(1, 1, 1)(2, 1, -3)^T = 0$$
$$(1, 1, 1)(4, -5, 1)^T = 0$$
$$(2, 1, -3)(4, -5, 1)^T = 0$$

Theorem 5.6.1. *If $\{\mathbf{v}_1, \ldots, \mathbf{v}_n\}$ is an orthogonal set of nonzero vectors in an inner product space V, then $\mathbf{v}_1, \mathbf{v}_2, \ldots, \mathbf{v}_n$ are linearly independent.*

PROOF. Suppose that $\mathbf{v}_1, \mathbf{v}_2, \ldots, \mathbf{v}_n$ are mutually orthogonal vectors which are linearly dependent. Then there exist scalars $\alpha_1, \alpha_2, \ldots, \alpha_n$ not all zero for which

$$\alpha_1\mathbf{v}_1 + \alpha_2\mathbf{v}_2 + \cdots + \alpha_n\mathbf{v}_n = \mathbf{0}$$

We may assume without loss of generality that $\alpha_1 \neq 0$. It follows that

$$\mathbf{v}_1 = \sum_{i=2}^{n} \beta_i\mathbf{v}_i \qquad \text{where} \quad \beta_i = \frac{-\alpha_i}{\alpha_1}$$

and hence

$$\begin{aligned} \|\mathbf{v}_1\|^2 &= \langle \mathbf{v}_1, \mathbf{v}_1 \rangle \\ &= \langle \mathbf{v}_1, \sum_{i=2}^{n} \beta_i\mathbf{v}_i \rangle \\ &= \sum_{i=2}^{n} \beta_i\langle \mathbf{v}_1, \mathbf{v}_i \rangle \\ &= 0 \end{aligned}$$

Thus $\mathbf{v}_1$ must be the zero vector. Therefore, if $\mathbf{v}_1, \mathbf{v}_2, \ldots, \mathbf{v}_n$ are nonzero and mutually orthogonal, they must be linearly independent.

Definition. An **orthonormal** set of vectors is an orthogonal set of unit vectors.

The set $\{\mathbf{u}_1, \mathbf{u}_2, \ldots, \mathbf{u}_n\}$ will be orthonormal if and only if

$$\langle \mathbf{u}_i, \mathbf{u}_j \rangle = \delta_{ij}$$

where

$$\delta_{ij} = \begin{cases} 1 & \text{if } i = j \\ 0 & \text{if } i \neq j \end{cases}$$

Given any orthogonal set of nonzero vectors $\{\mathbf{v}_1, \mathbf{v}_2, \ldots, \mathbf{v}_n\}$, it is possible to form an orthonormal set by defining

$$\mathbf{u}_i = \left(\frac{1}{\|\mathbf{v}_i\|}\right)\mathbf{v}_i \qquad \text{for} \quad i = 1, 2, \ldots, n$$

The reader may verify that $\{\mathbf{u}_1, \mathbf{u}_2, \ldots, \mathbf{u}_n\}$ will be an orthonormal set.

EXAMPLE 2. We saw in Example 1 that if $\mathbf{v}_1 = (1, 1, 1)^T$, $\mathbf{v}_2 = (2, 1, -3)^T$, and $\mathbf{v}_3 = (4, -5, 1)^T$, then $\{\mathbf{v}_1, \mathbf{v}_2, \mathbf{v}_3\}$ is an orthogonal set in R^3. To form an orthonormal set, let

$$\mathbf{u}_1 = \left(\frac{1}{\|\mathbf{v}_1\|}\right)\mathbf{v}_1 = \frac{1}{\sqrt{3}}(1, 1, 1)^T$$

$$\mathbf{u}_2 = \left(\frac{1}{\|\mathbf{v}_2\|}\right)\mathbf{v}_2 = \frac{1}{\sqrt{14}}(2, 1, -3)^T$$

$$\mathbf{u}_3 = \left(\frac{1}{\|\mathbf{v}_3\|}\right)\mathbf{v}_3 = \frac{1}{\sqrt{42}}(4, -5, 1)^T$$

EXAMPLE 3. In $C[-\pi, \pi]$ with inner product

$$\langle f, g\rangle = \int_{-\pi}^{\pi} f(x)g(x)\,dx$$

the set $\{1, \cos x, \sin x\}$ is an orthogonal set of vectors, since

$$\langle 1, \cos x\rangle = \int_{-\pi}^{\pi} \cos x\,dx = 0$$

$$\langle 1, \sin x\rangle = \int_{-\pi}^{\pi} \sin x\,dx = 0$$

$$\langle \cos x, \sin x\rangle = \int_{-\pi}^{\pi} \cos x \sin x\,dx = 0$$

To form an orthonormal set, we calculate the norms of the three vectors.

$$\|1\|^2 = \langle 1, 1\rangle = \int_{-\pi}^{\pi} 1\,dx = 2\pi$$

$$\|\cos x\|^2 = \langle \cos x, \cos x\rangle = \int_{-\pi}^{\pi} \cos^2 x\,dx = \pi$$

$$\|\sin x\|^2 = \langle \sin x, \sin x\rangle = \int_{-\pi}^{\pi} \sin^2 x\,dx = \pi$$

Thus, $\{1/\sqrt{2\pi}, (1/\sqrt{\pi})\cos x, (1/\sqrt{\pi})\sin x\}$ is an orthonormal set of vectors.

It follows from Theorem 5.6.1 that if $B = \{\mathbf{x}_1, \mathbf{x}_2, \ldots, \mathbf{x}_k\}$ is an orthonormal set in an inner product space V, then B is a basis for a subspace S of V. We say the B is an *orthonormal basis* for S. It is generally much easier to work with an orthonormal basis than with an ordinary basis. In particular, it is usually easier to calculate the coordinates of a given vector $\mathbf{x}$ with respect to an orthonormal basis. Once these coordinates have been determined, they can be used to compute $\|\mathbf{x}\|$.

Theorem 5.6.2. *Let $\{\mathbf{x}_1, \mathbf{x}_2, \ldots, \mathbf{x}_n\}$ be an orthonormal basis for an inner product space V. If*

$$\mathbf{x} = \sum_{i=1}^{n} c_i\mathbf{x}_i$$

then

$$c_i = \langle \mathbf{x}_i, \mathbf{x} \rangle$$

and

$$\|\mathbf{x}\|^2 = \sum_{i=1}^{n} c_i^2 \qquad \textit{(Parseval's formula)}$$

PROOF. Both of the assertions of the theorem can be proved by simple calculations.

$$\begin{aligned}
\langle \mathbf{x}_i, \mathbf{x} \rangle &= \langle \mathbf{x}_i, \sum_{j=1}^{n} c_j \mathbf{x}_j \rangle \\
&= \sum_{j=1}^{n} c_j \langle \mathbf{x}_i, \mathbf{x}_j \rangle \\
&= \sum_{j=1}^{n} c_j \delta_{ij} \\
&= c_i
\end{aligned}$$

The second assertion follows, since

$$\begin{aligned}
0 &= \left\| \mathbf{x} - \sum_{i=1}^{n} c_i \mathbf{x}_i \right\|^2 \\
&= \langle \mathbf{x} - \sum_{i=1}^{n} c_i \mathbf{x}_i, \mathbf{x} - \sum_{i=1}^{n} c_i \mathbf{x}_i \rangle \\
&= \|\mathbf{x}\|^2 - 2 \sum_{i=1}^{n} c_i \langle \mathbf{x}, \mathbf{x}_i \rangle + \|\mathbf{x}\|^2 \\
&= 2\|\mathbf{x}\|^2 - 2 \sum_{i=1}^{n} c_i^2
\end{aligned}$$

EXAMPLE 4. The vectors $\mathbf{u}_1 = (1/\sqrt{2}, 1/\sqrt{2})^T$ and $\mathbf{u}_2 = (1/\sqrt{2}, -1/\sqrt{2})^T$ form an orthonormal basis for R^2. Let $\mathbf{x} = (x_1, x_2)^T$ be any vector in R^2.

$$\mathbf{x}^T \mathbf{u}_1 = \frac{x_1 + x_2}{\sqrt{2}}$$

$$\mathbf{x}^T \mathbf{u}_2 = \frac{x_1 - x_2}{\sqrt{2}}$$

It follows from the theorem that

$$\mathbf{x} = \frac{x_1 + x_2}{\sqrt{2}} \mathbf{u}_1 + \frac{x_1 - x_2}{\sqrt{2}} \mathbf{u}_2$$

and

$$\|\mathbf{x}\|^2 = \left(\frac{x_1 + x_2}{\sqrt{2}} \right)^2 + \left(\frac{x_1 - x_2}{\sqrt{2}} \right)^2 = x_1^2 + x_2^2$$

Orthogonality plays an important role in solving least squares problems. Recall that if A is an $m \times n$ matrix of rank n, then the least squares problem $A\mathbf{x} = \mathbf{b}$ has a unique solution $\hat{\mathbf{x}}$ that is determined by solving the normal equations $A^TA\mathbf{x} = A^T\mathbf{b}$. The projection $\mathbf{p} = A\hat{\mathbf{x}}$ is the vector in $R(A)$ that is closest to $\mathbf{b}$. The least squares problem is especially easy to solve in the case where the column vectors of A form an orthonormal set in R^m.

Theorem 5.6.3. *If the column vectors of A form an orthonormal set of vectors in R^m, then $A^TA = I$ and the solution to the least squares problem is*

$$\hat{\mathbf{x}} = A^T\mathbf{b}$$

PROOF. The ijth entry of A^TA is formed from the ith row of A^T and the jth column of A. Thus the ijth entry is actually the scalar product of the ith and jth columns of A. Since the column vectors of A are orthonormal, it follows that

$$A^TA = (\delta_{ij}) = I$$

Consequently, the normal equations simplify to

$$\mathbf{x} = A^T\mathbf{b}$$

What if the columns of A are not orthonormal? In the next section we will learn a method for finding an orthonormal basis for $R(A)$. From this method we will obtain a factorization of A into a product QR, where Q has an orthonormal set of column vectors and R is upper triangular. With this factorization, the least squares problem can be solved quickly and accurately.

If one has an orthonormal basis for $R(A)$, then the projection $\mathbf{p} = A\hat{\mathbf{x}}$ can be determined in terms of the basis elements. Indeed, this is a special case of the more general least squares problem of finding the element $\mathbf{p}$ in a subspace S of an inner product space V that is closest to a given element $\mathbf{x}$ in V. This problem is easily solved if S has an orthonormal basis. We first prove the following theorem.

Theorem 5.6.4. *Let S be a subspace of an inner product space V and let $\mathbf{x} \in V$. Let $\{\mathbf{x}_1, \mathbf{x}_2, \ldots, \mathbf{x}_n\}$ be an orthonormal basis for S and let*

$$\mathbf{p} = \sum_{i=1}^{n} c_i\mathbf{x}_i \tag{1}$$

where

$$c_i = \langle \mathbf{x}, \mathbf{x}_i \rangle \quad \textit{for each } i \tag{2}$$

Then $\mathbf{p} - \mathbf{x} \in S^{\perp}$ (see Figure 5.6.1).

PROOF. We will show first that $(\mathbf{p}-\mathbf{x}) \perp \mathbf{x}_i$ for each i.

$$\begin{aligned}\langle \mathbf{x}_i, \mathbf{p}-\mathbf{x}\rangle &= \langle \mathbf{x}_i, \mathbf{p}\rangle - \langle \mathbf{x}_i, \mathbf{x}\rangle \\ &= \langle \mathbf{x}_i, \sum_{j=1}^{n} c_j \mathbf{x}_j\rangle - c_i \\ &= \sum_{j=1}^{n} c_j \langle \mathbf{x}_i, \mathbf{x}_j\rangle - c_i \\ &= 0\end{aligned}$$

So $\mathbf{p}-\mathbf{x}$ is orthogonal to all the $\mathbf{x}_i$'s. If $\mathbf{y} \in S$, then

$$\mathbf{y} = \sum_{i=1}^{n} \alpha_i \mathbf{x}_i$$

and hence

$$\begin{aligned}\langle \mathbf{p}-\mathbf{x}, \mathbf{y}\rangle &= \langle \mathbf{p}-\mathbf{x}, \sum_{i=1}^{n} \alpha_i \mathbf{x}_i\rangle \\ &= \sum_{i=1}^{n} \alpha_i \langle \mathbf{p}-\mathbf{x}, \mathbf{x}_i\rangle \\ &= 0\end{aligned}$$

If $\mathbf{x} \in S$, then the preceding result is trivial, since by Theorem 5.6.2, $\mathbf{p}-\mathbf{x} = \mathbf{0}$.

Theorem 5.6.5. *Under the hypothesis of Theorem* 5.6.4, $\mathbf{p}$ *is the element of* S *that is closest to* $\mathbf{x}$, *that is,*

$$\|\mathbf{y}-\mathbf{x}\| > \|\mathbf{p}-\mathbf{x}\|$$

for any $\mathbf{y} \neq \mathbf{p}$ *in* S.

PROOF

$$\|\mathbf{y}-\mathbf{x}\|^2 = \|(\mathbf{y}-\mathbf{p}) + (\mathbf{p}-\mathbf{x})\|^2$$

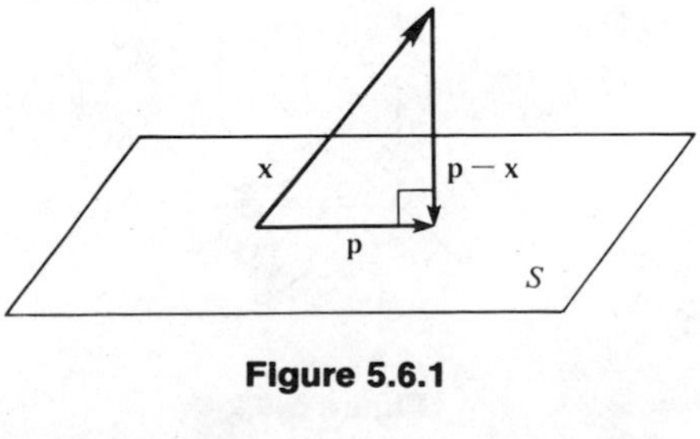

Figure 5.6.1

Since $\mathbf{y} - \mathbf{p} \in S$, it follows from Theorem 5.6.4 and the Pythagorean law that

$$\|\mathbf{y} - \mathbf{x}\|^2 = \|\mathbf{y} - \mathbf{p}\|^2 + \|\mathbf{p} - \mathbf{x}\|^2 > \|\mathbf{p} - \mathbf{x}\|^2$$

Therefore, $\|\mathbf{y} - \mathbf{x}\| > \|\mathbf{p} - \mathbf{x}\|$.

The vector $\mathbf{p}$ defined by (1) and (2) is said to be the *projection of* $\mathbf{x}$ *onto* S.

EXAMPLE 5. Let S be the set of all vectors in R^3 of the form $(x, y, 0)^T$. Find the vector $\mathbf{p}$ in S that is closest to $\mathbf{w} = (5, 3, 4)^T$ (see Figure 5.6.2).

SOLUTION. Let $\mathbf{u}_1 = (1, 0, 0)^T$ and $\mathbf{u}_2 = (0, 1, 0)^T$. Clearly, $\mathbf{u}_1$ and $\mathbf{u}_2$ form an orthonormal basis for S. Now

$$c_1 = \mathbf{w}^T\mathbf{u}_1 = 5$$
$$c_2 = \mathbf{w}^T\mathbf{u}_2 = 3$$

The vector $\mathbf{p}$ turns out to be exactly what we would expect.

$$\mathbf{p} = 5\mathbf{u}_1 + 3\mathbf{u}_2 = (5, 3, 0)^T$$

EXAMPLE 6. Find the best least squares approximation to e^x on the interval [0, 1] by a linear function.

SOLUTION. Let S be the subspace of all linear functions in $C[0, 1]$. Although the functions 1 and x span S, they are not orthogonal. We seek a function of the form $x - a$ that is orthogonal to 1.

$$\langle 1, x - a \rangle = \int_0^1 (x - a)\,dx = \tfrac{1}{2} - a$$

Thus $a = \frac{1}{2}$. Since $\|x - \frac{1}{2}\| = 1/\sqrt{12}$, it follows that

$$u_1(x) = 1 \qquad \text{and} \qquad u_2(x) = \sqrt{12}\left(x - \tfrac{1}{2}\right)$$

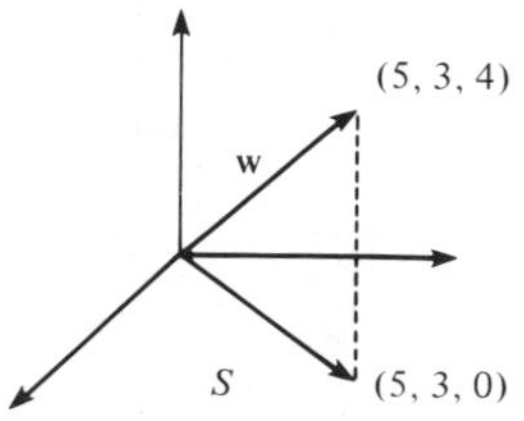

Figure 5.6.2

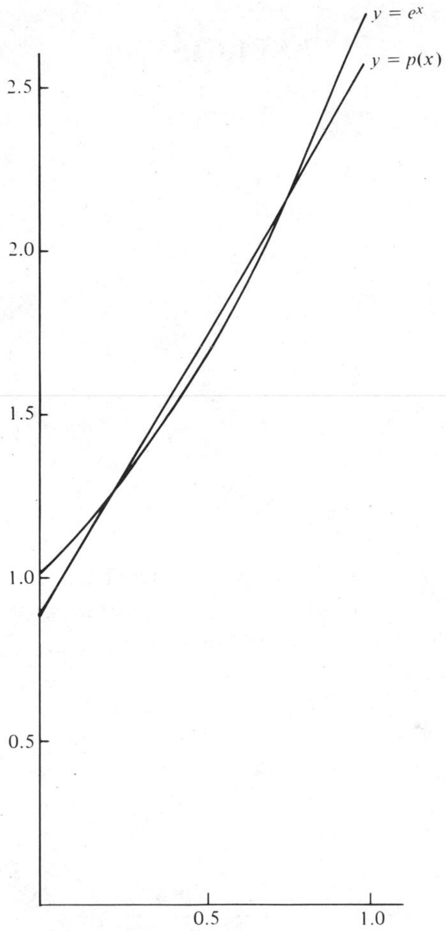

Figure 5.6.3

form an orthonormal basis for S.

Let

$$c_1 = \int_0^1 u_1(x)e^x\,dx = e - 1$$

$$c_2 = \int_0^1 u_2(x)e^x\,dx = \sqrt{3}\,(3 - e)$$

The projection

$$\begin{aligned} p(x) &= c_1u_1(x) + c_2u_2(x) \\ &= (e-1)\cdot 1 + \sqrt{3}\,(3-e)\left[\sqrt{12}\left(x - \tfrac{1}{2}\right)\right] \\ &= (4e - 10) + 6(3 - e)x \end{aligned}$$

is the best linear least squares approximation to e^x on [0, 1] (see Figure 5.6.3).

EXERCISES

1. Which of the following sets of vectors form an orthonormal basis for R^2?

(a) $\{(1, 0)^T, (0, 1)^T\}$

(b) $\{(\frac{3}{5}, \frac{4}{5})^T, (\frac{5}{13}, \frac{12}{13})^T\}$

(c) $\{(1, -1)^T, (1, 1)^T\}$

(d) $\left\{\left(\frac{\sqrt{3}}{2}, \frac{1}{2}\right)^T, \left(-\frac{1}{2}, \frac{\sqrt{3}}{2}\right)^T\right\}$

2. Let

$$\mathbf{x}_1 = \left(\frac{1}{3\sqrt{2}}, \frac{1}{3\sqrt{2}}, -\frac{4}{3\sqrt{2}}\right)^T, \quad \mathbf{x}_2 = \left(\frac{2}{3}, \frac{2}{3}, \frac{1}{3}\right)^T$$

$$\mathbf{x}_3 = \left(\frac{1}{\sqrt{2}}, -\frac{1}{\sqrt{2}}, 0\right)^T$$

(a) Show that $\{\mathbf{x}_1, \mathbf{x}_2, \mathbf{x}_3\}$ is an orthonormal basis for R^3.

(b) Let $\mathbf{x} = (1, 1, 1)^T$. Write $\mathbf{x}$ as a linear combination of $\mathbf{x}_1$, $\mathbf{x}_2$, and $\mathbf{x}_3$ and compute $\|\mathbf{x}\|$ using Theorem 5.6.2.

3. Let S be the subspace of R^3 spanned by the vectors $\mathbf{x}_2$ and $\mathbf{x}_3$ of Exercise 2. Let $\mathbf{x} = (1, 2, 2)^T$. Find the projection $\mathbf{p}$ of $\mathbf{x}$ onto S. Show that $(\mathbf{p} - \mathbf{x}) \perp \mathbf{x}_2$ and $(\mathbf{p} - \mathbf{x}) \perp \mathbf{x}_3$.

4. Let

$$S = \left\{\frac{1}{\sqrt{2\pi}}, \frac{1}{\sqrt{\pi}}\cos x, \frac{1}{\sqrt{\pi}}\cos 2x, \frac{1}{\sqrt{\pi}}\cos 3x, \frac{1}{\sqrt{\pi}}\cos 4x\right\}$$

(a) Show that S is an orthonormal set in $C[-\pi, \pi]$.

(b) Use trigonometric identities to write the function $\sin^4 x$ as a linear combination of elements of S.

(c) Use part (b) and Theorem 5.6.2 to find the values of the following integrals.

(i) $\int_{-\pi}^{\pi} \sin^4 x\, dx$ (ii) $\int_{-\pi}^{\pi} \sin^4 x \cos x\, dx$

(iii) $\int_{-\pi}^{\pi} \sin^4 x \cos 2x\, dx$ (iv) $\int_{-\pi}^{\pi} \sin^4 x \cos 3x\, dx$

(v) $\int_{-\pi}^{\pi} \sin^4 x \cos 4x\, dx$

5. Let V be the subspace of $C[-\pi, \pi]$ spanned by the vectors in the set S from Exercise 4. Find the best least squares approximation to $f(x) = |x|$ on $[-\pi, \pi]$ by elements from V.

6. Let

$$A = \begin{bmatrix} \frac{1}{\sqrt{3}} & \frac{2}{\sqrt{6}} \\ \frac{1}{\sqrt{3}} & -\frac{1}{\sqrt{6}} \\ \frac{1}{\sqrt{3}} & -\frac{1}{\sqrt{6}} \end{bmatrix}$$

(a) Show that the column vectors of A form an orthonormal set in R^3.
(b) Solve the least squares problem $A\mathbf{x} = \mathbf{b}$ for each of the following choices of $\mathbf{b}$.
(i) $\mathbf{b} = (3, 0, 0)^T$ (ii) $\mathbf{b} = (1, 2, 3)^T$
(iii) $\mathbf{b} = (1, 1, 2)^T$

7. Let $\{\mathbf{x}_1, \mathbf{x}_2, \ldots, \mathbf{x}_k, \mathbf{x}_{k+1}, \ldots, \mathbf{x}_n\}$ be an orthonormal basis for an inner product space V. Let S_1 be the subspace of V spanned by $\mathbf{x}_1, \ldots, \mathbf{x}_k$ and let S_2 be the subspace spanned by $\mathbf{x}_{k+1}, \mathbf{x}_{k+2}, \ldots, \mathbf{x}_n$. Show that $S_1 \perp S_2$.

8. Let $\mathbf{x}$ be an element of the inner product space V in Exercise 7 and let $\mathbf{p}_1$ and $\mathbf{p}_2$ be the projections of $\mathbf{x}$ onto S_1 and S_2, respectively.
(a) Show that $\mathbf{x} = \mathbf{p}_1 + \mathbf{p}_2$.
(b) Show that if $\mathbf{x} \in S_1^\perp$, then $\mathbf{p}_1 = \mathbf{0}$ and hence $S_1^\perp = S_2$.

9. Let S be a subspace of an inner product space V. Let $\{\mathbf{x}_1, \ldots, \mathbf{x}_n\}$ be an orthogonal basis for S and let $\mathbf{x} \in V$. Show that the best least squares approximation to $\mathbf{x}$ by elements of S is given by

$$\mathbf{p} = \sum_{i=1}^{n} \frac{\langle \mathbf{x}, \mathbf{x}_i \rangle}{\langle \mathbf{x}_i, \mathbf{x}_i \rangle} \mathbf{x}_i$$

7. THE GRAM–SCHMIDT ORTHOGONALIZATION PROCESS

In this section we will learn a process for constructing an orthonormal basis for an n-dimensional inner product space V. Starting with a given basis $\{\mathbf{x}_1, \mathbf{x}_2, \ldots, \mathbf{x}_n\}$, the method involves using projections to construct an orthonormal basis $\{\mathbf{u}_1, \mathbf{u}_2, \ldots, \mathbf{u}_n\}$.

We will construct the $\mathbf{u}_i$'s so that $\mathcal{S}(\mathbf{u}_1, \ldots, \mathbf{u}_k) = \mathcal{S}(\mathbf{x}_1, \ldots, \mathbf{x}_k)$ for $k = 1, \ldots, n$. To begin the process, let

$$(1) \qquad \mathbf{u}_1 = \left(\frac{1}{\|\mathbf{x}_1\|}\right)\mathbf{x}_1$$

$\mathcal{S}(\mathbf{u}_1) = \mathcal{S}(\mathbf{x}_1)$, since $\mathbf{u}_1$ is a unit vector in the direction of $\mathbf{x}_1$. Let $\mathbf{p}_1$ denote the projection of $\mathbf{x}_2$ onto $\mathcal{S}(\mathbf{x}_1) = \mathcal{S}(\mathbf{u}_1)$.

$$\mathbf{p}_1 = \langle \mathbf{x}_2, \mathbf{u}_1 \rangle \mathbf{u}_1$$

By Theorem 5.6.4,

$$(\mathbf{x}_2 - \mathbf{p}_1) \perp \mathbf{u}_1$$

Note that $\mathbf{x}_2 - \mathbf{p}_1 \neq \mathbf{0}$, since

$$(2) \qquad \mathbf{x}_2 - \mathbf{p}_1 = \frac{-\langle \mathbf{x}_2, \mathbf{u}_1 \rangle}{\|\mathbf{x}_1\|}\mathbf{x}_1 + \mathbf{x}_2$$

and $\mathbf{x}_1$ and $\mathbf{x}_2$ are linearly independent. Let

$$(3) \qquad \mathbf{u}_2 = \frac{1}{\|\mathbf{x}_2 - \mathbf{p}_1\|}(\mathbf{x}_2 - \mathbf{p}_1)$$

$\mathbf{u}_2$ is a unit vector orthogonal to $\mathbf{u}_1$. It follows from (1), (2), and (3) that $\mathcal{S}(\mathbf{u}_1, \mathbf{u}_2) \subset \mathcal{S}(\mathbf{x}_1, \mathbf{x}_2)$. Since $\mathbf{u}_1$ and $\mathbf{u}_2$ are linearly independent, it follows that $\{\mathbf{u}_1, \mathbf{u}_2\}$ is an orthonormal basis for $\mathcal{S}(\mathbf{x}_1, \mathbf{x}_2)$, and hence $\mathcal{S}(\mathbf{x}_1, \mathbf{x}_2) = \mathcal{S}(\mathbf{u}_1, \mathbf{u}_2)$.

To construct $\mathbf{u}_3$, continue in the same manner. Let $\mathbf{p}_2$ be the projection of $\mathbf{x}_3$ onto $\mathcal{S}(\mathbf{x}_1, \mathbf{x}_2) = \mathcal{S}(\mathbf{u}_1, \mathbf{u}_2)$,

$$\mathbf{p}_2 = \langle \mathbf{x}_3, \mathbf{u}_1 \rangle \mathbf{u}_1 + \langle \mathbf{x}_3, \mathbf{u}_2 \rangle \mathbf{u}_2$$

and set

$$\mathbf{u}_3 = \frac{1}{\|\mathbf{x}_3 - \mathbf{p}_2\|}(\mathbf{x}_3 - \mathbf{p}_2)$$

and so on (see Figure 5.7.1).

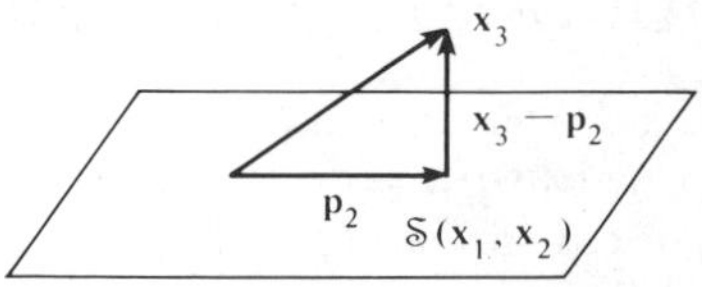

Figure 5.7.1

Theorem 5.7.1 (The Gram–Schmidt Process). *Let* $\{\mathbf{x}_1, \mathbf{x}_2, \ldots, \mathbf{x}_n\}$ *be a basis for the inner product space* V. *Let*

$$\mathbf{u}_1 = \left(\frac{1}{\|\mathbf{x}_1\|}\right)\mathbf{x}_1$$

and define $\mathbf{u}_2, \ldots, \mathbf{u}_n$ *recursively by*

$$\mathbf{u}_{k+1} = \frac{1}{\|\mathbf{x}_{k+1} - \mathbf{p}_k\|}(\mathbf{x}_{k+1} - \mathbf{p}_k) \qquad \text{for} \quad k = 1, \ldots, n-1$$

where

$$\mathbf{p}_k = \langle \mathbf{x}_{k+1}, \mathbf{u}_1\rangle \mathbf{u}_1 + \langle \mathbf{x}_{k+1}, \mathbf{u}_2\rangle \mathbf{u}_2 + \cdots + \langle \mathbf{x}_{k+1}, \mathbf{u}_k\rangle \mathbf{u}_k$$

is the projection of $\mathbf{x}_{k+1}$ *onto* $\mathcal{S}(\mathbf{u}_1, \mathbf{u}_2, \ldots, \mathbf{u}_k)$. *The set* $\{\mathbf{u}_1, \mathbf{u}_2, \ldots, \mathbf{u}_n\}$ *is an orthonormal basis for* V.

PROOF. We will argue inductively. Clearly, $\mathcal{S}(\mathbf{u}_1) = \mathcal{S}(\mathbf{x}_1)$. Suppose that $\mathbf{u}_1, \mathbf{u}_2, \ldots, \mathbf{u}_k$ have been constructed so that $\{\mathbf{u}_1, \mathbf{u}_2, \ldots, \mathbf{u}_k\}$ is an orthonormal set and

$$\mathcal{S}(\mathbf{u}_1, \mathbf{u}_2, \ldots, \mathbf{u}_k) = \mathcal{S}(\mathbf{x}_1, \mathbf{x}_2, \ldots, \mathbf{x}_k)$$

Since $\mathbf{p}_k$ is a linear combination of the $\mathbf{u}_i$'s, $1 \leqslant i \leqslant k$, it follows that $\mathbf{p}_k \in \mathcal{S}(\mathbf{x}_1, \ldots, \mathbf{x}_k)$ and $\mathbf{x}_{k+1} - \mathbf{p}_k \in \mathcal{S}(\mathbf{x}_1, \ldots, \mathbf{x}_{k+1})$.

$$\mathbf{x}_{k+1} - \mathbf{p}_k = \mathbf{x}_{k+1} - \sum_{i=1}^{k} c_i \mathbf{x}_i$$

Since $\mathbf{x}_1, \ldots, \mathbf{x}_{k+1}$ are linearly independent, it follows that $\mathbf{x}_{k+1} - \mathbf{p}_k$ is nonzero and by Theorem 5.6.4 it is orthogonal to each $\mathbf{u}_i$, $1 \leqslant i \leqslant k$. Thus $\{\mathbf{u}_1, \mathbf{u}_2, \ldots, \mathbf{u}_{k+1}\}$ is an orthonormal set of vectors in $\mathcal{S}(\mathbf{x}_1, \ldots, \mathbf{x}_{k+1})$. Since $\mathbf{u}_1, \ldots, \mathbf{u}_{k+1}$ are linearly independent, they form a basis for $\mathcal{S}(\mathbf{x}_1, \ldots, \mathbf{x}_{k+1})$, and consequently $\mathcal{S}(\mathbf{u}_1, \ldots, \mathbf{u}_{k+1}) = \mathcal{S}(\mathbf{x}_1, \ldots, \mathbf{x}_{k+1})$. It follows by induction that $\{\mathbf{u}_1, \mathbf{u}_2, \ldots, \mathbf{u}_n\}$ is an orthonormal basis for V.

EXAMPLE 1. Find an orthonormal basis for P_3 if the inner product on P_3 is defined by

$$\langle p, q\rangle = \sum_{i=1}^{3} p(x_i)q(x_i)$$

where $x_1 = -1$, $x_2 = 0$, and $x_3 = 1$.

SOLUTION. Starting with the basis $\{1, x, x^2\}$, the Gram–Schmidt process can be used to generate an orthonormal basis.

$$\|1\|^2 = \langle 1, 1\rangle = 3$$

so

$$\mathbf{u}_1 = \left(\frac{1}{\|1\|}\right)1 = \frac{1}{\sqrt{3}}$$

Set

$$p_1 = \langle x, \frac{1}{\sqrt{3}}\rangle \frac{1}{\sqrt{3}} = \left(-1\cdot\frac{1}{\sqrt{3}} + 0\cdot\frac{1}{\sqrt{3}} + 1\cdot\frac{1}{\sqrt{3}}\right)\frac{1}{\sqrt{3}} = 0$$

Therefore,

$$x - p_1 = x \quad \text{and} \quad \|x - p_1\|^2 = \langle x, x\rangle = 2$$

Hence

$$\mathbf{u}_2 = \frac{1}{\sqrt{2}}x$$

Finally,

$$p_2 = \langle x^2, \frac{1}{\sqrt{3}}\rangle \frac{1}{\sqrt{3}} + \langle x^2, \frac{1}{\sqrt{2}}x\rangle \frac{1}{\sqrt{2}}x = \frac{2}{3}$$

$$\|x^2 - p_2\|^2 = \langle x^2 - \tfrac{2}{3}, x^2 - \tfrac{2}{3}\rangle = \tfrac{2}{3}$$

and hence

$$\mathbf{u}_3 = \frac{\sqrt{6}}{2}\left(x^2 - \frac{2}{3}\right)$$

Orthogonal polynomials will be studied in more detail in Section 8.

EXAMPLE 2. Let

$$A = \begin{bmatrix} 1 & 1 & 2 \\ 1 & 2 & 3 \\ 1 & 2 & 1 \\ 1 & 1 & 6 \end{bmatrix}$$

Find an orthonormal basis for the column space of A.

SOLUTION. Let $\mathbf{x}_1$, $\mathbf{x}_2$, and $\mathbf{x}_3$ represent the respective column vectors of A. Since these vectors are linearly independent, they form a basis for a three-dimensional subspace of R^4. The Gram–Schmidt process may be used to form an orthonormal basis as follows:

(4) $$\mathbf{u}_1 = \left(\frac{1}{\|\mathbf{x}_1\|}\right)\mathbf{x}_1 = \frac{1}{2}\mathbf{x}_1 = \left(\frac{1}{2}, \frac{1}{2}, \frac{1}{2}, \frac{1}{2}\right)^T$$

$$\mathbf{p}_1 = \langle \mathbf{x}_2, \mathbf{u}_1\rangle \mathbf{u}_1 = 3\mathbf{u}_1$$

(5) $$\mathbf{x}_2 - \mathbf{p}_1 = \mathbf{x}_2 - 3\mathbf{u}_1 = \left(-\frac{1}{2}, \frac{1}{2}, \frac{1}{2}, -\frac{1}{2}\right)^T$$

$$(6)\qquad \mathbf{u}_2 = \left(\frac{1}{\|\mathbf{x}_2 - \mathbf{p}_1\|}\right)(\mathbf{x}_2 - \mathbf{p}_1)$$

$$= 1\left(-\frac{1}{2}, \frac{1}{2}, \frac{1}{2}, -\frac{1}{2}\right)^T$$

$$p_2 = \langle \mathbf{x}_3, \mathbf{u}_1 \rangle \mathbf{u}_1 + \langle \mathbf{x}_3, \mathbf{u}_2 \rangle \mathbf{u}_2$$

$$= 6\mathbf{u}_1 - 2\mathbf{u}_2$$

$$(7)\qquad \mathbf{x}_3 - \mathbf{p}_2 = \mathbf{x}_3 - 6\mathbf{u}_1 + 2\mathbf{u}_2 = (-2, 1, -1, 2)^T$$

$$(8)\qquad \mathbf{u}_3 = \frac{1}{\|\mathbf{x}_3 - \mathbf{p}_2\|}(\mathbf{x}_3 - \mathbf{p}_2) = \frac{1}{\sqrt{10}}(-2, 1, -1, 2)^T$$

The Gram–Schmidt process applied to the column vectors of a matrix gives rise to a useful factorization.

Theorem 5.7.2 (QR Factorization). *If A is an $m \times n$ matrix of rank n, then A can be factored into a product QR, where Q is an $m \times n$ matrix with orthonormal columns and R is $n \times n$ matrix that is upper triangular and invertible.*

PROOF. Let $\mathbf{x}_1, \mathbf{x}_2, \ldots, \mathbf{x}_n$ denote the respective column vectors of A. Let $\mathbf{p}_1, \ldots, \mathbf{p}_{n-1}$ be defined as in Theorem 5.7.1 and let $\{\mathbf{u}_1, \mathbf{u}_2, \ldots, \mathbf{u}_n\}$ be the orthonormal basis of $R(A)$ derived from the Gram–Schmidt process. Define

$$\alpha_{11} = \|\mathbf{x}_1\|$$

$$\alpha_{kk} = \|\mathbf{x}_k - \mathbf{p}_{k-1}\| \qquad \text{for} \quad k = 2, \ldots, n$$

and

$$\alpha_{ik} = \langle \mathbf{x}_k, \mathbf{u}_i \rangle \qquad \text{for} \quad i = 1, \ldots, k-1 \text{ and } k = 2, \ldots, n$$

By the Gram–Schmidt process,

$$\alpha_{11}\mathbf{u}_1 = \mathbf{x}_1$$

$$(9)\qquad \alpha_{kk}\mathbf{u}_k = \mathbf{x}_k - \alpha_{1k}\mathbf{u}_1 - \alpha_{2k}\mathbf{u}_2 - \cdots - \alpha_{k-1,k}\mathbf{u}_{k-1}$$
$$\text{for} \quad k = 2, \ldots, n$$

System (9) may be rewritten in the form

$$\mathbf{x}_1 = \alpha_{11}\mathbf{u}_1$$

$$\mathbf{x}_2 = \alpha_{12}\mathbf{u}_1 + \alpha_{22}\mathbf{u}_2$$

$$\vdots$$

$$\mathbf{x}_n = \alpha_{1n}\mathbf{u}_1 + \cdots + \alpha_{nn}\mathbf{u}_n$$

If each of these equations is transposed and the resulting system is written as a matrix equation, then

$$A^T = \begin{bmatrix} \mathbf{x}_1^T \\ \mathbf{x}_2^T \\ \vdots \\ \mathbf{x}_n^T \end{bmatrix} = \begin{bmatrix} \alpha_{11} & 0 & \cdots & 0 \\ \alpha_{12} & \alpha_{22} & \cdots & 0 \\ \vdots & & & \\ \alpha_{1n} & \alpha_{2n} & \cdots & \alpha_{nn} \end{bmatrix} \begin{bmatrix} \mathbf{u}_1^T \\ \mathbf{u}_2^T \\ \vdots \\ \mathbf{u}_n^T \end{bmatrix}$$

Transposing once more, we obtain

$$A = (\mathbf{u}_1 \quad \mathbf{u}_2 \quad \cdots \quad \mathbf{u}_n) \begin{bmatrix} \alpha_{11} & \alpha_{12} & \cdots & \alpha_{1n} \\ 0 & \alpha_{22} & \cdots & \alpha_{2n} \\ \vdots & & & \\ 0 & 0 & \cdots & \alpha_{nn} \end{bmatrix} = QR$$

The columns of Q are the orthonormal vectors $\mathbf{u}_1, \mathbf{u}_2, \ldots, \mathbf{u}_n$ and R is nonsingular, since $\alpha_{kk} \neq 0$ for $k = 1, \ldots, n$.

EXAMPLE 3. Factor the matrix A from Example 2 into a product QR.

SOLUTION. It follows from (4) that

$$2\mathbf{u}_1 = \mathbf{x}_1$$

It follows from (5) and (6) that

$$1\mathbf{u}_2 = \mathbf{x}_2 - 3\mathbf{u}_1$$

By (7) and (8),

$$\sqrt{10}\,\mathbf{u}_3 = \mathbf{x}_3 - 6\mathbf{u}_1 + 2\mathbf{u}_2$$

Solving these equations for $\mathbf{x}_1$, $\mathbf{x}_2$, and $\mathbf{x}_3$ yields

$$\begin{aligned} \mathbf{x}_1 &= 2\mathbf{u}_1 \\ \mathbf{x}_2 &= 3\mathbf{u}_1 + \mathbf{u}_2 \\ \mathbf{x}_3 &= 6\mathbf{u}_1 - 2\mathbf{u}_2 + \sqrt{10}\,\mathbf{u}_3 \end{aligned}$$

Let

$$Q = (\mathbf{u}_1 \quad \mathbf{u}_2 \quad \mathbf{u}_3) = \begin{bmatrix} \frac{1}{2} & -\frac{1}{2} & \frac{-2}{\sqrt{10}} \\ \frac{1}{2} & \frac{1}{2} & \frac{1}{\sqrt{10}} \\ \frac{1}{2} & \frac{1}{2} & \frac{-1}{\sqrt{10}} \\ \frac{1}{2} & -\frac{1}{2} & \frac{2}{\sqrt{10}} \end{bmatrix}$$

and

$$R = \begin{bmatrix} 2 & 3 & 6 \\ 0 & 1 & -2 \\ 0 & 0 & \sqrt{10} \end{bmatrix}$$

The reader may verify that $A = QR$.

We saw in Section 6 that if the columns of an $m \times n$ matrix A form an orthonormal set, then the least squares solution to $A\mathbf{x} = \mathbf{b}$ is simply $\hat{\mathbf{x}} = A^T\mathbf{b}$. If A has rank n but its column vectors do not form an orthonormal set in R^m, then the QR factorization can be used to solve the least squares problem.

Theorem 5.7.3. *If A is an $m \times n$ matrix of rank n, then the solution to the least squares problem $A\mathbf{x} = \mathbf{b}$ is given by $\hat{\mathbf{x}} = R^{-1}Q^T\mathbf{b}$, where Q and R are the matrices obtained from the factorization given in Theorem 5.7.2. The solution $\hat{\mathbf{x}}$ may be obtained by using back substitution to solve $R\mathbf{x} = Q^T\mathbf{b}$.*

PROOF. Let $\hat{\mathbf{x}}$ be the solution to the least squares problem $A\mathbf{x} = \mathbf{b}$ guaranteed by Theorem 5.5.1. Thus $\hat{\mathbf{x}}$ is the solution to the normal equations

$$A^TA\mathbf{x} = A^T\mathbf{b}$$

If A is factored into a product QR, these equations become

$$(QR)^TQR\mathbf{x} = (QR)^T\mathbf{b}$$

or

$$R^T(Q^TQ)R\mathbf{x} = R^TQ^T\mathbf{b}$$

Since Q has orthonormal columns, it follows that $Q^TQ = I$ and hence

$$R^TR\mathbf{x} = R^TQ^T\mathbf{b}$$

Since R^T is invertible, this simplifies to

$$R\mathbf{x} = Q^T\mathbf{b} \qquad \text{or} \qquad \mathbf{x} = R^{-1}Q^T\mathbf{b}$$

EXAMPLE 4. Find the least squares solution to

$$\begin{bmatrix} 1 & 1 & 2 \\ 1 & 2 & 3 \\ 1 & 2 & 1 \\ 1 & 1 & 6 \end{bmatrix} \begin{bmatrix} x_1 \\ x_2 \\ x_3 \end{bmatrix} = \begin{bmatrix} 3 \\ -4 \\ 2 \\ 1 \end{bmatrix}$$

SOLUTION. The coefficient matrix of this system was factored in Example 3. Using that factorization, we have

$$Q^T\mathbf{b} = \begin{bmatrix} \frac{1}{2} & \frac{1}{2} & \frac{1}{2} & \frac{1}{2} \\ -\frac{1}{2} & \frac{1}{2} & \frac{1}{2} & -\frac{1}{2} \\ -\frac{2}{\sqrt{10}} & \frac{1}{\sqrt{10}} & -\frac{1}{\sqrt{10}} & \frac{2}{\sqrt{10}} \end{bmatrix} \begin{bmatrix} 3 \\ -4 \\ 2 \\ 1 \end{bmatrix} = \begin{bmatrix} 1 \\ -3 \\ -\sqrt{10} \end{bmatrix}$$

$$R\mathbf{x} = \begin{bmatrix} 2 & 3 & 6 \\ 0 & 1 & -2 \\ 0 & 0 & \sqrt{10} \end{bmatrix} \begin{bmatrix} x_1 \\ x_2 \\ x_3 \end{bmatrix} = \begin{bmatrix} 1 \\ -3 \\ -\sqrt{10} \end{bmatrix}$$

This system is easily solved by back substitution. The solution we get is $(x_1, x_2, x_3) = (11, -5, -1)$.

Of particular interest are $n \times n$ matrices whose column vectors form an orthonormal set in R^n. Such matrices are called *orthogonal* matrices.

***Theorem* 5.7.4.** *If Q is an orthogonal matrix, then Q is invertible and $Q^{-1} = Q^T$.*

PROOF. Q has rank n, since its column vectors are linearly independent. Therefore, Q^{-1} exists. It follows from Theorem 5.6.3 that

$$Q^T = Q^TI = Q^T(QQ^{-1}) = (Q^TQ)Q^{-1} = IQ^{-1} = Q^{-1}$$

EXAMPLE 5. For any fixed θ, the matrix

$$Q = \begin{pmatrix} \cos\theta & -\sin\theta \\ \sin\theta & \cos\theta \end{pmatrix}$$

is orthogonal and

$$Q^{-1} = \begin{pmatrix} \cos\theta & \sin\theta \\ -\sin\theta & \cos\theta \end{pmatrix}$$

The matrix Q in Example 5 can be thought of as a linear transformation from R^2 onto R^2 that has the effect of rotating each vector by an angle θ while leaving the length of the vector unchanged. Similarly, Q^{-1} can be thought of as a rotation by the angle $-\theta$ (see Figure 5.7.2).

In general, inner products are preserved under multiplication by an orthogonal matrix (i.e., $\langle \mathbf{x}, \mathbf{y} \rangle = \langle Q\mathbf{x}, Q\mathbf{y} \rangle$). This is true, since

$$\langle Q\mathbf{x}, Q\mathbf{y} \rangle = (Q\mathbf{y})^TQ\mathbf{x} = \mathbf{y}^TQ^TQ\mathbf{x} = \mathbf{y}^T\mathbf{x} = \langle \mathbf{x}, \mathbf{y} \rangle$$

In particular, if $\mathbf{x} = \mathbf{y}$, then $\|Q\mathbf{x}\|^2 = \|\mathbf{x}\|^2$ and hence $\|Q\mathbf{x}\| = \|\mathbf{x}\|$. Multiplication by an orthogonal matrix preserves the lengths of vectors.

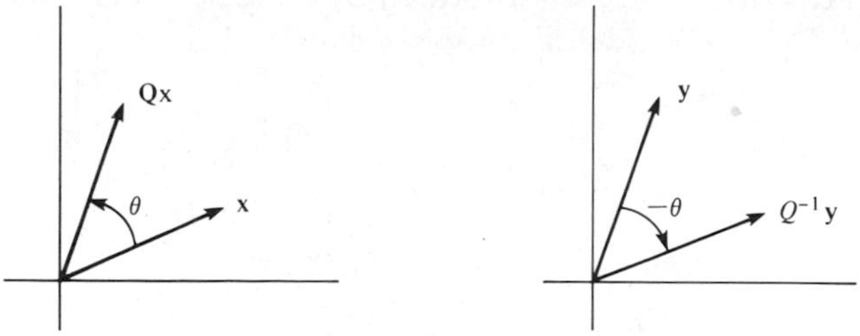

Figure 5.7.2

Permutation Matrices

A *permutation matrix* is a matrix formed from the identity matrix by reordering its columns. Clearly, then, permutation matrices are orthogonal matrices. If P is the permutation matrix formed by reordering the columns of I in the order $(k_1, \ldots, k_n)$, then $P = (\mathbf{e}_{k_1}, \ldots, \mathbf{e}_{k_n})$. If A is an $m \times n$ matrix, then

$$AP = (A\mathbf{e}_{k_1}, \ldots, A\mathbf{e}_{k_n}) = (\mathbf{a}_{k_1}, \ldots, \mathbf{a}_{k_n})$$

Postmultiplication of A by P reorders the columns of A in the order $(k_1, \ldots, k_n)$. For example, if

$$A = \begin{pmatrix} 1 & 2 & 3 \\ 1 & 2 & 3 \end{pmatrix} \quad \text{and} \quad P = \begin{bmatrix} 0 & 1 & 0 \\ 0 & 0 & 1 \\ 1 & 0 & 0 \end{bmatrix}$$

then

$$AP = \begin{pmatrix} 3 & 1 & 2 \\ 3 & 1 & 2 \end{pmatrix}$$

Since $P = (\mathbf{e}_{k_1}, \ldots, \mathbf{e}_{k_n})$ is orthogonal, it follows that

$$P^{-1} = P^T = \begin{bmatrix} \mathbf{e}_{k_1}^T \\ \vdots \\ \mathbf{e}_{k_n}^T \end{bmatrix}$$

The k_1 column of P^T will be $\mathbf{e}_1$, the k_2 column will be $\mathbf{e}_2$, and so on. Thus P^T is a permutation matrix. The matrix P^T can be formed directly from I by reordering its rows in the order $(k_1, k_2, \ldots, k_n)$. In general, a permutation matrix can be formed from I by either reordering its rows or its columns.

If Q is the permutation matrix formed by reordering the rows of I in the order $(k_1, k_2, \ldots, k_n)$ and B is an $n \times r$ matrix, then

$$QB = \begin{bmatrix} \mathbf{e}_{k_1}^T \\ \vdots \\ \mathbf{e}_{k_n}^T \end{bmatrix} B = \begin{bmatrix} \mathbf{e}_{k_1}^T B \\ \vdots \\ \mathbf{e}_{k_n}^T B \end{bmatrix}$$

Thus QB is the matrix formed by reordering the rows of B in the order $(k_1, k_2, \ldots, k_n)$. For example, if

$$Q = \begin{bmatrix} 0 & 0 & 1 \\ 1 & 0 & 0 \\ 0 & 1 & 0 \end{bmatrix} \quad \text{and} \quad B = \begin{bmatrix} 1 & 1 \\ 2 & 2 \\ 3 & 3 \end{bmatrix}$$

then

$$QB = \begin{bmatrix} 3 & 3 \\ 1 & 1 \\ 2 & 2 \end{bmatrix}$$

In general, if P is an $n \times n$ permutation matrix, then premultiplication of an $n \times r$ matrix B by P reorders the rows of B and postmultiplication of an $m \times n$ matrix A by P reorders the columns of A.

EXERCISES

1. Consider the vector space $C[-1, 1]$ with inner product defined by

$$\langle f, g \rangle = \int_{-1}^{1} f(x)g(x)\,dx$$

Find an orthonormal basis for the subspace spanned by 1, x, and x^2.

2. Let

$$A = \begin{bmatrix} 2 & 1 \\ 1 & 1 \\ 2 & 1 \end{bmatrix} \quad \text{and} \quad \mathbf{b} = \begin{bmatrix} 12 \\ 6 \\ 18 \end{bmatrix}$$

(a) Use the Gram–Schmidt process to find an orthonormal basis for the column space of A.

(b) Factor A into a product QR, where Q has an orthonormal set of column vectors and R is upper triangular.

(c) Solve the least squares problem

$$A\mathbf{x} = \mathbf{b}$$

3. Show that $\mathbf{u}_1 = (0, 1, 0)^T$ and $\mathbf{u}_2 = (1/\sqrt{2}, 0, 1/\sqrt{2})^T$ also form an orthonormal basis for the column space of the matrix A in Exercise 2.

4. Repeat Exercise 2 using

$$A = \begin{bmatrix} 3 & -1 \\ 4 & 2 \\ 0 & 2 \end{bmatrix} \quad \text{and} \quad \mathbf{b} = \begin{bmatrix} 0 \\ 20 \\ 10 \end{bmatrix}$$

5. Let Q be the matrix in Example 5 and let $\mathbf{x} = (r \cos \alpha, r \sin \alpha)^T$. Verify that

$$Q\mathbf{x} = (r \cos(\alpha + \theta), r \sin(\alpha + \theta))^T$$

6. Let $\mathbf{u} = (\cos \theta/2, \sin \theta/2)^T$ and let $Q = I - 2\mathbf{u}\mathbf{u}^T$.
(a) Show that Q is orthogonal.
(b) Show that Q is a combination of a rotation and a reflection.

7. Prove that an $n \times n$ matrix Q is orthogonal if and only if $Q^TQ = I$.

8. Prove that the transpose of an orthogonal matrix is an orthogonal matrix.

9. Let Q be an $n \times n$ orthogonal matrix. Use mathematical induction to prove each of the following:
(a) $(Q^m)^{-1} = (Q^T)^m = (Q^m)^T$ for any positive integer m
(b) $\|Q^m\mathbf{x}\| = \|\mathbf{x}\|$ for any $\mathbf{x} \in R^n$

10. Let $\mathbf{u}$ be a unit vector in R^n and let $H = I - 2\mathbf{u}\mathbf{u}^T$. Show that H is both orthogonal and symmetric and hence is its own inverse.

8. ORTHOGONAL POLYNOMIALS

We have already seen how polynomials can be used for data fitting and for approximating continuous functions. Since both of these problems are least squares problems, they can be simplified by selecting an orthogonal basis for the class of approximating polynomials. This leads us to the concept of orthogonal polynomials.

In this section we will study families of orthogonal polynomials associated with various inner products on $C[a, b]$. We will see that the polynomials in each of these classes satisfy a three-term recursion relation. This recursion relation is particularly useful in computer applications.

Certain families of orthogonal polynomials have important applications in many areas of mathematics. We will refer to these polynomials as classical polynomials and examine them in more detail. In particular, the classical polynomials are solutions to certain classes of second-order linear differential equations that arise in the solution of many partial differential equations from mathematical physics.

Orthogonal Sequences

Since the proof of Theorem 5.7.1 was by induction, the Gram–Schmidt process is valid for a denumerable set. Thus, if $\mathbf{x}_1, \mathbf{x}_2, \ldots$ is a sequence of vectors in an inner product space V and $\mathbf{x}_1, \mathbf{x}_2, \ldots \mathbf{x}_n$ are linearly independent for each n, then the Gram–Schmidt process may be used to form a sequence $\mathbf{u}_1, \mathbf{u}_2, \ldots,$ where $\{\mathbf{u}_1, \mathbf{u}_2, \ldots\}$ is an orthonormal set and $\mathcal{S}(\mathbf{x}_1, \mathbf{x}_2 \ldots, \mathbf{x}_n) = \mathcal{S}(\mathbf{u}_1, \mathbf{u}_2, \ldots, \mathbf{u}_n)$ for each n. In particular, from the sequence $1, x, x^2, \ldots$ it is possible to construct an "orthonormal sequence" $p_0(x), p_1(x), \ldots$.

Let P be the vector space of all polynomials and define the inner product $\langle\, ,\, \rangle$ on P by

$$(1) \qquad \langle p, q\rangle = \int_a^b p(x)q(x)w(x)\, dx$$

where $w(x)$ is a weight function. The interval can be taken as either open or closed and may be finite or infinite. If, however,

$$\int_a^b p(x)w(x)\, dx \text{ is improper, we require that it converge}$$

for every $p \in P$.

Definition. Let $p_0(x), p_1(x), \ldots$ be a sequence of polynomials with $\deg p_i(x) = i$ for each i. If $\langle p_i(x), p_j(x)\rangle = 0$ whenever $i \neq j$, then $\{p_n(x)\}$ is said to be a **sequence of orthogonal polynomials.** If $\langle p_i, p_j\rangle = \delta_{ij}$, then $\{p_n(x)\}$ is said to be a **sequence of orthonormal polynomials.**

Theorem 5.8.1. *If $p_0, p_1, \ldots$ is a sequence of orthogonal polynomials, then*

(i) $p_0, \ldots, p_{n-1}$ *form a basis for* P_n.

(ii) $p_n \in P_n^{\perp}$ (i.e., p_n *is orthogonal to every polynomial of degree less than* n).

PROOF. It follows from Theorem 5.6.1 that $p_0, p_1, \ldots, p_{n-1}$ are linearly independent in P_n. Since $\dim P_n = n$, these n vectors must form a basis for P_n. Let $p(x)$ be any polynomial of degree less than n. Then

$$p(x) = \sum_{i=0}^{n-1} c_i p_i(x)$$

and hence

$$\langle p_n, p\rangle = \langle p_n, \sum_{i=0}^{n-1} c_i p_i\rangle = \sum_{i=0}^{n-1} c_i\langle p_n, p_i\rangle = 0$$

Therefore, $p_n \in P_n^{\perp}$.

If $\{p_0, p_1, \ldots, p_{n-1}\}$ is an orthogonal set in P_n and

$$u_i = \left(\frac{1}{\|p_i\|}\right)p_i \qquad \text{for} \quad i = 0, \ldots, n-1$$

then $\{u_0, \ldots, u_{n-1}\}$ is an orthonormal basis for P_n. Hence, if $p \in P_n$, then

$$\begin{aligned} p &= \sum_{i=0}^{n-1} \langle p, u_i\rangle u_i \\ &= \sum_{i=0}^{n-1} \left\langle p, \left(\frac{1}{\|p_i\|}\right)p_i\right\rangle\left(\frac{1}{\|p_i\|}\right)p_i \\ &= \sum_{i=0}^{n-1} \frac{\langle p, p_i\rangle}{\langle p_i, p_i\rangle} p_i \end{aligned}$$

Similarly, if $f \in C[a, b]$, then the best least squares approximation to f by elements of P_n is given by

$$p = \sum_{i=0}^{n-1} \frac{\langle f, p_i\rangle}{\langle p_i, p_i\rangle} p_i$$

where $p_0, p_1, \ldots, p_{n-1}$ are orthogonal polynomials.

Another nice feature of sequences of orthogonal polynomials is that they satisfy a three-term recursion relation.

Theorem 5.8.2. *Let $p_0, p_1, \ldots$ be a sequence of orthogonal polynomials. Let a_i denote the lead coefficient of p_i for each i and define $p_{-1}(x)$ to be the zero polynomial. Then*

$$\alpha_{n+1}p_{n+1}(x) = (x - \beta_n)p_n(x) - \alpha_n\gamma_n p_{n-1}(x)$$

where

$$\alpha_n = \frac{a_{n-1}}{a_n}, \qquad \beta_n = \frac{\langle p_n, xp_n\rangle}{\langle p_n, p_n\rangle}, \qquad \gamma_n = \frac{\langle p_n, p_n\rangle}{\langle p_{n-1}, p_{n-1}\rangle}$$

PROOF. Since $p_0, p_1, \ldots, p_{n+1}$ form a basis for P_{n+2}, we can write

(2) $$xp_n(x) = \sum_{k=0}^{n+1} c_{nk}p_k(x)$$

where

$$c_{nk} = \frac{\langle xp_n, p_k \rangle}{\langle p_k, p_k \rangle} \tag{3}$$

For any inner product defined by (1),

$$\langle xf, g \rangle = \langle f, xg \rangle$$

In particular,

$$\langle xp_n, p_k \rangle = \langle p_n, xp_k \rangle$$

It follows from Theorem 5.8.1 that if $k < n - 1$, then

$$c_{nk} = \frac{\langle xp_n, p_k \rangle}{\langle p_k, p_k \rangle} = \frac{\langle p_n, xp_k \rangle}{\langle p_k, p_k \rangle} = 0$$

Therefore, (2) simplifies to

$$xp_n(x) = c_{n,n-1}p_{n-1}(x) + c_{n,n}p_n(x) + c_{n,n+1}p_{n+1}(x)$$

This can be rewritten in the form

$$c_{n,n+1}p_{n+1}(x) = (x - c_{n,n})p_n(x) - c_{n,n-1}p_{n-1}(x) \tag{4}$$

Comparing the lead coefficients of the polynomials on each side of (4), we see that

$$c_{n,n+1}a_{n+1} = a_n$$

or

$$c_{n,n+1} = \frac{a_n}{a_{n+1}} = \alpha_{n+1} \tag{5}$$

It also follows from (4) that

$$\begin{aligned} c_{n,n+1}\langle p_n, p_{n+1} \rangle &= \langle p_n, (x - c_{n,n})p_n \rangle - c_{n,n-1}\langle p_n, p_{n-1} \rangle \\ 0 &= \langle p_n, xp_n \rangle - c_{nn}\langle p_n, p_n \rangle \end{aligned}$$

and hence

$$c_{nn} = \frac{\langle p_n, xp_n \rangle}{\langle p_n, p_n \rangle} = \beta_n$$

It follows from (3) that

$$\begin{aligned} \langle p_{n-1}, p_{n-1} \rangle c_{n,n-1} &= \langle xp_n, p_{n-1} \rangle \\ &= \langle p_n, xp_{n-1} \rangle \\ &= \langle p_n, p_n \rangle c_{n-1,n} \end{aligned}$$

and hence by (5) we have

$$c_{n,n-1} = \frac{\langle p_n, p_n \rangle}{\langle p_{n-1}, p_{n-1} \rangle}\alpha_n = \gamma_n\alpha_n$$

In generating a sequence of orthogonal polynomials by the recursion relation in Theorem 5.8.2, we are free to choose any nonzero lead coefficient a_{n+1} we want at each step. This is reasonable, since any nonzero multiple of a particular p_{n+1} will also be orthogonal to $p_0, \ldots, p_n$. If we were to choose our a_i's to be 1, for example, then the recursion relation would simplify to

$$p_{n+1}(x) = (x - \beta_n)p_n(x) - \gamma_n p_{n-1}(x)$$

Let us now look at some examples. Because of their importance we will consider the classical polynomials beginning with the simplest, the Legendre polynomials.

Legendre Polynomials

The Legendre polynomials are orthogonal with respect to the inner product

$$\langle p, q \rangle = \int_{-1}^{1} p(x)q(x)\,dx$$

Let $P_n(x)$ denote the Legendre polynomial of degree n. If we choose the lead coefficients so that $P_n(1) = 1$ for each n, then the recursion formula for the Legendre polynomials is

$$(n+1)P_{n+1}(x) = (2n+1)xP_n(x) - nP_{n-1}(x)$$

By the use of this formula, the sequence of Legendre polynomials is easily generated.

$$\begin{aligned} P_0(x) &= 1 \\ P_1(x) &= x \\ P_2(x) &= \tfrac{1}{2}(3x^2 - 1) \\ P_3(x) &= \tfrac{1}{2}(5x^2 - 3x) \\ P_4(x) &= \tfrac{1}{8}(35x^4 - 30x^2 + 3) \end{aligned}$$

Tchebycheff Polynomials

These polynomials are orthogonal with respect to the inner product,

$$\langle p, q \rangle = \int_{-1}^{1} p(x)q(x)(1 - x^2)^{-1/2}\,dx$$

It is customary to normalize the lead coefficients so that $a_0 = 1$ and $a_k = 2^{k-1}$ for $k = 1, 2, \ldots$. The Tchebycheff polynomials are denoted by $T_n(x)$ and have the interesting property that

$$T_n(\cos\theta) = \cos n\theta \qquad (6)$$

This property, together with the trigonometric identity

$$\cos(n+1)\theta = 2\cos\theta\cos n\theta - \cos(n-1)\theta$$

can be used to derive the recursion relation

$$T_{n+1}(x) = 2xT_n(x) - T_{n-1}(x)$$

for $n \geqslant 1$.

The Legendre and the Tchebycheff polynomials are both special cases of the *Jacobi polynomials*. The Jacobi polynomials $P_n^{(\lambda,\mu)}$ are orthogonal with respect to the inner product,

$$\langle p, q\rangle = \int_{-1}^{1} p(x)q(x)(1-x)^{\lambda}(1+x)^{\mu}\,dx$$

where $\lambda, \mu > -1$.

The *Hermite polynomials* are defined on the interval $(-\infty, \infty)$. They are orthogonal with respect to the inner product.

$$\langle p, q\rangle = \int_{-\infty}^{\infty} p(x)q(x)e^{-x^2}\,dx$$

The recursion relation for Hermite polynomials is given by

$$H_{n+1}(x) = 2xH_n - 2nH_{n-1}(x)$$

The *Laguerre polynomials* are defined on the interval $(0, \infty)$ and are orthogonal with respect to the inner product,

$$\langle p, q\rangle = \int_{0}^{\infty} p(x)q(x)x^{\lambda}e^{-x}\,dx$$

where $\lambda > -1$. The recursion relation for the Laguerre polynomials is given by

$$(n+1)L_{n+1}^{(\lambda)}(x) = (2n+\lambda+1-x)L_n^{(\lambda)}(x) - (n+\lambda)L_{n-1}^{(\lambda)}(x)$$

The Tchebycheff, Hermite, and Laguerre polynomials are compared in Table 1.

TABLE 1

Tchebycheff	*Hermite*	*Laguerre* ($\lambda = 0$)
$T_{n+1} = 2xT_n - T_{n-1},\ n > 1$	$H_{n+1} = 2xH_n - 2nH_{n-1}$	$(n+1)L_{n+1}^{(0)} = (2n+1-x)L_n^{(0)} - nL_{n-1}^{(0)}$
$T_0 = 1$	$H_0 = 1$	$L_0^{(0)} = 1$
$T_1 = x$	$H_1 = 2x$	$L_1^{(0)} = 1 - x$
$T_2 = 2x^2 - 1$	$H_2 = 4x^2 - 2$	$L_2^{(0)} = \frac{1}{2}(x^2 - 4x + 2)$
$T_3 = 4x^3 - 3x$	$H_3 = 8x^3 - 12x$	$L_3^{(0)} = \frac{1}{6}(-x^3 + 9x^2 - 18x + 6)$

EXERCISES

1. Use the recursion formulas to calculate T_4, T_5 and H_4, H_5.

2. Let $p_0(x)$, $p_1(x)$, and $p_2(x)$ be orthogonal with respect to the inner product,

$$\langle p(x), q(x) \rangle = \int_{-1}^{1} \frac{p(x)q(x)}{1+x^2}\, dx$$

Use Theorem 5.8.2 to calculate $p_1(x)$ and $p_2(x)$ if all the polynomials have lead coefficient 1.

3. Show that the Tchebycheff polynomials have the following properties:
(a) $2T_m(x)T_n(x) = T_{m+n}(x) + T_{m-n}(x)$, for $m > n$
(b) $T_m(T_n(x)) = T_{mn}(x)$

4. Find the best quadratic least squares approximation to e^x on $[-1, 1]$ with respect to the inner product,

$$\langle f, g \rangle = \int_{-1}^{1} f(x)g(x)\, dx$$

5. Let $p_0, p_1, \ldots$ be a sequence of orthogonal polynomials and let a_n denote the lead coefficient of p_n. Prove that

$$\|p_n\|^2 = a_n \langle x^n, p_n \rangle$$

6. Let $T_n(x)$ denote the Tchebycheff polynomial of degree n and define

$$U_{n-1}(x) = \frac{1}{n} T_n'(x)$$

for $n = 1, 2, \ldots$.
(a) Compute $U_0(x)$, $U_1(x)$, and $U_2(x)$
(b) If $x = \cos\theta$, show that

$$U_{n-1}(x) = \frac{\sin n\theta}{\sin\theta}$$

7. Let $U_{n-1}(x)$ be defined as in Exercise 6 for $n \geqslant 1$ and define $U_{-1}(x) = 0$. Show that:
(a) $T_n(x) = U_n(x) - xU_{n-1}(x)$, for $n \geqslant 0$
(b) $U_n(x) = 2xU_{n-1}(x) - U_{n-2}(x)$, for $n \geqslant 1$

8. Show the U_i's defined in Exercise 6 are orthogonal with respect to the inner product

$$\langle p, q \rangle = \int_{-1}^{1} p(x)q(x)(1 - x^2)^{1/2}\, dx$$

The U_i's are called Tchebycheff polynomials of the second kind.

9. For $n = 0, 1, 2$ show that the Legendre polynomial $P_n(x)$ satisfies the second-order equation

$$(1 - x^2)y'' - 2xy' + n(n+1)y = 0$$

10. Use mathematical induction to prove each of the following.
(a) $H_n'(x) = 2nH_{n-1}(x)$, $n = 0, 1, \ldots$
(b) $H_n''(x) - 2xH_n'(x) + 2nH_n(x) = 0$, $n = 0, 1, \ldots$

6

Eigenvalues

In Section 1 we will be concerned with the equation $A\mathbf{x} = \lambda\mathbf{x}$. This equation occurs in many applications of linear algebra. If the equation has a nonzero solution $\mathbf{x}$, then λ is said to be an *eigenvalue* of A and $\mathbf{x}$ is said to be an *eigenvector* belonging to λ. One of the main applications of eigenvalues is in the solution of systems of linear differential equations. This application is presented in Section 2.

If A is an $n \times n$ matrix, we can think of A as representing a linear transformation from R^n into itself. Eigenvalues and eigenvectors provide the key to understanding how the operator works. For example, if $\lambda > 0$, the effect of the operator on any eigenvector belonging to λ is simply a stretching or a shrinking by a constant factor. Indeed, the effect of the operator is easily determined on any linear combination of eigenvectors. In particular, if it is possible to find a basis of eigenvectors for R^n, the operator can be represented by a diagonal matrix D with respect to that basis and the matrix A can be factored into a product SDS^{-1}. In Section 3 we will see how this is done and look at a number of applications.

In Section 4 we consider matrices with complex entries. In this setting we will be concerned with matrices whose eigenvectors can be used to form an orthonormal basis for C^n (the vector space of all n-tuples of complex numbers). Section 5 deals with the application of eigenvalues to quadratic equations in several variables, and in Section 6 we consider applications to finding maximum and minimum points of functions of several variables. Finally, in Section 7 we study matrices with nonnegative entries and some applications to economics.

1. EIGENVALUES AND EIGENVECTORS

In this section we will be concerned with the problem of finding a scalar λ such that the $n \times n$ system

$$A\mathbf{x} = \lambda\mathbf{x}$$

has a nonzero solution.

Definition. Let A be an $n \times n$ matrix. A scalar λ is said to be an **eigenvalue** or a **characteristic value** of A if there exists a nonzero vector $\mathbf{x}$ in R^n such that $A\mathbf{x} = \lambda\mathbf{x}$. The vector $\mathbf{x}$ is said to be an **eigenvector** or a **characteristic vector** belonging to λ.

The equation $A\mathbf{x} = \lambda\mathbf{x}$ can be written in the form

$$(A - \lambda I)\mathbf{x} = \mathbf{0} \tag{1}$$

Thus λ is an eigenvalue of A if and only if (1) has a nontrivial solution. The set of solutions to (1) is $N(A - \lambda I)$, which is a subspace of R^n. Thus, if λ is an eigenvalue of A, then $N(A - \lambda I) \neq \{\mathbf{0}\}$ and any nonzero vector in $N(A - \lambda I)$ is an eigenvector belonging to λ. The subspace $N(A - \lambda I)$ is called the *eigenspace* corresponding to the eigenvalue λ.

Equation (1) will have a nontrivial solution if and only if $A - \lambda I$ is singular or, equivalently,

$$|A - \lambda I| = 0 \tag{2}$$

If the determinant in (2) is expanded, we obtain an nth-degree polynomial in the variable λ,

$$p(\lambda) = |A - \lambda I|$$

This polynomial is called the *characteristic polynomial* and equation (2) is called the *characteristic equation* for the matrix A. The roots of the characteristic polynomial are the eigenvalues of A. If we count roots according to multiplicity, the characteristic polynomial will have exactly n roots. Thus A will have n eigenvalues, some of which may be repeated and some of which may be complex numbers. To take care of the latter case, it will be necessary to expand our field of scalars to the complex numbers and to allow complex entries for our vectors and matrices.

We have now established a number of equivalent conditions for λ to be an eigenvalue of A.

> Let A be an $n \times n$ matrix and λ be a scalar. The following statements are equivalent.
>
> (a) λ is an eigenvalue of A.
> (b) $(A - \lambda I)\mathbf{x} = \mathbf{0}$ has a nontrivial solution.
> (c) $N(A - \lambda I) \neq \{\mathbf{0}\}$
> (d) $A - \lambda I$ is singular.
> (e) $|A - \lambda I| = 0$

We will now use statement (e) to determine the eigenvalues in a number of examples.

EXAMPLE 1. Find the eigenvalues of the matrix

$$A = \begin{pmatrix} 3 & 2 \\ 3 & -2 \end{pmatrix}$$

SOLUTION. The characteristic equation is

$$\begin{vmatrix} 3-\lambda & 2 \\ 3 & -2-\lambda \end{vmatrix} = 0$$

or

$$\lambda^2 - \lambda - 12 = 0$$

Thus the eigenvalues of A are $\lambda_1 = 4$ and $\lambda_2 = -3$. To find the eigenvectors belonging to $\lambda_1 = 4$, we must determine the nullspace of $A - 4I$.

$$A - 4I = \begin{pmatrix} -1 & 2 \\ 3 & -6 \end{pmatrix}$$

Solving $(A - 4I)\mathbf{x} = \mathbf{0}$, we get

$$\mathbf{x} = (2x_2, x_2)^T$$

Thus any nonzero multiple of $(2, 1)^T$ is an eigenvector belonging to λ_1 and $\{(2, 1)^T\}$ is a basis for the eigenspace corresponding to λ_1. Similarly, to find eigenvectors for λ_2, we must solve $(A + 3I)\mathbf{x} = \mathbf{0}$. In this case $\{(-1, 3)^T\}$ is a basis for $N(A + 3I)$ and any nonzero multiple of $(-1, 3)^T$ is an eigenvector belonging to λ_2.

EXAMPLE 2. Let

$$A = \begin{bmatrix} 3 & -1 & -2 \\ 2 & 0 & -2 \\ 2 & -1 & -1 \end{bmatrix}$$

Find the eigenvalues of A and the corresponding eigenspaces.

SOLUTION

$$\begin{vmatrix} 3-\lambda & -1 & -2 \\ 2 & -\lambda & -2 \\ 2 & -1 & -1-\lambda \end{vmatrix} = -\lambda(\lambda-1)^2$$

Thus the characteristic polynomial has roots $\lambda_1 = 0$, $\lambda_2 = \lambda_3 = 1$. The eigenspace corresponding to $\lambda_1 = 0$ is $N(A)$. The reader may verify that $\{(1, 1, 1)^T\}$ is a basis for $N(A)$. To find the eigenspace corresponding to $\lambda = 1$, one must solve the system $(A - I)\mathbf{x} = \mathbf{0}$.

$$A - I = \begin{bmatrix} 2 & -1 & -2 \\ 2 & -1 & -2 \\ 2 & -1 & -2 \end{bmatrix}$$

A vector $\mathbf{x} = (x_1, x_2, x_3)^T$ will be in $N(A - I)$, provided that $x_2 = 2x_1 - 2x_3$. Thus $\{(1, 2, 0)^T, (0, -2, 1)^T\}$ is a basis for $N(A - I)$.

EXAMPLE 3. Find the eigenvalues and the corresponding eigenspaces of the matrix

$$A = \begin{pmatrix} 1 & 2 \\ -2 & 1 \end{pmatrix}$$

SOLUTION

$$\begin{vmatrix} 1-\lambda & 2 \\ -2 & 1-\lambda \end{vmatrix} = (1-\lambda)^2 + 4$$

The roots of the characteristic polynomial are $\lambda_1 = 1 + 2i$, $\lambda_2 = 1 - 2i$.

$$A - \lambda_1 I = \begin{pmatrix} -2i & 2 \\ -2 & -2i \end{pmatrix} = -2\begin{pmatrix} i & -1 \\ 1 & i \end{pmatrix}$$

It follows that $\{(1, i)^T\}$ is a basis for the eigenspace corresponding to $\lambda_1 = 1 + 2i$. Similarly,

$$A - \lambda_2 I = \begin{pmatrix} 2i & 2 \\ -2 & 2i \end{pmatrix} = 2\begin{pmatrix} i & 1 \\ -1 & i \end{pmatrix}$$

and $\{(1, -i)^T\}$ is a basis for $N(A - \lambda_2 I)$.

Let A be an $n \times n$ matrix and let $p(\lambda)$ be the characteristic polynomial of A.

$$(3)\qquad p(\lambda) = |A - \lambda I| = \begin{vmatrix} a_{11}-\lambda & a_{12} & \cdots & a_{1n} \\ a_{21} & a_{22}-\lambda & & a_{2n} \\ \vdots & & & \\ a_{n1} & a_{n2} & & a_{nn}-\lambda \end{vmatrix}$$

It follows from the definition of the determinant that the only term in the expansion of $|A - \lambda I|$ involving more than $n-2$ of the diagonal elements will be

$$(4)\qquad (a_{11}-\lambda)(a_{22}-\lambda)\cdots(a_{nn}-\lambda)$$

When this is expanded, the coefficient of λ^n will be $(-1)^n$. Thus, if $\lambda_1, \ldots, \lambda_n$ are the eigenvalues of A, then

$$(5)\qquad \begin{aligned} p(\lambda) &= (-1)^n(\lambda-\lambda_1)(\lambda-\lambda_2)\cdots(\lambda-\lambda_n) \\ &= (\lambda_1-\lambda)(\lambda_2-\lambda)\cdots(\lambda_n-\lambda) \end{aligned}$$

It follows from (3) and (5) that

$$\boxed{\lambda_1 \cdot \lambda_2 \cdots \lambda_n = p(0) = |A|}$$

From (4) we also see that the coefficient of $(-\lambda)^{n-1}$ is $\sum_{i=1}^{n} a_{ii}$. If we determine this same coefficient using (5), we obtain $\sum_{i=1}^{n} \lambda_i$. It follows that

$$\boxed{\sum_{i=1}^{n} \lambda_i = \sum_{i=1}^{n} a_{ii}}$$

The sum of the diagonal elements of A is called the *trace* of A and is denoted by tr A.

EXAMPLE 4. If

$$A = \begin{pmatrix} 5 & -18 \\ 1 & -1 \end{pmatrix}$$

then

$$|A| = -5 + 18 = 13 \qquad \text{and} \qquad \operatorname{tr} A = 5 - 1 = 4$$

The characteristic polynomial of A is given by

$$\begin{vmatrix} 5-\lambda & -18 \\ 1 & -1-\lambda \end{vmatrix} = \lambda^2 - 4\lambda + 13$$

and hence the eigenvalues of A are $\lambda_1 = 2 + 3i$ and $\lambda_2 = 2 - 3i$. Note that

$$\lambda_1 + \lambda_2 = 4 = \operatorname{tr} A$$

and

$$\lambda_1\lambda_2 = 13 = |A|$$

In the examples we have looked at so far, n has always been less than 4. For larger n it is more difficult to find the roots of the characteristic polynomial. In Chapter 7 we will learn numerical methods for computing eigenvalues. (These methods will not involve the characteristic polynomial at all.) If the eigenvalues of A have been computed using some numerical method, one way to check their accuracy is to compare their sum to the trace of A.

EXERCISES

1. Find the eigenvalues and the corresponding eigenspaces for each of the following matrices.

(a) $\begin{pmatrix} 3 & 2 \\ 4 & 1 \end{pmatrix}$ (b) $\begin{pmatrix} 1 & 1 \\ -2 & 3 \end{pmatrix}$

(c) $\begin{pmatrix} 1 & 1 & 1 \\ 0 & 2 & 1 \\ 0 & 0 & 1 \end{pmatrix}$ (d) $\begin{pmatrix} 4 & -5 & 1 \\ 1 & 0 & -1 \\ 0 & 1 & -1 \end{pmatrix}$

(e) $\begin{pmatrix} -2 & 0 & 1 \\ 1 & 0 & -1 \\ 0 & 1 & -1 \end{pmatrix}$ (f) $\begin{pmatrix} 2 & 0 & 0 & 0 \\ 0 & 2 & 0 & 0 \\ 0 & 0 & 3 & 0 \\ 0 & 0 & 0 & 4 \end{pmatrix}$

2. Show that the eigenvalues of a triangular matrix are the diagonal elements of the matrix.

3. Let λ be an eigenvalue of A and let $\mathbf{x}$ be an eigenvector belonging to λ. Use mathematical induction to show that λ^m is an eigenvalue of A^m and $\mathbf{x}$ is an eigenvector of A^m belonging to λ^m for $m = 1, 2, \ldots$.

4. Let A be an $n \times n$ matrix and let $B = A - \alpha I$ for some scalar α. How do the eigenvalues of A and B compare? Explain.

5. Show that A and A^T have the same eigenvalues. Do they necessarily have the same eigenvectors? Explain.

6. Let λ_1 and λ_2 be distinct real eigenvalues of A. Let $\mathbf{x}$ be an eigenvector of A belonging to λ_1 and let $\mathbf{y}$ be an eigenvector of A^T belonging to λ_2. Show that $\mathbf{x}$ and $\mathbf{y}$ are orthogonal.

7. Let A be an $n \times n$ matrix and let λ be an eigenvalue of A. If $A - \lambda I$ has a rank k, what is the dimension of the eigenspace corresponding to λ? Explain.

8. Let A and B be $n \times n$ matrices. Show that AB and BA have the same eigenvalues.

9. Let λ a nonzero eigenvalue of A and let $\mathbf{x}$ be an eigenvector belonging to λ. Show that $A^m\mathbf{x}$ is also an eigenvector belonging to λ for $m = 1, 2, \ldots$.

10. Let A be an $n \times n$ matrix. Show that a vector $\mathbf{x}$ in R^n is an eigenvector belonging to A if and only if the subspace S of R^n spanned by $\mathbf{x}$ and $A\mathbf{x}$ has dimension 1.

11. Let $p(\lambda) = (-1)^n(\lambda^n - a_{n-1}\lambda^{n-1} - \cdots - a_1\lambda - a_0)$ be a polynomial of degree $n \geqslant 1$ and let

$$C = \begin{bmatrix} a_{n-1} & a_{n-2} & \cdots & a_1 & a_0 \\ 1 & 0 & & 0 & 0 \\ 0 & 1 & & 0 & 0 \\ \vdots & & & & \\ 0 & 0 & & 1 & 0 \end{bmatrix}$$

Show that $p(\lambda)$ is the characteristic polynomial of C. The matrix C is called the *companion matrix* of $p(\lambda)$.

12. Let $A = (a_{ij})$ be an $n \times n$ matrix with eigenvalues $\lambda_1, \ldots, \lambda_n$. Show that

$$\lambda_j = a_{jj} + \sum_{i \neq j} (a_{ii} - \lambda_i) \qquad \text{for } j = 1, \ldots, n$$

13. Let A be a 2×2 matrix and let $p(\lambda) = \lambda^2 + b\lambda + c$ be the characteristic polynomial of A. Show that $b = -\operatorname{tr} A$ and $c = |A|$.

14. Show that the matrix

$$A = \begin{pmatrix} \cos\theta & -\sin\theta \\ \sin\theta & \cos\theta \end{pmatrix}$$

will have complex eigenvalues if θ is not a multiple of π. Give a geometric interpretation of this result.

15. Let A be a matrix with real entries and let $\lambda = a + bi$ be a complex eigenvalue of A.
    (a) Show that $\bar{\lambda} = a - bi$ is also an eigenvalue of A.
    (b) If $\mathbf{z}$ is an eigenvector belonging to λ, show that $\bar{\mathbf{z}}$ is an eigenvector belonging to $\bar{\lambda}$.

16. Let $\mathbf{x}_1, \ldots, \mathbf{x}_r$ be eigenvectors of an $n \times n$ matrix A and let S be the subspace of R^n spanned by $\mathbf{x}_1, \mathbf{x}_2, \ldots, \mathbf{x}_r$. Show that S is invariant under A, (i.e., show that $A\mathbf{x} \in S$ whenever $\mathbf{x} \in S$).

17. Let A be a matrix whose columns all add up to a fixed constant δ. Show that δ in an eigenvalue of A.

2. SYSTEMS OF LINEAR DIFFERENTIAL EQUATIONS

Eigenvalues play an important role in the solution of systems of linear differential equations. In this section we will see how they are used in the solution of systems of linear differential equations with constant coefficients. We begin by considering systems of first-order equations of the form

$$\begin{aligned} y_1' &= a_{11}y_1 + a_{12}y_2 + \cdots + a_{1n}y_n \\ y_2' &= a_{21}y_1 + a_{22}y_2 + \cdots + a_{2n}y_n \\ &\vdots \\ y_n' &= a_{n1}y_1 + a_{n2}y_2 + \cdots + a_{nn}y_n \end{aligned}$$

where $y_i = f_i(t)$ is a function in $C^1[a, b]$ for each i. If we let $\mathbf{Y} = (y_1, y_2, \ldots, y_n)^T$ and $\mathbf{Y}' = (y_1', y_2', \ldots, y_n')^T$, the system can be written in the form

$$\mathbf{Y}' = A\mathbf{Y}$$

$\mathbf{Y}$ and $\mathbf{Y}'$ are both vector functions of t. Let us consider the simplest case first. When $n = 1$, the system is simply

$$y' = ay$$

Clearly, any function of the form

$$y(t) = ce^{at} \qquad (c \text{ an arbitrary constant})$$

satisfies this equation. To generalize to the n-dimensional case, let us assume there is a solution of the form

$$\mathbf{Y} = \begin{bmatrix} x_1 e^{\lambda t} \\ x_2 e^{\lambda t} \\ \vdots \\ x_n e^{\lambda t} \end{bmatrix} = e^{\lambda t}\mathbf{x}$$

where $\mathbf{x} = (x_1, x_2, \ldots, x_n)^T$. It follows that

$$\mathbf{Y}' = \lambda e^{\lambda t}\mathbf{x} = \lambda \mathbf{Y}$$

If we choose λ to be an eigenvalue of A and $\mathbf{x}$ is an eigenvector belonging to λ, then

$$A\mathbf{Y} = e^{\lambda t}A\mathbf{x} = \lambda e^{\lambda t}\mathbf{x} = \lambda \mathbf{Y} = \mathbf{Y}'$$

and hence $\mathbf{Y}$ is a solution to the system. Thus, if λ is an eigenvalue of A and $\mathbf{x}$ is an eigenvector belonging to λ, then $e^{\lambda t}\mathbf{x}$ is a solution of the system $\mathbf{Y}' = A\mathbf{Y}$. This will be true whether λ is real or complex. Note that if $\mathbf{Y}_1$ and $\mathbf{Y}_2$ are both solutions to $\mathbf{Y}' = A\mathbf{Y}$, then $\alpha\mathbf{Y}_1 + \beta\mathbf{Y}_2$ is also a solution, since

$$\begin{aligned}(\alpha\mathbf{Y}_1 + \beta\mathbf{Y}_2)' &= \alpha\mathbf{Y}_1' + \beta\mathbf{Y}_2' \\ &= \alpha A\mathbf{Y}_1 + \beta A\mathbf{Y}_2 \\ &= A(\alpha\mathbf{Y}_1 + \beta\mathbf{Y}_2)\end{aligned}$$

It follows by induction that if $\mathbf{Y}_1, \ldots, \mathbf{Y}_n$ are solutions to $\mathbf{Y}' = A\mathbf{Y}$, then any linear combination $c_1\mathbf{Y}_1 + \cdots + c_n\mathbf{Y}_n$ will also be a solution.

In general, the solutions to an $n \times n$ first-order system of the form

$$\mathbf{Y}' = A\mathbf{Y}$$

will form an n-dimensional subspace of the vector space of all continuous vector-valued functions. If in addition we require that $\mathbf{Y}(t)$ take on a prescribed value when $t = 0$, the problem will have a unique solution (see [23], p. 228). A problem of the form

$$\mathbf{Y}' = A\mathbf{Y} \qquad \mathbf{Y}(0) = \mathbf{Y}_0$$

is called an *initial value problem*.

EXAMPLE 1. Solve the system

$$\begin{aligned} y_1' &= 3y_1 + 4y_2 \\ y_2' &= 3y_1 + 2y_2 \end{aligned}$$

SOLUTION

$$A = \begin{pmatrix} 3 & 4 \\ 3 & 2 \end{pmatrix}$$

The eigenvalues of A are $\lambda_1 = 6$ and $\lambda_2 = -1$. Solving $(A - \lambda I)\mathbf{x} = \mathbf{0}$ with $\lambda = \lambda_1$ and $\lambda = \lambda_2$, we see that $\mathbf{x}_1 = (4, 3)^T$ is an eigenvector belonging to λ_1 and $\mathbf{x}_2 = (1, -1)^T$ is an eigenvector belonging to λ_2. Thus any vector function of the form

$$\begin{aligned}\mathbf{Y} &= c_1 e^{\lambda_1 t}\mathbf{x}_1 + c_2 e^{\lambda_2 t}\mathbf{x}_2 \\ &= \begin{pmatrix} 4c_1 e^{6t} + c_2 e^{-t} \\ 3c_1 e^{6t} - c_2 e^{-t} \end{pmatrix}\end{aligned}$$

is a solution to the system.

In Example 1 suppose that we require that $y_1 = 6$ and $y_2 = 1$ when $t = 0$. Thus

$$\mathbf{Y}(0) = \begin{pmatrix} 4c_1 + c_2 \\ 3c_1 - c_2 \end{pmatrix} = \begin{pmatrix} 6 \\ 1 \end{pmatrix}$$

and it follows that $c_1 = 1$ and $c_2 = 2$. Hence the solution to the initial value problem is given by

$$\begin{aligned}\mathbf{Y} &= e^{6t}\mathbf{x}_1 + 2e^{-t}\mathbf{x}_2 \\ &= \begin{pmatrix} 4e^{6t} + 2e^{-t} \\ 3e^{6t} - 2e^{-t} \end{pmatrix}\end{aligned}$$

Complex Eigenvalues

Let A be a real $n \times n$ matrix with a complex eigenvalue $\lambda = a + bi$ and let $\mathbf{x}$ be an eigenvector belonging to λ. The vector $\mathbf{x}$ can be split up into its real and imaginary parts.

$$\mathbf{x} = \begin{bmatrix} \operatorname{Re} x_1 + i \operatorname{Im} x_1 \\ \operatorname{Re} x_2 + i \operatorname{Im} x_2 \\ \vdots \\ \operatorname{Re} x_n + i \operatorname{Im} x_n \end{bmatrix} = \begin{bmatrix} \operatorname{Re} x_1 \\ \operatorname{Re} x_2 \\ \vdots \\ \operatorname{Re} x_n \end{bmatrix} + i \begin{bmatrix} \operatorname{Im} x_1 \\ \operatorname{Im} x_2 \\ \vdots \\ \operatorname{Im} x_n \end{bmatrix} = \operatorname{Re} \mathbf{x} + i \operatorname{Im} \mathbf{x}$$

$\bar{\lambda} = a - bi$ is also an eigenvalue of A and since

$$A\bar{\mathbf{x}} = \bar{A}\bar{\mathbf{x}} = \overline{A\mathbf{x}} = \overline{\lambda\mathbf{x}} = \bar{\lambda}\bar{\mathbf{x}}$$

it follows that $\bar{\mathbf{x}} = (\text{Re } \mathbf{x} - i \text{ Im } \mathbf{x})$ is an eigenvector belonging to $\bar{\lambda}$. Thus $e^{\lambda t}\mathbf{x}$ and $e^{\bar{\lambda} t}\bar{\mathbf{x}}$ are both solutions to the first-order system $\mathbf{Y}' = A\mathbf{Y}$.

$$\begin{aligned} e^{\lambda t}\mathbf{x} &= e^{(a+ib)t}\mathbf{x} \\ &= e^{at}(\cos bt + i \sin bt)(\text{Re } \mathbf{x} + i \text{ Im } \mathbf{x}) \\ e^{\bar{\lambda} t}\bar{\mathbf{x}} &= e^{(a-ib)t}\bar{\mathbf{x}} \\ &= e^{at}(\cos bt - i \sin bt)(\text{Re } \mathbf{x} - i \text{ Im } \mathbf{x}) \end{aligned}$$

Let

$$\mathbf{Y}_1 = \tfrac{1}{2}(e^{\lambda t}\mathbf{x} + e^{\bar{\lambda} t}\bar{\mathbf{x}}) = \text{Re}(e^{\lambda t}\mathbf{x})$$

and

$$\mathbf{Y}_2 = \tfrac{1}{2i}(e^{\lambda t}\mathbf{x} - e^{\bar{\lambda} t}\bar{\mathbf{x}}) = \text{Im}(e^{\lambda t}\mathbf{x})$$

The vector functions $\mathbf{Y}_1$ and $\mathbf{Y}_2$ are real-valued solutions to $\mathbf{Y}' = A\mathbf{Y}$.

$$\begin{aligned} \mathbf{Y}_1 &= e^{at}\big[(\cos bt) \text{ Re } \mathbf{x} - (\sin bt) \text{ Im } \mathbf{x}\big] \\ \mathbf{Y}_2 &= e^{at}\big[(\cos bt) \text{ Im } \mathbf{x} + (\sin bt) \text{ Re } \mathbf{x}\big] \end{aligned}$$

EXAMPLE 2. Solve the system

$$\begin{aligned} y_1' &= y_1 + y_2 \\ y_2' &= -2y_1 + 3y_2 \end{aligned}$$

SOLUTION. Let

$$A = \begin{pmatrix} 1 & 1 \\ -2 & 3 \end{pmatrix}$$

The eigenvalues of A are $\lambda = 2 + i$ and $\bar{\lambda} = 2 - i$ with eigenvectors $\mathbf{x} = (1, 1 + i)^T$ and $\bar{\mathbf{x}} = (1, 1 - i)^T$, respectively.

$$\begin{aligned} e^{\lambda t}\mathbf{x} &= \begin{pmatrix} e^{2t}(\cos t + i \sin t) \\ e^{2t}(\cos t + i \sin t)(1 + i) \end{pmatrix} \\ &= \begin{pmatrix} e^{2t} \cos t + ie^{2t} \sin t \\ e^{2t}(\cos t - \sin t) + ie^{2t}(\cos t + \sin t) \end{pmatrix} \end{aligned}$$

Let

$$\mathbf{Y}_1 = \text{Re}(e^{\lambda t}\mathbf{x}) = \begin{pmatrix} e^{2t} \cos t \\ e^{2t}(\cos t - \sin t) \end{pmatrix}$$

and

$$\mathbf{Y}_2 = \text{Im}(e^{\lambda t}\mathbf{x}) = \begin{pmatrix} e^{2t} \sin t \\ e^{2t}(\cos t + \sin t) \end{pmatrix}$$

any linear combination

$$\mathbf{Y} = c_1\mathbf{Y}_1 + c_2\mathbf{Y}_2$$

will be a solution to the system.

If the $n \times n$ coefficient matrix A of the system $\mathbf{Y}' = A\mathbf{Y}$ has n linearly independent eigenvectors, the general solution can be obtained by the methods that have been presented. The case when A has less than n linearly independent eigenvectors is more complicated and consequently will not be covered in this book.

Higher-Order Systems

Given a second-order system of the form

$$\mathbf{Y}'' = A_1\mathbf{Y} + A_2\mathbf{Y}'$$

we may translate it into a first-order system by setting

$$\begin{aligned} y_{n+1}(t) &= y_1'(t) \\ y_{n+2}(t) &= y_2'(t) \\ &\vdots \\ y_{2n}(t) &= y_n'(t) \end{aligned}$$

If we let

$$\mathbf{Y}_1 = \mathbf{Y} = (y_1, y_2, \ldots, y_n)^T$$

and

$$\mathbf{Y}_2 = \mathbf{Y}' = (y_{n+1}, \ldots, y_{2n})^T$$

then

$$\mathbf{Y}_1' = O\mathbf{Y}_1 + I\mathbf{Y}_2$$

and

$$\mathbf{Y}_2' = A_1\mathbf{Y}_1 + A_2\mathbf{Y}_2$$

The equations can be combined to give the $2n \times 2n$ first-order system

$$\begin{pmatrix} \mathbf{Y}_1' \\ \mathbf{Y}_2' \end{pmatrix} = \begin{pmatrix} O & I \\ A_1 & A_2 \end{pmatrix} \begin{pmatrix} \mathbf{Y}_1 \\ \mathbf{Y}_2 \end{pmatrix}$$

If the values of $\mathbf{Y}_1 = \mathbf{Y}$ and $\mathbf{Y}_2 = \mathbf{Y}'$ are specified when $t = 0$, then the initial value problem will have a unique solution.

EXAMPLE 3. Solve the initial value problem

$$\begin{aligned} y_1'' &= 2y_1 + y_2 + y_1' + y_2' \\ y_2'' &= -5y_1 + 2y_2 + 5y_1' - y_2' \\ y_1(0) = y_2(0) &= y_1'(0) = 4, \qquad y_2'(0) = -4 \end{aligned}$$

SOLUTION. Set $y_3 = y_1'$ and $y_4 = y_2'$. This gives the first-order system

$$\begin{aligned} y_1' &= \qquad\qquad\qquad y_3 \\ y_2' &= \qquad\qquad\qquad\qquad y_4 \\ y_3' &= 2y_1 + y_2 + y_3 + y_4 \\ y_4' &= -5y_1 + 2y_2 + 5y_3 - y_4 \end{aligned}$$

The coefficient matrix for this system,

$$A = \begin{bmatrix} 0 & 0 & 1 & 0 \\ 0 & 0 & 0 & 1 \\ 2 & 1 & 1 & 1 \\ -5 & 2 & 5 & -1 \end{bmatrix}$$

has eigenvalues

$$\lambda_1 = 1, \qquad \lambda_2 = -1, \qquad \lambda_3 = 3, \qquad \lambda_4 = -3$$

Corresponding to these eigenvalues are the eigenvectors

$$\mathbf{x}_1 = (1, -1, 1, -1)^T, \qquad \mathbf{x}_2 = (1, 5, -1, -5)^T,$$

$$\mathbf{x}_3 = (1, 1, 3, 3)^T, \qquad \mathbf{x}_4 = (1, -5, -3, 15)^T$$

Thus the solution will be of the form

$$c_1\mathbf{x}_1e^t + c_2\mathbf{x}_2e^{-t} + c_3\mathbf{x}_3e^{3t} + c_4\mathbf{x}_4e^{-3t}$$

We can use the initial conditions to find c_1, c_2, c_3, and c_4. For $t = 0$, we have

$$c_1\mathbf{x}_1 + c_2\mathbf{x}_2 + c_3\mathbf{x}_3 + c_4\mathbf{x}_4 = (4, 4, 4, -4)^T$$

or, equivalently,

$$\begin{bmatrix} 1 & 1 & 1 & 1 \\ -1 & 5 & 1 & -5 \\ 1 & -1 & 3 & -3 \\ -1 & -5 & 3 & 15 \end{bmatrix} \begin{bmatrix} c_1 \\ c_2 \\ c_3 \\ c_4 \end{bmatrix} = \begin{bmatrix} 4 \\ 4 \\ 4 \\ -4 \end{bmatrix}$$

The solution to this system is $\mathbf{c} = (2, 1, 1, 0)^T$ and hence the solution to the initial value problem is

$$\mathbf{Y} = 2\mathbf{x}_1e^t + \mathbf{x}_2e^{-t} + \mathbf{x}_3e^{3t}$$

Therefore,

$$\begin{bmatrix} y_1 \\ y_2 \\ y_1' \\ y_2' \end{bmatrix} = \begin{bmatrix} 2e^t + e^{-t} + e^{3t} \\ -2e^t + 5e^{-t} + e^{3t} \\ 2e^t - e^{-t} + 3e^{3t} \\ -2e^t - 5e^{-t} + 3e^{3t} \end{bmatrix}$$

In general, if we have an mth-order system of the form

$$\mathbf{Y}^{(m)} = A_1\mathbf{Y} + A_2\mathbf{Y}' + \cdots + A_m\mathbf{Y}^{(m-1)}$$

where each A_i is an $n \times n$ matrix, we can transform it into a first-order system by setting

$$\mathbf{Y}_1 = \mathbf{Y}, \mathbf{Y}_2 = \mathbf{Y}'_1, \ldots, \mathbf{Y}_m = \mathbf{Y}'_{m-1}$$

We will end up with a system of the form

$$\begin{bmatrix} \mathbf{Y}'_1 \\ \mathbf{Y}'_2 \\ \vdots \\ \mathbf{Y}'_{m-1} \\ \mathbf{Y}'_m \end{bmatrix} = \begin{bmatrix} O & I & O & \cdots & O \\ O & O & I & \cdots & O \\ \vdots & & & & \\ O & O & O & \cdots & I \\ A_1 & A_2 & A_3 & \cdots & A_m \end{bmatrix} \begin{bmatrix} \mathbf{Y}_1 \\ \mathbf{Y}_2 \\ \vdots \\ \mathbf{Y}_{m-1} \\ \mathbf{Y}_m \end{bmatrix}$$

If in addition we require that $\mathbf{Y}, \mathbf{Y}', \ldots, \mathbf{Y}^{(m-1)}$ take on specified values when $t = 0$, there will be exactly one solution to the problem.

If the system is simply of the form $\mathbf{Y}^{(m)} = A\mathbf{Y}$, it is usually not necessary to introduce new variables. In this case one need only calculate the mth roots of the eigenvalues of A. If λ is an eigenvalue of A, $\mathbf{x}$ is an eigenvector belonging to λ, σ is an mth root of λ, and $\mathbf{Y} = e^{\sigma t}\mathbf{x}$, then

$$\mathbf{Y}^{(m)} = \sigma^m e^{\sigma t}\mathbf{x} = \lambda\mathbf{Y}$$

and

$$A\mathbf{Y} = e^{\sigma t}A\mathbf{x} = \lambda e^{\sigma t}\mathbf{x} = \lambda\mathbf{Y}$$

Therefore, $\mathbf{Y} = e^{\sigma t}\mathbf{x}$ is a solution to the system.

Application

In Figure 6.2.1 two masses are adjoined by springs and the ends A and B are fixed. The masses are free to move horizontally. We will assume that the three springs are uniform and that initially the system is in the equilibrium position. A force is exerted on the system to set the masses in motion. The horizontal displacements of the masses at time t will be denoted by $x_1(t)$ and $x_2(t)$, respectively. We will assume that there are no retarding forces such as friction. Then the only forces acting on mass m_1 at time t will be from springs 1 and 2. The force from spring 1 will be $-kx_1$ and the force from spring 2 will be $k(x_2 - x_1)$. By Newton's second law,

$$m_1 x_1''(t) = -kx_1 + k(x_2 - x_1)$$

Similarly, the only forces acting on the second mass will be from springs 2 and 3. Using Newton's second law again, we get

$$m_2 x_2''(t) = -k(x_2 - x_1) - kx_2$$

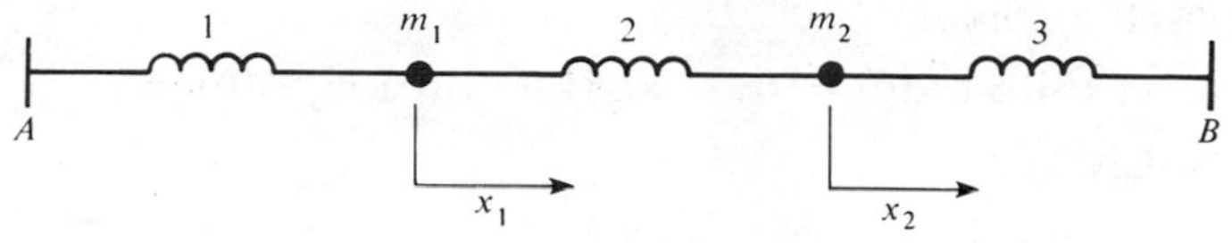

Figure 6.2.1

Thus we end up with the second-order system

$$x_1'' = -\frac{k}{m_1}(2x_1 - x_2)$$

$$x_2'' = -\frac{k}{m_2}(-x_1 + 2x_2)$$

Suppose now that $m_1 = m_2 = 1$, $k = 1$, and that the initial velocity of both masses is $+2$ units per second. To determine the displacements x_1 and x_2 as functions of t, we write the system in the form

(1) $$\mathbf{X}'' = A\mathbf{X}$$

The coefficient matrix

$$A = \begin{pmatrix} -2 & 1 \\ 1 & -2 \end{pmatrix}$$

has eigenvalues $\lambda_1 = -1$ and $\lambda_2 = -3$. Corresponding to λ_1, we have the eigenvector $\mathbf{v}_1 = (1, 1)^T$ and $\sigma_1 = \pm i$. Thus $e^{it}\mathbf{v}_1$ and $e^{-it}\mathbf{v}_1$ are both solutions to (1). It follows that

$$\tfrac{1}{2}(e^{it} + e^{-it})\mathbf{v}_1 = (\text{Re } e^{it})\mathbf{v}_1 = (\cos t)\mathbf{v}_1$$

and

$$\frac{1}{2i}(e^{it} - e^{-it})\mathbf{v}_1 = (\text{Im } e^{it})\mathbf{v}_1 = (\sin t)\mathbf{v}_1$$

are both solutions to (1). Similarly, for $\lambda_2 = -3$, we have the eigenvector $\mathbf{v}_2 = (1, -1)^T$ and $\sigma_2 = \pm\sqrt{3}\, i$. It follows that

$$(\text{Re } e^{\sqrt{3}\, it})\mathbf{v}_2 = (\cos \sqrt{3}\, t)\mathbf{v}_2$$

and

$$(\text{Im } e^{\sqrt{3}\, it})\mathbf{v}_2 = (\sin \sqrt{3}\, t)\mathbf{v}_2$$

are also solutions to (1). Thus the general solution will be of the form

$$\begin{aligned} \mathbf{X}(t) &= c_1(\cos t)\mathbf{v}_1 + c_2(\sin t)\mathbf{v}_1 + c_3(\cos \sqrt{3}\, t)\mathbf{v}_2 + c_4(\sin \sqrt{3}\, t)\mathbf{v}_2 \\ &= \begin{bmatrix} c_1 \cos t + c_2 \sin t + c_3 \cos \sqrt{3}\, t + c_4 \sin \sqrt{3}\, t \\ c_1 \cos t + c_2 \sin t - c_3 \cos \sqrt{3}\, t - c_4 \sin \sqrt{3}\, t \end{bmatrix} \end{aligned}$$

At time $t = 0$, we have

$$x_1(0) = x_2(0) = 0 \qquad \text{and} \qquad x_1'(0) = x_2'(0) = 2$$

It follows that

$$\begin{aligned} c_1 + c_3 &= 0 \qquad \text{and} \qquad & c_2 + \sqrt{3}\, c_4 &= 2 \\ c_1 - c_3 &= 0 & c_2 - \sqrt{3}\, c_4 &= 2 \end{aligned}$$

and hence

$$c_1 = c_3 = c_4 = 0 \qquad \text{and} \qquad c_2 = 2$$

Therefore, the solution to the initial value problem is simply

$$\mathbf{X}(t) = \begin{pmatrix} 2 \sin t \\ 2 \sin t \end{pmatrix}$$

The masses will oscillate with frequency 1 and amplitude 2.

EXERCISES

1. Find the general solution to each of the following systems.

(a) $\begin{aligned} y_1' &= y_1 + y_2 \\ y_2' &= -2y_1 + 4y_2 \end{aligned}$ (b) $\begin{aligned} y_1' &= y_1 - y_2 \\ y_2' &= y_1 + y_2 \end{aligned}$

(c) $\begin{aligned} y_1'' &= -2y_2 \\ y_2'' &= y_1 + 3y_2 \end{aligned}$

2. Solve each of the following initial value problems.

(a) $\begin{aligned} y_1' &= -y_1 + 2y_2 \\ y_2' &= 2y_1 - y_2 \end{aligned}$ $\quad y_1(0) = 3, y_2(0) = 1$

(b) $\begin{aligned} y_1' &= y_1 - 2y_2 \\ y_2' &= 2y_1 + y_2 \end{aligned}$ $\quad y_1(0) = 1, y_2(0) = -2$

(c) $\begin{aligned} y_1' &= 2y_1 - 6y_3 \\ y_2' &= y_1 - 3y_3 \\ y_3' &= y_2 - 2y_3 \end{aligned}$ $\quad y_1(0) = y_2(0) = y_3(0) = 2$

3. In the application problem, assume that the solutions are of the form $x_1 = a_1 \sin \sigma t$, $x_2 = a_2 \sin \sigma t$. Substitute these expressions into the system and solve for the frequency σ and the amplitudes a_1 and a_2.

4. Solve the application problem with the initial conditions $x_1(0) = x_2(0) = 1$, $x_1'(0) = 4$, and $x_2'(0) = 2$.

5. Two masses are connected by springs as shown in the diagram. Both springs have the same spring constant and the end of the first spring is fixed. If y_1 and y_2 represent the displacements from the equilibrium position, derive a system of second-order linear differential equations that describes the motion of the system.

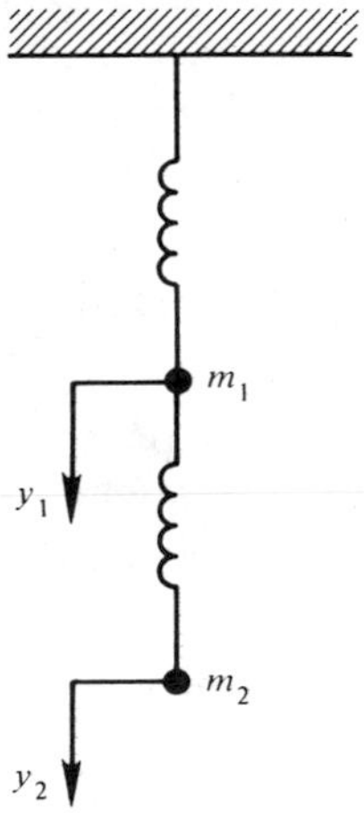

6. Three masses are connected by a series of springs between two fixed points as shown in the figure. Assume that the springs all have the same spring constant and let $x_1(t)$, $x_2(t)$, and $x_3(t)$ represent the displacements of the respective masses at time t.

(a) Derive a system of second-order linear differential equations which describes the motion of this system.

(b) Solve the system if $m_1 = m_2 = m_3 = 1$, $k = 1$, $x_1(0) = x_2(0) = x_3(0) = 1$, and $x_1'(0) = x_2'(0) = x_3'(0) = 0$.

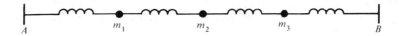

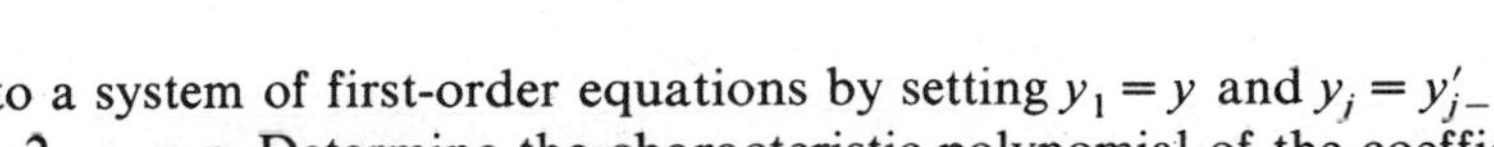

7. Transform the nth-order equation

$$y^{(n)} = a_0 y + a_1 y' + \cdots + a_{n-1} y^{(n-1)}$$

into a system of first-order equations by setting $y_1 = y$ and $y_j = y_{j-1}'$ for $j = 2, \ldots, n$. Determine the characteristic polynomial of the coefficient matrix of this system.

8. Solve the initial value problem

$$\begin{aligned} y_1'' &= -2y_2 + y_1' + 2y_2' \\ y_2'' &= 2y_1 + 2y_1' - y_2' \end{aligned}$$

$$y_1(0) = 1, \quad y_2(0) = 0, \quad y_1'(0) = -3, \quad y_2'(0) = 2$$

9. Let A be an $n \times n$ matrix with n linearly independent eigenvectors and let B be the $2n \times 2n$ matrix given by

$$B = \begin{pmatrix} O & I \\ A & O \end{pmatrix}$$

How are the eigenvalues of A and B related? Explain.

3. DIAGONALIZATION

In this section we will consider the problem of factoring an $n \times n$ matrix A into a product of the form SDS^{-1}, where D is diagonal. We will give a necessary and sufficient condition for the existence of such a factorization and we look at a number of examples. We begin by showing that eigenvectors belonging to distinct eigenvalues are linearly independent.

Theorem 6.3.1. *If $\lambda_1, \lambda_2, \ldots, \lambda_k$ are distinct eigenvalues of the $n \times n$ matrix A with corresponding eigenvectors $\mathbf{x}_1, \mathbf{x}_2, \ldots, \mathbf{x}_k$, then $\mathbf{x}_1, \ldots, \mathbf{x}_k$ are linearly independent.*

PROOF. Let r be the dimension of the subspace of R^n spanned by $\mathbf{x}_1, \ldots, \mathbf{x}_k$ and suppose that $r < k$. We may assume (reordering the $\mathbf{x}_i$'s and λ_i's if necessary) that $\mathbf{x}_1, \ldots, \mathbf{x}_r$ are linearly independent. Since $\mathbf{x}_1, \mathbf{x}_2, \ldots, \mathbf{x}_r, \mathbf{x}_{r+1}$ are linearly dependent, there exist scalars $c_1, \ldots, c_r, c_{r+1}$ not all zero such that

$$(1) \qquad c_1\mathbf{x}_1 + \cdots + c_r\mathbf{x}_r + c_{r+1}\mathbf{x}_{r+1} = \mathbf{0}$$

Note that c_{r+1} must be nonzero; otherwise, $\mathbf{x}_1, \ldots, \mathbf{x}_r$ would be dependent. So $c_{r+1}\mathbf{x}_{r+1} \neq \mathbf{0}$ and hence $c_1, \ldots, c_r$ cannot all be zero. Multiplying (1) by A, we get

$$c_1A\mathbf{x}_1 + \cdots + c_rA\mathbf{x}_r + c_{r+1}A\mathbf{x}_{r+1} = \mathbf{0}$$

or

$$(2) \qquad c_1\lambda_1\mathbf{x}_1 + \cdots + c_r\lambda_r\mathbf{x}_r + c_{r+1}\lambda_{r+1}\mathbf{x}_{r+1} = \mathbf{0}$$

Subtracting λ_{r+1} times (1) from (2) gives

$$c_1(\lambda_1 - \lambda_{r+1})\mathbf{x}_1 + \cdots + c_r(\lambda_r - \lambda_{r+1})\mathbf{x}_r = \mathbf{0}$$

This contradicts the independence of $\mathbf{x}_1, \ldots, \mathbf{x}_r$. Therefore, r must equal k.

Definition. An $n \times n$ matrix A is said to be **diagonalizable** if there exists a nonsingular matrix S and a diagonal matrix D such that

$$S^{-1}AS = D$$

We say that S *diagonalizes* A.

Theorem 6.3.2. *An $n \times n$ matrix A is diagonalizable if and only if A has n linearly independent eigenvectors.*

PROOF. Suppose that A has n linearly independent eigenvectors $\mathbf{x}_1, \mathbf{x}_2, \ldots, \mathbf{x}_n$. Let λ_i be the eigenvalue of A corresponding to $\mathbf{x}_i$ for each i. (Some of the λ_i's may be equal.) Let S be the matrix whose jth column vector is $\mathbf{x}_j$ for $j = 1, \ldots, n$. It follows that $A\mathbf{x}_j = \lambda_j\mathbf{x}_j$ is the jth column vector of AS. Thus

$$\begin{aligned} AS &= (A\mathbf{x}_1, A\mathbf{x}_2, \ldots, A\mathbf{x}_n) \\ &= (\lambda_1\mathbf{x}_1, \lambda_2\mathbf{x}_2, \ldots, \lambda_n\mathbf{x}_n) \\ &= (\mathbf{x}_1, \mathbf{x}_2, \ldots, \mathbf{x}_n)\begin{bmatrix} \lambda_1 & & & \\ & \lambda_2 & & \\ & & \ddots & \\ & & & \lambda_n \end{bmatrix} \\ &= SD \end{aligned}$$

Since S has n linearly independent column vectors, it follows that S is nonsingular and hence

$$D = S^{-1}SD = S^{-1}AS$$

Conversely, suppose that A is diagonalizable. Then there exists a nonsingular matrix S such that $AS = SD$. If $\mathbf{x}_1, \mathbf{x}_2, \ldots, \mathbf{x}_n$ are the column vectors of S, then

$$A\mathbf{x}_j = \lambda_j\mathbf{x}_j \qquad (\lambda_j = d_{jj})$$

for each j. Thus for each j, λ_j is an eigenvalue of A and $\mathbf{x}_j$ is an eigenvector belonging to λ_j. Since the column vectors of S are linearly independent, it follows that A has n linearly independent eigenvectors.

Remarks

1. If A is diagonalizable, then the column vectors of the diagonalizing matrix S are eigenvectors of A and the diagonal elements of D are the corresponding eigenvalues of A.
2. If A has n distinct eigenvalues, then A is diagonalizable.
3. The diagonalizing matrix S is not unique. Reordering the columns of a given diagonalizing matrix S or multiplying them by nonzero scalars will produce a new diagonalizing matrix.
4. If A is diagonalizable, then A can be factored into a product SDS^{-1}.

It follows from remark (4) that

$$A^2 = (SDS^{-1})(SDS^{-1}) = SD^2S^{-1}$$

and in general

$$A^k = SD^kS^{-1}$$

$$= S\begin{bmatrix} (\lambda_1)^k & & & \\ & (\lambda_2)^k & & \\ & & \ddots & \\ & & & (\lambda_n)^k \end{bmatrix} S^{-1}$$

Once we have a factorization $A = SDS^{-1}$, it is then easy to compute powers of A.

EXAMPLE 1. Let

$$A = \begin{pmatrix} 2 & -3 \\ 2 & -5 \end{pmatrix}$$

The eigenvalues of A are $\lambda_1 = 1$ and $\lambda_2 = -4$. Corresponding to λ_1 and λ_2 we have the eigenvectors $\mathbf{x}_1 = (3, 1)^T$ and $\mathbf{x}_2 = (1, 2)^T$. Let

$$S = \begin{pmatrix} 3 & 1 \\ 1 & 2 \end{pmatrix}$$

It follows that

$$S^{-1}AS = \tfrac{1}{5}\begin{pmatrix} 2 & -1 \\ -1 & 3 \end{pmatrix}\begin{pmatrix} 2 & -3 \\ 2 & -5 \end{pmatrix}\begin{pmatrix} 3 & 1 \\ 1 & 2 \end{pmatrix}$$

$$= \begin{pmatrix} 1 & 0 \\ 0 & -4 \end{pmatrix}$$

EXAMPLE 2. Let

$$A = \begin{bmatrix} 3 & -1 & -2 \\ 2 & 0 & -2 \\ 2 & -1 & -1 \end{bmatrix}$$

We saw in Example 2 of Section 1 that the eigenvalues of A are $\lambda_1 = 0$, $\lambda_2 = 1$, $\lambda_3 = 1$. Corresponding to $\lambda_1 = 0$, we have the eigenvector $(1, 1, 1)^T$, and corresponding to $\lambda = 1$, we have the eigenvectors $(1, 2, 0)^T$ and $(0, -2, 1)^T$. Let

$$S = \begin{bmatrix} 1 & 1 & 0 \\ 1 & 2 & -2 \\ 1 & 0 & 1 \end{bmatrix}$$

It follows that

$$S^{-1}AS = \begin{bmatrix} -2 & 1 & 2 \\ 3 & -1 & -2 \\ 2 & -1 & -1 \end{bmatrix}\begin{bmatrix} 3 & -1 & -2 \\ 2 & 0 & -2 \\ 2 & -1 & -1 \end{bmatrix}\begin{bmatrix} 1 & 1 & 0 \\ 1 & 2 & -2 \\ 1 & 0 & 1 \end{bmatrix} = \begin{bmatrix} 0 & 0 & 0 \\ 0 & 1 & 0 \\ 0 & 0 & 1 \end{bmatrix}$$

Note that

$$A^k = SD^kS^{-1} = SDS^{-1} = A$$

for any $k \geqslant 1$.

If an $n \times n$ matrix A has fewer than n linearly independent eigenvectors, we say that A is *defective*. A defective matrix is not diagonalizable.

EXAMPLE 3. Let

$$A = \begin{pmatrix} 1 & 1 \\ 0 & 1 \end{pmatrix}$$

The eigenvalues of A are both equal to 1. Any eigenvector corresponding to $\lambda = 1$ must be a multiple of $\mathbf{x}_1 = (1, 0)^T$. Thus A is defective and cannot be diagonalized.

Application 1

Recall Application 3 from Section 3 of the first chapter. In a certain town 30 percent of the married women get divorced each year and 20 percent of the single women get married each year. There are 8000 married women and 2000 single women and the total population remains constant. Find the number of married women and single women after 5 years. What will be the long-range prospects if these percentages of marriages and divorces continue indefinitely into the future?

SOLUTION. To find the number of married and single women after 1 year, we multiplied the vector $\mathbf{x} = (8000, 2000)^T$ by

$$A = \begin{pmatrix} 0.7 & 0.2 \\ 0.3 & 0.8 \end{pmatrix}$$

The number of married and single women after 5 years will be $A^5\mathbf{x}$. To compute $A^5\mathbf{x}$, we will factor A into a product SDS^{-1}. The eigenvalues of A are $\lambda_1 = 1$ and $\lambda_2 = \frac{1}{2}$. $\mathbf{x}_1 = (2, 3)^T$ is an eigenvector belonging to λ_1 and $\mathbf{x}_2 = (1, -1)^T$ is an eigenvector belonging to λ_2.

Let

$$S = \begin{pmatrix} 2 & 1 \\ 3 & -1 \end{pmatrix}$$

It follows that

$$A = SDS^{-1} = \frac{1}{5}\begin{pmatrix} 2 & 1 \\ 3 & -1 \end{pmatrix}\begin{pmatrix} 1 & 0 \\ 0 & \frac{1}{2} \end{pmatrix}\begin{pmatrix} 1 & 1 \\ 3 & -2 \end{pmatrix}$$

Thus

$$\begin{aligned} A^5\mathbf{x} &= \frac{1}{5}\begin{pmatrix} 2 & 1 \\ 3 & -1 \end{pmatrix}\begin{pmatrix} 1 & 0 \\ 0 & \frac{1}{2} \end{pmatrix}^5\begin{pmatrix} 1 & 1 \\ 3 & -2 \end{pmatrix}\begin{pmatrix} 8000 \\ 2000 \end{pmatrix} \\ &= \begin{pmatrix} 4125 \\ 5875 \end{pmatrix} \end{aligned}$$

After 5 years there will be 4125 married women and 5875 single women. After n years the number of married and single women is given by $A^n\mathbf{x}$. To find the long-range trend, we take the limit as n approaches infinity.

$$\lim_{n\to\infty} A^n\mathbf{x} = \lim_{n\to\infty} SD^nS^{-1}\mathbf{x}$$

As $n\to\infty$, $D^n \to \begin{pmatrix} 1 & 0 \\ 0 & 0 \end{pmatrix}$. Thus

$$\lim_{n\to\infty} A^n\mathbf{x} = \tfrac{1}{5}\begin{pmatrix} 2 & 1 \\ 3 & -1 \end{pmatrix}\begin{pmatrix} 1 & 0 \\ 0 & 0 \end{pmatrix}\begin{pmatrix} 1 & 1 \\ 3 & -2 \end{pmatrix}\begin{pmatrix} 8000 \\ 2000 \end{pmatrix} = \begin{pmatrix} 4000 \\ 6000 \end{pmatrix}$$

In the long run, 40 percent of the women will be married and 60 percent will be single.

Application 2. Sex-Linked Genes

Sex-linked genes are genes that are located on the X chromosome. For example, the gene for blue-green color blindness is a recessive sex-linked gene. To devise a mathematical model to describe color blindness in a given population, it is necessary to divide the population into two classes, males and females. Let $x_1^{(0)}$ be the proportion of genes for color blindness in the male population and let $x_2^{(0)}$ be the proportion in the female population. [Since color blindness is recessive, the actual proportion of color blind females will be less than $x_2^{(0)}$.] Since the male receives one X chromosome from the mother, the proportion $x_1^{(1)}$ of color blind males in the next generation will be the same as the proportion of recessive genes in the present generation of females. Since the female receives an X chromosome from each parent, the proportion $x_2^{(1)}$ of recessive genes in the next generation of females will be the average of $x_1^{(0)}$ and $x_2^{(0)}$. Thus

$$\begin{aligned} x_2^{(0)} &= x_1^{(1)} \\ \tfrac{1}{2}x_1^{(0)} + \tfrac{1}{2}x_2^{(0)} &= x_2^{(1)} \end{aligned}$$

If $x_1^{(0)} = x_2^{(0)}$, the proportions will not change in future generations. Let us assume that $x_1^{(0)} \neq x_2^{(0)}$ and write the system as a matrix equation.

$$\begin{pmatrix} 0 & 1 \\ \frac{1}{2} & \frac{1}{2} \end{pmatrix}\begin{pmatrix} x_1^{(0)} \\ x_2^{(0)} \end{pmatrix} = \begin{pmatrix} x_1^{(1)} \\ x_2^{(1)} \end{pmatrix}$$

Let A denote the coefficient matrix and let $\mathbf{x}^{(n)} = (x_1^{(n)}, x_2^{(n)})^T$ denote the proportion of colorblind genes in the male and female populations of the $(n + 1)$st generation. Thus

$$\mathbf{x}^{(n)} = A^n\mathbf{x}^{(0)}$$

To compute A^n, we note that A has eigenvalues 1 and $-\frac{1}{2}$ and consequently can be factored into a product

$$A = \begin{pmatrix} 1 & -2 \\ 1 & 1 \end{pmatrix} \begin{pmatrix} 1 & 0 \\ 0 & -\frac{1}{2} \end{pmatrix} \begin{bmatrix} \frac{1}{3} & \frac{2}{3} \\ -\frac{1}{3} & \frac{1}{3} \end{bmatrix}$$

Thus

$$\begin{aligned} \mathbf{x}^{(n)} &= \begin{pmatrix} 1 & -2 \\ 1 & 1 \end{pmatrix} \begin{pmatrix} 1 & 0 \\ 0 & -\frac{1}{2} \end{pmatrix}^n \begin{bmatrix} \frac{1}{3} & \frac{2}{3} \\ -\frac{1}{3} & \frac{1}{3} \end{bmatrix} \begin{pmatrix} x_1^{(0)} \\ x_2^{(0)} \end{pmatrix} \\ &= \frac{1}{3} \begin{bmatrix} 1-\left(-\frac{1}{2}\right)^{n-1} & 2+\left(-\frac{1}{2}\right)^{n-1} \\ 1-\left(-\frac{1}{2}\right)^{n} & 2+\left(-\frac{1}{2}\right)^{n} \end{bmatrix} \begin{pmatrix} x_1^{(0)} \\ x_2^{(0)} \end{pmatrix} \end{aligned}$$

and hence

$$\begin{aligned} \lim_{n\to\infty} \mathbf{x}^{(n)} &= \frac{1}{3} \begin{pmatrix} 1 & 2 \\ 1 & 2 \end{pmatrix} \begin{pmatrix} x_1^{(0)} \\ x_2^{(0)} \end{pmatrix} \\ &= \begin{bmatrix} \dfrac{x_1^{(0)} + 2x_2^{(0)}}{3} \\ \dfrac{x_1^{(0)} + 2x_2^{(0)}}{3} \end{bmatrix} \end{aligned}$$

The proportions of genes for color blindness in the male and female populations will tend to the same value as the number of generations increases. If the proportion of colorblind men is p, and over a number of generations no outsiders have entered the population, there is justification for assuming that the proportion of genes for color blindness in the female population is also p. Since color blindness is recessive, we would expect the proportion of colorblind women to be about p^2. Thus, if 1 percent of the male population is colorblind, we would expect about 0.01 percent of the female population to be colorblind.

EXERCISES

1. In each of the following, factor the matrix A into a product SDS^{-1}, where D is diagonal.

(a) $A = \begin{pmatrix} 6 & -6 \\ 2 & -1 \end{pmatrix}$ (b) $A = \begin{pmatrix} -8 & 5 \\ -14 & 9 \end{pmatrix}$

(c) $A = \begin{bmatrix} 2 & 2 & 1 \\ 0 & 1 & 2 \\ 0 & 0 & -1 \end{bmatrix}$ (d) $A = \begin{bmatrix} 3 & 0 & 0 \\ -2 & 2 & 3 \\ 1 & 1 & 4 \end{bmatrix}$

2. Let A be a nonsingular $n \times n$ matrix with distinct eigenvalues $\lambda_1, \ldots, \lambda_n$. Show that the eigenvalues of A^{-1} are $1/\lambda_1, \ldots, 1/\lambda_n$.

3. For each of the matrices in Exercise 1, compute A^2, A^3, and A^{-1}.

4. Let

$$A = \begin{pmatrix} 4 & 1 & 1 \\ 0 & 4 & 5 \\ 0 & 3 & 6 \end{pmatrix}$$

Find a matrix B such that $B^2 = A$.

5. Let $B = S^{-1}AS$ for some nonsingular matrix S. Show that B and A have the same eigenvalues.

6. Let A be an $n \times n$ matrix with positive real eigenvalues $\lambda_1 > \lambda_2 > \cdots > \lambda_n$. Let $\mathbf{x}_i$ be an eigenvector belonging to λ_i for each i and let $\mathbf{x} = \alpha_1 \mathbf{x}_1 + \cdots + \alpha_n \mathbf{x}_n$.
(a) Show that $A^m \mathbf{x} = \sum_{i=1}^{n} \alpha_i \lambda_i^m \mathbf{x}_i$.
(b) If $\lambda_1 = 1$, show that $\lim_{m \to \infty} A^m \mathbf{x} = \alpha_1 \mathbf{x}_1$.

7. Show that any 3×3 matrix of the form

$$\begin{pmatrix} a & 1 & 0 \\ 0 & a & 1 \\ 0 & 0 & b \end{pmatrix}$$

is defective.

8. The town of Midvale maintains a constant population of 300,000 people from year to year. A political science study estimated that there were 150,000 independents, 90,000 Democrats, and 60,000 Republicans in the town. It was also estimated that each year 20 percent of the independents become Democrats and 10 percent become Republicans. Similarly, 20 percent of the Democrats become independents and 10 percent become Republicans while 10 percent of the Republicans defect to the Democrats and 10 percent become independents each year. Let

$$\mathbf{x} = \begin{pmatrix} 150{,}000 \\ 90{,}000 \\ 60{,}000 \end{pmatrix}$$

and let $\mathbf{x}^{(1)}$ be a vector representing the number of people in each group after 1 year.
(a) Find a matrix A such that $A\mathbf{x} = \mathbf{x}^{(1)}$.
(b) Show that $\lambda_1 = 1.0$, $\lambda_2 = 0.5$, and $\lambda_3 = 0.7$ are eigenvalues of A and factor A into a product SDS^{-1}, where D is diagonal.
(c) Which group will dominate in the long run? Justify your answer by computing $\lim_{n \to \infty} A^n \mathbf{x}$.

4. COMPLEX MATRICES

Let C^n denote the vector space of all n-tuples of complex numbers. The set C of all complex numbers will be taken as our field of scalars. We have already seen that a matrix A with real entries may have complex eigenvalues and eigenvectors. In this section we will study matrices with complex entries and look at the complex analogues of symmetric and orthogonal matrices.

If $\alpha = a + bi$ is a complex scalar, the length of α is given by

$$|\alpha| = \sqrt{\bar{\alpha}\alpha} = \sqrt{a^2 + b^2}$$

The length of a vector $\mathbf{z} = (z_1, z_2, \ldots, z_n)^T$ in C^n is given by

$$\begin{aligned}\|\mathbf{z}\| &= \left(|z_1|^2 + |z_2|^2 + \cdots + |z_n|^2\right)^{1/2} \\ &= \left(\bar{z}_1 z_1 + \bar{z}_2 z_2 + \cdots + \bar{z}_n z_n\right)^{1/2} \\ &= (\bar{\mathbf{z}}^T \mathbf{z})^{1/2}\end{aligned}$$

As a notational convenience, we write $\mathbf{z}^H$ for the transpose of $\bar{\mathbf{z}}$. Thus

$$\bar{\mathbf{z}}^T = \mathbf{z}^H \qquad \text{and} \qquad \|\mathbf{z}\| = (\mathbf{z}^H\mathbf{z})^{1/2}$$

Definition. Let V be a vector space over the complex numbers. An **inner product** on V is an operation that assigns to each pair of vectors $\mathbf{z}$ and $\mathbf{w}$ in V a complex number $\langle \mathbf{z}, \mathbf{w} \rangle$ satisfying the following conditions.

(i) $\langle \mathbf{z}, \mathbf{z} \rangle \geq 0$ with equality if and only if $\mathbf{z} = \mathbf{0}$

(ii) $\langle \mathbf{z}, \mathbf{w} \rangle = \overline{\langle \mathbf{w}, \mathbf{z} \rangle}$ for all $\mathbf{z}$ and $\mathbf{w}$ in V

(iii) $\langle \alpha\mathbf{z} + \beta\mathbf{w}, \mathbf{u} \rangle = \alpha\langle \mathbf{z}, \mathbf{u} \rangle + \beta\langle \mathbf{w}, \mathbf{u} \rangle$

Note that for a complex inner product space $\langle \mathbf{z}, \mathbf{w} \rangle = \overline{\langle \mathbf{w}, \mathbf{z} \rangle}$ rather than $\langle \mathbf{w}, \mathbf{z} \rangle$. If we make the proper modifications to allow for this, the theorems on real inner product spaces in Chapter 5, Section 6, will all be valid for complex inner product spaces. In particular, let us recall Theorem 5.6.2; if $\{\mathbf{u}_1, \ldots, \mathbf{u}_n\}$ is an orthonormal basis for a real inner product space V and

$$\mathbf{x} = \sum_{i=1}^{n} c_i \mathbf{u}_i$$

then

$$c_i = \langle \mathbf{u}_i, \mathbf{x} \rangle = \langle \mathbf{x}, \mathbf{u}_i \rangle \qquad \text{and} \qquad \|\mathbf{x}\|^2 = \sum_{i=1}^{n} c_i^2$$

In the case of a complex inner product space, if $\{\mathbf{w}_1, \ldots, \mathbf{w}_n\}$ is an orthonormal basis and

$$\mathbf{z} = \sum_{i=1}^{n} c_i \mathbf{w}_i$$

then

$$c_i = \langle \mathbf{z}, \mathbf{w}_i \rangle, \ \bar{c}_i = \langle \mathbf{w}_i, \mathbf{z} \rangle \qquad \text{and} \qquad \|\mathbf{z}\|^2 = \sum_{i=1}^{n} c_i \bar{c}_i$$

We can define an inner product on C^n by

$$\langle \mathbf{z}, \mathbf{w} \rangle = \mathbf{w}^H \mathbf{z} \tag{1}$$

for all $\mathbf{z}$ and $\mathbf{w}$ in C^n. We leave it to the reader to verify that (1) actually does define an inner product on C^n. The complex inner product space C^n is quite similar to the real inner product space R^n. The main difference is that in the complex case it is necessary to conjugate before transposing when taking an inner product.

R^n	C^n
$\langle \mathbf{x}, \mathbf{y} \rangle = \mathbf{y}^T \mathbf{x}$	$\langle \mathbf{z}, \mathbf{w} \rangle = \mathbf{w}^H \mathbf{z}$
$\mathbf{x}^T \mathbf{y} = \mathbf{y}^T \mathbf{x}$	$\mathbf{z}^H \mathbf{w} = \overline{\mathbf{w}^H \mathbf{z}}$
$\|\mathbf{x}\|^2 = \mathbf{x}^T \mathbf{x}$	$\|\mathbf{z}\|^2 = \mathbf{z}^H \mathbf{z}$

Let $M = (m_{ij})$ be an $m \times n$ matrix with $m_{ij} = a_{ij} + ib_{ij}$ for each i and j. We may write M in the form

$$M = A + iB$$

where $A = (a_{ij})$ and $B = (b_{ij})$ have real entries. We define the conjugate of M by

$$\overline{M} = A - iB$$

Thus $\overline{M}$ is the matrix formed by conjugating each of the entries of M. The transpose of $\overline{M}$ will be denoted by M^H.

Definition. A matrix M is said to be **Hermitian** if $M = M^H$.

EXAMPLE 1. The matrix

$$M = \begin{pmatrix} 3 & 2-i \\ 2+i & 4 \end{pmatrix}$$

is Hermitian, since

$$M^H = \begin{bmatrix} \overline{3} & \overline{2-i} \\ \overline{2+i} & \overline{4} \end{bmatrix}^T = \begin{pmatrix} 3 & 2-i \\ 2+i & 4 \end{pmatrix} = M$$

If M is a matrix with real entries, then $M^H = M^T$. In particular, if M is a real symmetric matrix, then M is Hermitian. Thus we may view Hermitian matrices as the complex analogue of real symmetric matrices. Hermitian matrices have many nice properties, as we shall see in the following theorem.

Theorem 6.4.1. *The eigenvalues of an Hermitian matrix are all real. Furthermore, eigenvectors belonging to distinct eigenvalues are orthogonal.*

PROOF. Let A be an Hermitian matrix. Let λ be an eigenvalue of A and let $\mathbf{x}$ be an eigenvector belonging to λ. If $\alpha = \mathbf{x}^H A\mathbf{x}$, then

$$\bar{\alpha} = \alpha^H = (\mathbf{x}^H A\mathbf{x})^H = \mathbf{x}^H A\mathbf{x} = \alpha$$

Thus α is real. It follows that

$$\alpha = \mathbf{x}^H A\mathbf{x} = \mathbf{x}^H \lambda\mathbf{x} = \lambda\|\mathbf{x}\|^2$$

and hence

$$\lambda = \frac{\alpha}{\|\mathbf{x}\|^2}$$

is real. If $\mathbf{x}_1$ and $\mathbf{x}_2$ are eigenvectors belonging to distinct eigenvalues λ_1 and λ_2, respectively, then

$$(A\mathbf{x}_1)^H\mathbf{x}_2 = \mathbf{x}_1^H A^H\mathbf{x}_2 = \mathbf{x}_1^H A\mathbf{x}_2 = \lambda_2\mathbf{x}_1^H\mathbf{x}_2$$

and

$$(A\mathbf{x}_1)^H\mathbf{x}_2 = (\mathbf{x}_2^H A\mathbf{x}_1)^H = (\lambda_1\mathbf{x}_2^H\mathbf{x}_1)^H = \lambda_1\mathbf{x}_1^H\mathbf{x}_2$$

Thus

$$\lambda_1\mathbf{x}_1^H\mathbf{x}_2 = \lambda_2\mathbf{x}_1^H\mathbf{x}_2$$

and since $\lambda_1 \neq \lambda_2$, it follows that

$$\langle\mathbf{x}_2, \mathbf{x}_1\rangle = \mathbf{x}_1^H\mathbf{x}_2 = 0$$

Definition. An $n \times n$ matrix U is said to be *unitary* if its column vectors form an orthonormal set in C^n.

Thus U is unitary if and only if $U^H U = I$. If U is unitary, then, since the column vectors are orthonormal, U must have rank n. It follows that

$$U^{-1} = IU^{-1} = U^H UU^{-1} = U^H$$

A real unitary matrix is an orthogonal matrix.

Corollary 6.4.2. *If the eigenvalues of a Hermitian matrix A are distinct, then there exists a unitary matrix U that diagonalizes A.*

PROOF. Let $\mathbf{x}_i$ be an eigenvector belonging to λ_i for each eigenvalue λ_i of A. Let $\mathbf{u}_i = (1/\|\mathbf{x}_i\|)\mathbf{x}_i$. Thus $\mathbf{u}_i$ is a unit eigenvector belonging to λ_i for each i. It follows from Theorem 6.4.1 that $\{\mathbf{u}_1, \ldots, \mathbf{u}_n\}$ is an orthonormal set in C^n. Let U be the matrix whose ith column vector is $\mathbf{u}_i$ for each i; U is unitary and U diagonalizes A.

EXAMPLE 2. Let

$$A = \begin{pmatrix} 2 & 1-i \\ 1+i & 1 \end{pmatrix}$$

Find a unitary matrix U that diagonalizes A.

SOLUTION. The eigenvalues of A are $\lambda_1 = 3$ and $\lambda_2 = 0$ with corresponding eigenvectors $\mathbf{x}_1 = (1-i, 1)^T$ and $\mathbf{x}_2 = (-1, 1+i)^T$. Let

$$\mathbf{u}_1 = \frac{1}{\|\mathbf{x}_1\|}\mathbf{x}_1 = \frac{1}{\sqrt{3}}(1-i, 1)^T$$

and

$$\mathbf{u}_2 = \frac{1}{\|\mathbf{x}_2\|}\mathbf{x}_2 = \frac{1}{\sqrt{3}}(-1, 1+i)^T$$

Thus

$$U = \frac{1}{\sqrt{3}}\begin{pmatrix} 1-i & -1 \\ 1 & 1+i \end{pmatrix}$$

and

$$\begin{aligned} U^HAU &= \tfrac{1}{3}\begin{pmatrix} 1+i & 1 \\ -1 & 1-i \end{pmatrix}\begin{pmatrix} 2 & 1-i \\ 1+i & 1 \end{pmatrix}\begin{pmatrix} 1-i & -1 \\ 1 & 1+i \end{pmatrix} \\ &= \begin{pmatrix} 3 & 0 \\ 0 & 0 \end{pmatrix} \end{aligned}$$

Actually, Corollary 6.4.2 is valid even if the eigenvalues of A are not distinct. To show this, we will first prove the following theorem.

Theorem 6.4.3 (Schur's Theorem). *For each $n \times n$ matrix A, there exists a unitary matrix U such that U^HAU is upper triangular.*

PROOF. The proof is by induction on n. The result is obvious if $n = 1$. Assume that the hypothesis holds for all $k \times k$ matrices and let A be a $(k+1) \times (k+1)$ matrix. Let λ_1 be an eigenvalue of A and let $\mathbf{w}_1$ be a unit eigenvector belonging to λ_1. Using the Gram–Schmidt process, construct $\mathbf{w}_2, \ldots, \mathbf{w}_{k+1}$ such that $\{\mathbf{w}_1, \ldots, \mathbf{w}_{k+1}\}$ is an orthonormal basis for C^{k+1}. Let W be the matrix whose ith column vector is $\mathbf{w}_i$ for $i = 1, \ldots, k+1$. Thus, by construction, W is unitary. The first column of W^HAW will be $W^HA\mathbf{w}_1$.

$$W^HA\mathbf{w}_1 = \lambda_1 W^H\mathbf{w}_1 = \lambda_1\mathbf{e}_1$$

Thus W^HAW is a matrix of the form

$$\left(\begin{array}{c|cccc} \lambda_1 & \times & \times & \cdots & \times \\ \hline 0 & & & & \\ \vdots & & & M & \\ 0 & & & & \end{array}\right)$$

where M is a $k \times k$ matrix. By the induction hypothesis there exists a $k \times k$ unitary matrix V_1 such that $V_1^HMV_1 = T_1$, where T_1 is triangular. Let

$$V = \left(\begin{array}{c|ccc} 1 & 0 & \cdots & 0 \\ \hline 0 & & & \\ \vdots & & V_1 & \\ 0 & & & \end{array}\right)$$

Here V is unitary and

$$\begin{aligned} V^HW^HAWV &= \left(\begin{array}{c|ccc} \lambda_1 & \times & \cdots & \times \\ \hline 0 & & & \\ \vdots & & V_1^HMV_1 & \\ 0 & & & \end{array}\right) \\ &= \left(\begin{array}{c|ccc} \lambda_1 & \times & \cdots & \times \\ \hline 0 & & & \\ \vdots & & T_1 & \\ 0 & & & \end{array}\right) \\ &= T \end{aligned}$$

Let $U = WV$. The matrix U is unitary, since

$$U^HU = (WV)^HWV = V^HW^HWV = I$$

Thus $U^HAU = T$, and this completes the proof.

The factorization $A = UTU^H$ is often referred to as the *Schur decomposition* of A. In the case that A is Hermitian, the matrix T will be diagonal.

Corollary 6.4.4. *If A is Hermitian, there exists a unitary matrix U that diagonalizes A.*

PROOF. By Theorem 6.4.3, there is a unitary matrix U such that $U^HAU = T$, where T is upper triangular.

$$T^H = (U^HAU)^H = U^HA^HU = U^HAU = T$$

Therefore, T is Hermitian and consequently must be diagonal.

It follows from Corollary 6.4.4 that each Hermitian matrix A can be factored into a product UDU^H, where U is unitary and D is diagonal. Since U diagonalizes A, it follows that the diagonal elements of D are the eigenvalues of A and the column vectors of U are eigenvectors of A. Thus A cannot be defective. It has a complete set of eigenvectors which form an orthonormal basis for C^n. This is in a sense the ideal situation. We have seen how to express a vector as a linear combination of orthonormal basis elements (Theorem 5.6.2), and the action of A on any linear combination of eigenvectors can be easily determined. Thus, if A has an orthonormal set of eigenvectors $\{\mathbf{u}_1, \ldots, \mathbf{u}_n\}$ and $\mathbf{x} = \alpha_1\mathbf{u}_1 + \cdots + \alpha_n\mathbf{u}_n$, then

$$A\mathbf{x} = \alpha_1\lambda_1\mathbf{u}_1 + \cdots + \alpha_n\lambda_n\mathbf{u}_n$$

Furthermore,

$$\alpha_i = \langle \mathbf{x}, \mathbf{u}_i \rangle = \mathbf{u}_i^H\mathbf{x}$$

Thus

$$A\mathbf{x} = \lambda_1(\mathbf{u}_1^H\mathbf{x})\mathbf{u}_1 + \cdots + \lambda_n(\mathbf{u}_n^H\mathbf{x})\mathbf{u}_n$$

There are non-Hermitian matrices that possess complete orthonormal sets of eigenvectors. For example, skew symmetric and skew Hermitian matrices have this property. (A is *skew Hermitian* if $A^H = -A$.) In general, if A is any matrix with a complete orthonormal set of eigenvectors, then $A = UDU^H$, where U is unitary and D is a diagonal matrix (whose diagonal elements may be complex). Thus, in general, $D^H \neq D$ and, consequently, $A^H = UD^HU^H \neq A$. However,

$$AA^H = UDU^HUD^HU^H = UDD^HU^H$$

and

$$A^HA = UD^HU^HUDU^H = UD^HDU^H$$

Since

$$D^HD = DD^H = \begin{pmatrix} |\lambda_1|^2 & & & \\ & |\lambda_2|^2 & & \\ & & \ddots & \\ & & & |\lambda_n|^2 \end{pmatrix}$$

it follows that

$$AA^H = A^HA$$

Definition. A matrix A is said to be **normal** if $AA^H = A^HA$.

We have shown that if a matrix has a complete orthonormal set of eigenvectors, it is normal. The converse is also true.

Theorem 6.4.5. *A matrix A is normal if and only if A possesses a complete orthonormal set of eigenvectors.*

PROOF. In view of the preceding remarks, we need only show that a normal matrix A has a complete orthonormal set of eigenvectors. By Theorem 6.4.3 there exists a unitary matrix U and a triangular matrix T such that $T = U^HAU$. We claim that T is also normal.

$$T^HT = U^HA^HUU^HAU = U^HA^HAU$$

and

$$TT^H = U^HAUU^HA^HU = U^HAA^HU$$

Since $A^HA = AA^H$, it follows that $T^HT = TT^H$. Comparing the diagonal elements of TT^H and T^HT, we see that

$$\begin{aligned} |t_{11}|^2 + |t_{12}|^2 + |t_{13}|^2 + \cdots + |t_{1n}|^2 &= |t_{11}|^2 \\ |t_{22}|^2 + |t_{23}|^2 + \cdots + |t_{2n}|^2 &= |t_{12}|^2 + |t_{22}|^2 \\ &\vdots \\ |t_{nn}|^2 &= |t_{1n}|^2 + |t_{2n}|^2 + |t_{3n}|^2 + \cdots + |t_{nn}|^2 \end{aligned}$$

It follows that $t_{ij} = 0$ whenever $i \neq j$. Thus U diagonalizes A and the column vectors of U are eigenvectors of A.

EXERCISES

1. Which of the following matrices are Hermitian? Normal?

(a) $\begin{pmatrix} 1-i & 2 \\ 2 & 3 \end{pmatrix}$ (b) $\begin{pmatrix} 1 & 2-i \\ 2+i & -1 \end{pmatrix}$

(c) $\begin{bmatrix} \frac{1}{\sqrt{2}} & -\frac{1}{\sqrt{2}} \\ \frac{1}{\sqrt{2}} & \frac{1}{\sqrt{2}} \end{bmatrix}$ (d) $\begin{bmatrix} \frac{1}{\sqrt{2}}i & \frac{1}{\sqrt{2}} \\ \frac{1}{\sqrt{2}} & -\frac{1}{\sqrt{2}}i \end{bmatrix}$

(e) $\begin{bmatrix} 0 & i & 1 \\ i & 0 & -2+i \\ -1 & 2+i & 0 \end{bmatrix}$ (f) $\begin{bmatrix} 3 & 1+i & i \\ 1-i & 1 & 3 \\ -i & 3 & 1 \end{bmatrix}$

2. Find a unitary diagonalizing matrix for each of the following.

(a) $\begin{pmatrix} 2 & 1 \\ 1 & 2 \end{pmatrix}$ (b) $\begin{pmatrix} 1 & 3+i \\ 3-i & 4 \end{pmatrix}$

(c) $\begin{bmatrix} 2 & i & 0 \\ -i & 2 & 0 \\ 0 & 0 & 2 \end{bmatrix}$ (d) $\begin{bmatrix} 2 & 1 & 1 \\ 1 & 3 & -2 \\ 1 & -2 & 3 \end{bmatrix}$

(e) $\begin{bmatrix} 1 & 1 & 1 \\ 1 & 1 & 1 \\ 1 & 1 & 1 \end{bmatrix}$

3. Show that any basis $\{\mathbf{x}_1, \ldots, \mathbf{x}_n\}$ for R^n is also a basis for C^n.

4. Let A be an $m \times n$ matrix and B be an $n \times r$ matrix. Show that $(AB)^H = B^H A^H$.

5. Show that

$$\langle \mathbf{z}, \mathbf{w} \rangle = \mathbf{w}^H \mathbf{z}$$

defines an inner product on C^n.

6. Let $\{\mathbf{u}_1, \ldots, \mathbf{u}_n\}$ be an orthonormal basis for a complex inner product space V and let $\mathbf{z}$ and $\mathbf{w}$ be elements of V. Show that

$$\langle \mathbf{z}, \mathbf{w} \rangle = \sum_{i=1}^{n} (\langle \mathbf{z}, \mathbf{u}_i \rangle \cdot \langle \mathbf{u}_i, \mathbf{w} \rangle)$$

7. Given

$$A = \begin{bmatrix} 4 & 0 & 0 \\ 0 & 1 & i \\ 0 & -i & 1 \end{bmatrix}$$

find a matrix B such that $B^H B = A$.

8. Let U be a unitary matrix. Prove:
(a) U is normal.
(b) $\|U\mathbf{x}\| = \|\mathbf{x}\|$ for all $\mathbf{x} \in C^n$.
(c) If λ is an eigenvalue of U, then $|\lambda| = 1$.

9. Let $\mathbf{u}$ be a unit vector in C^n and define $U = I - 2\mathbf{u}\mathbf{u}^H$. Show that U is both unitary and Hermitian and consequently is its own inverse.

10. Show that $M = A + iB$ (A and B real matrices) is skew Hermitian if and only if A is skew symmetric and B is symmetric.

11. Let $p(x) = -x^3 + cx^2 + (c+3)x + 1$, where c is a real number. Let C denote the companion matrix of $p(x)$,

$$C = \begin{bmatrix} c & c+3 & 1 \\ 1 & 0 & 0 \\ 0 & 1 & 0 \end{bmatrix}$$

and let

$$A = \begin{bmatrix} -1 & 2 & -c-3 \\ 1 & -1 & c+2 \\ -1 & 1 & -c-1 \end{bmatrix}$$

(a) Compute $A^{-1}CA$.

(b) Use the result from part (a) to prove that $p(x)$ will have only real roots regardless of the value of c.

5. QUADRATIC FORMS

By this time the reader should be well aware of the important role that matrices play in the study of linear equations. In this section we will see that matrices also play an important role in the study of quadratic equations. With each quadratic equation we can associate a vector function $f(\mathbf{x}) = \mathbf{x}^T A\mathbf{x}$. Such a vector function is called a "quadratic form." Quadratic forms arise in a wide variety of applied problems. They are particularly important in the study of optimization theory.

Definition. A **quadratic equation** in two variables x and y is an equation of the form

(1) $$ax^2 + 2bxy + cy^2 + dx + ey + f = 0$$

Equation (1) may be rewritten in the form

(2) $$(x \quad y)\begin{pmatrix} a & b \\ b & c \end{pmatrix}\begin{pmatrix} x \\ y \end{pmatrix} + (d \quad e)\begin{pmatrix} x \\ y \end{pmatrix} + f = 0$$

Let

$$\mathbf{x} = \begin{pmatrix} x \\ y \end{pmatrix} \quad \text{and} \quad A = \begin{pmatrix} a & b \\ b & c \end{pmatrix}$$

The term

$$\mathbf{x}^T A\mathbf{x} = ax^2 + 2bxy + cy^2$$

is called the **quadratic form** associated with (1).

Conic Sections

The graph of an equation of the form (1) is called a *conic section*. [If there are no ordered pairs (x, y) that satisfy (1), we say that the equation represents an imaginary conic.] If the graph of (1) consists of a single point, a line, or a pair of lines, we say that (1) represents a degenerate conic. Of more interest are the nondegenerate conics. Graphs of nondegenerate conics turn out to be circles, ellipses, parabolas, or hyperbolas. The graph of a conic is particularly easy to sketch when its equation can be put into one of the following standard forms:

(i) $$x^2 + y^2 = r^2$$ (circle)

(ii) $$\frac{x^2}{\alpha^2} + \frac{y^2}{\beta^2} = 1$$ (ellipse)

(iii) $$\frac{x^2}{\alpha^2} - \frac{y^2}{\beta^2} = 1 \quad \text{or} \quad \frac{y^2}{\alpha^2} - \frac{x^2}{\beta^2} = 1$$ (hyperbola)

(iv) $$x^2 = \alpha y \quad \text{or} \quad y^2 = \alpha x$$ (parabola)

where α, β, and r are nonzero real numbers. Note that the circle is a special case of the ellipse ($\alpha = \beta = r$). A conic section is said to be in *standard position* if its equation can be put into one of these four standard forms. The graphs of (i), (ii), and (iii) will all be symmetric to both coordinate axes and the origin. We say that these curves are centered at the origin. A parabola in standard position will have its vertex at the origin and will be symmetric to one of the axes.

What about conics that are not in standard position? Let us consider the following cases.

Case (1). The conic section has been translated horizontally from the standard position. This occurs when the x^2 and x terms in (1) both have nonzero coefficients.

Case (2). The conic section has been translated vertically from the standard position. This occurs when the y^2 and y terms in (1) have nonzero coefficients (i.e., $c \neq 0$ and $e \neq 0$).

Case (3). The conic section has been rotated from the standard position by an angle θ which is not a multiple of 90°. This occurs when the coefficient of the xy term is nonzero (i.e., $b \neq 0$).

In general, we may have any one or combination of these three cases. In order to graph a conic that is not in standard position, we usually find a new set of axes x' and y' such that the conic section is in standard position with respect to these new axes. This is not difficult if the conic has only been translated horizontally and vertically, in which case the new axes can

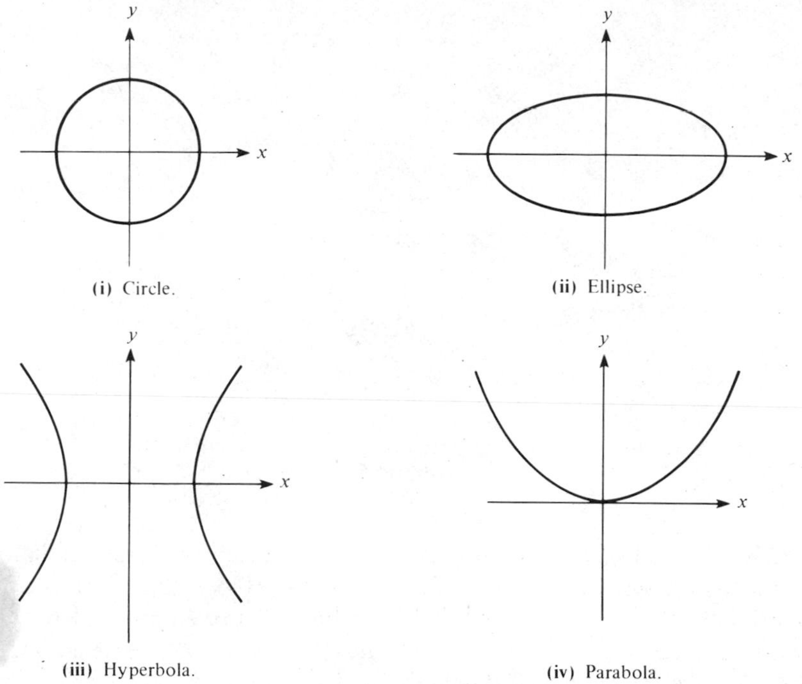

Figure 6.5.1

be found by completing the squares. The following example illustrates how this is done.

EXAMPLE 1. Sketch the graph of the equation

$$9x^2 - 18x + 4y^2 + 16y - 11 = 0$$

SOLUTION. To see how to choose our new axis system, we complete the squares.

$$9(x^2 - 2x + 1) + 4(y^2 + 4y + 4) - 11 = 9 + 16$$

This equation can be simplified to the form

$$\frac{(x-1)^2}{2^2} + \frac{(y+2)^2}{3^2} = 1$$

If we let

$$x' = x - 1 \quad \text{and} \quad y' = y + 2$$

the equation becomes

$$\frac{(x')^2}{2^2} + \frac{(y')^2}{3^2} = 1$$

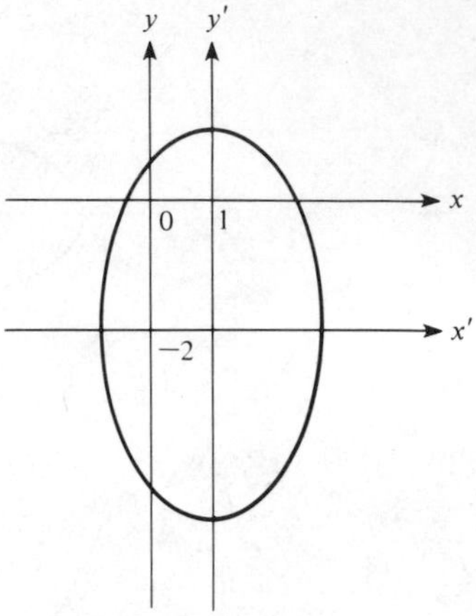

Figure 6.5.2

which is in standard form with respect to the variables x' and y'. Thus the graph, as shown in Figure 6.5.2, will be an ellipse that is in standard position in the $x'y'$ axis system. The center of the ellipse will be at the origin of the $x'y'$ plane [i.e., at the point $(x, y) = (1, -2)$]. The equation of the x' axis is simply $y' = 0$, which is the equation of the line $y = -2$ in the xy plane. Similarly, the y' axis coincides with the line $x = 1$.

There is little problem if the center or vertex of the conic section has been translated. If, however, the conic section has also been rotated from the standard position, it is necessary to change coordinates so that the equation in terms of the new coordinates x' and y' involves no $x'y'$ term. Let $\mathbf{x} = (x, y)^T$ and $\mathbf{x}' = (x', y')^T$. Since the new coordinates differ from the old coordinates by a rotation, we have

$$\mathbf{x} = Q\mathbf{x}'$$

where

$$Q = \begin{pmatrix} \cos\theta & \sin\theta \\ -\sin\theta & \cos\theta \end{pmatrix}$$

for some angle θ. Equation (2) then becomes

$$(\mathbf{x}')^T(Q^TAQ)\mathbf{x}' + (d' \quad e')\mathbf{x}' + f = 0 \tag{3}$$

where $(d' \quad e') = (d \quad e)Q$. This equation will involve no $x'y'$ term if and only if Q^TAQ is diagonal. Since A is symmetric, it is possible to find a pair of orthonormal eigenvectors $\mathbf{q}_1 = (x_1, -y_1)$ and $\mathbf{q}_2 = (y_1, x_1)^T$. Thus if we set $\cos\theta = x_1$ and $\sin\theta = y_1$, then

$$Q = (\mathbf{q}_1 \quad \mathbf{q}_2) = \begin{pmatrix} x_1 & y_1 \\ -y_1 & x_1 \end{pmatrix}$$

diagonalizes A and (3) simplifies to

$$\lambda_1(x')^2 + \lambda_2(y')^2 + d'x' + e'y' + f = 0$$

EXAMPLE 2. Consider the conic section

$$3x^2 + 2xy + 3y^2 - 8 = 0$$

This equation can be written in the form

$$(x \quad y)\begin{pmatrix} 3 & 1 \\ 1 & 3 \end{pmatrix}\begin{pmatrix} x \\ y \end{pmatrix} = 8$$

The matrix

$$\begin{pmatrix} 3 & 1 \\ 1 & 3 \end{pmatrix}$$

has eigenvalues $\lambda = 2$ and $\lambda = 4$ with corresponding unit eigenvectors

$$\left(\frac{1}{\sqrt{2}}, -\frac{1}{\sqrt{2}}\right)^T \quad \text{and} \quad \left(\frac{1}{\sqrt{2}}, \frac{1}{\sqrt{2}}\right)^T$$

Let

$$Q = \begin{bmatrix} \frac{1}{\sqrt{2}} & \frac{1}{\sqrt{2}} \\ -\frac{1}{\sqrt{2}} & \frac{1}{\sqrt{2}} \end{bmatrix} = \begin{pmatrix} \cos 45^\circ & \sin 45^\circ \\ -\sin 45^\circ & \cos 45^\circ \end{pmatrix}$$

and set

$$\begin{pmatrix} x \\ y \end{pmatrix} = \begin{bmatrix} \frac{1}{\sqrt{2}} & \frac{1}{\sqrt{2}} \\ -\frac{1}{\sqrt{2}} & \frac{1}{\sqrt{2}} \end{bmatrix}\begin{pmatrix} x' \\ y' \end{pmatrix}$$

Thus

$$Q^TAQ = \begin{pmatrix} 2 & 0 \\ 0 & 4 \end{pmatrix}$$

and the equation of the conic becomes

$$2(x')^2 + 4(y')^2 = 8$$

or

$$\frac{(x')^2}{4} + \frac{(y')^2}{2} = 1$$

(see Figure 6.5.3).

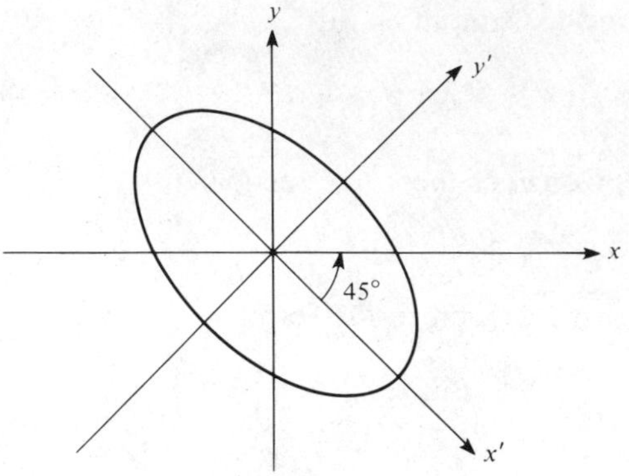

Figure 6.5.3

EXAMPLE 3. Given the quadratic equation

$$3x^2 + 2xy + 3y^2 + 8\sqrt{2}\,y - 4 = 0$$

find a change of coordinates so that the resulting equation represents a conic in standard position.

SOLUTION. The xy term is eliminated in the same manner as in Example 2. In this case, using the rotation matrix,

$$Q = \begin{pmatrix} \frac{1}{\sqrt{2}} & \frac{1}{\sqrt{2}} \\ -\frac{1}{\sqrt{2}} & \frac{1}{\sqrt{2}} \end{pmatrix}$$

the equation becomes

$$2(x')^2 + 4(y')^2 + (0 \quad 8\sqrt{2}\,)Q\begin{pmatrix} x' \\ y' \end{pmatrix} = 4$$

or

$$(x')^2 - 4x' + 2(y')^2 + 4y' = 2$$

If we complete the square, we get

$$(x' - 2)^2 + 2(y' + 1)^2 = 8$$

If we set $x'' = x' - 2$ and $y'' = y' + 1$ (Figure 6.5.4), then the equation simplifies to

$$\frac{(x'')^2}{8} + \frac{(y'')^2}{4} = 1$$

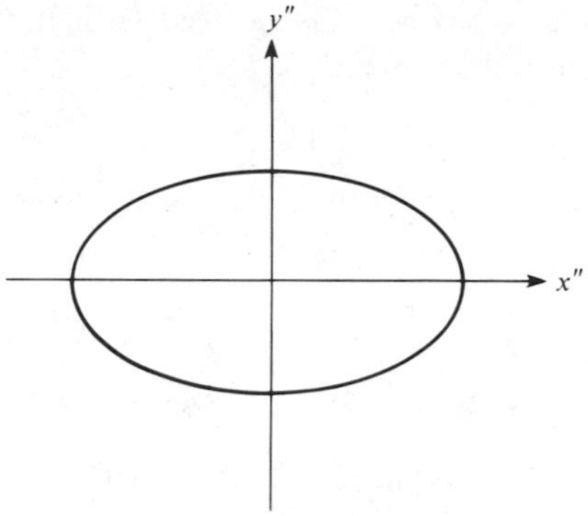

Figure 6.5.4

To summarize, a quadratic equation in the variables x and y can be written in the form

$$\mathbf{x}^T A\mathbf{x} + B\mathbf{x} + f = 0$$

where $\mathbf{x} = (x, y)^T$, A is a 2×2 symmetric matrix, B is a 1×2 matrix, and f is a scalar. If A is nonsingular, then by rotating and translating the axes, it is possible to rewrite the equation in the form

$$\lambda_1(x')^2 + \lambda_2(y')^2 + f' = 0 \tag{4}$$

where λ_1 and λ_2 are the eigenvalues of A. If $f' \neq 0$, then (4) represents a nondegenerate conic that will either be an ellipse or a hyperbola, depending on whether λ_1 and λ_2 agree in sign or differ in sign. If A is singular and exactly one of its eigenvalues is zero, the quadratic equation can be reduced to either

$$\lambda_1(x')^2 + e'y' + f' = 0 \quad \text{or} \quad \lambda_2(y')^2 + d'x' + f' = 0$$

These equations will represent parabolas, provided that e' and d' are nonzero.

There is no reason to limit ourselves to two variables. We could just as well have quadratic equations and quadratic forms in any number of variables. Indeed, a *quadratic equation in n variables* $x_1, \ldots, x_n$ is one of the form

$$\mathbf{x}^T A\mathbf{x} + B\mathbf{x} + \alpha = 0 \tag{5}$$

where $\mathbf{x} = (x_1, \ldots, x_n)^T$, A is an $n \times n$ symmetric matrix, B is a $1 \times n$ matrix, and α is a scalar. The vector function

$$f(\mathbf{x}) = \mathbf{x}^T A\mathbf{x} = \sum_{i=1}^{n}\left(\sum_{j=1}^{n} a_{ij}x_j\right)x_i$$

is the *quadratic form in n variables* associated with the quadratic equation.

In the case of three unknowns, if

$$\mathbf{x} = \begin{pmatrix} x \\ y \\ z \end{pmatrix}, \qquad A = \begin{bmatrix} a & d & e \\ d & b & f \\ e & f & c \end{bmatrix}, \qquad B = \begin{bmatrix} g \\ h \\ i \end{bmatrix}$$

then (5) becomes

(6) $ax^2 + by^2 + cz^2 + 2dxy + 2exz + 2fyz + gx + hy + iz + \alpha = 0$

The graph of a quadratic equation in three variables is called a *quadric surface*.

There are four basic types of nondegenerate quadric surfaces:

(i) Ellipsoids
(ii) Hyperboloids (of one or two sheets)
(iii) Cones
(iv) Paraboloids (either elliptic or hyperbolic)

In order to see what type of surface is represented by a quadratic equation

$$\mathbf{x}^T A \mathbf{x} + B\mathbf{x} + \alpha = 0$$

we generally transform the equation into standard form. As before, this can be done using a rotation and a translation. In this case, by a rotation we mean a transformation $\mathbf{x} = Q\mathbf{x}'$, where Q is orthogonal and $|Q| = 1$. Since the matrix A of the quadratic form is symmetric, it is diagonalizable. If A is nonsingular, the equation can be transformed into one of the form

(7) $$\lambda_1(x')^2 + \lambda_2(y')^2 + \lambda_3(z')^2 + \alpha = 0$$

If (7) represents a nondegenerate quadric surface, the graph will be

(i) An ellipsoid if λ_1, λ_2, and λ_3 all have the same sign
(ii) A cone if λ_1, λ_2, and λ_3 do not all have the same sign and $\alpha = 0$
(iii) A hyperboloid if λ_1, λ_2, and λ_3 do not all have the same sign and $\alpha \neq 0$. (The number of sheets will depend on how many of the eigenvalues agree with α in sign.)

If exactly one of the eigenvalues of A is zero, the equation will be a paraboloid if it can be transformed into one of the following forms:

$$x' = \lambda_2(y')^2 + \lambda_3(z')^2$$
$$y' = \lambda_1(x')^2 + \lambda_3(z')^2$$
$$z' = \lambda_1(x')^2 + \lambda_2(y')^2$$

If the two nonzero eigenvalues agree in sign, the paraboloid will be elliptic; otherwise, it will be hyperbolic. If A has 0 as a multiple eigenvalue, the quadric surface will be degenerate.

EXAMPLE 4. Given the quadric surface

$$2(x^2+y^2-xy-xz-yz)-1=0$$

find a rotation of axes such that the transformed equation will be in standard form.

SOLUTION. The equation of the surface can be written in the form $\mathbf{x}^T A\mathbf{x}=1$, where

$$\mathbf{x}=\begin{pmatrix} x \\ y \\ z \end{pmatrix} \quad \text{and} \quad A=\begin{bmatrix} 2 & -1 & -1 \\ -1 & 2 & -1 \\ -1 & -1 & 0 \end{bmatrix}$$

The eigenvalues of A are $\lambda_1=2$, $\lambda_2=-1$, and $\lambda_3=3$, and the corresponding unit eigenvectors are

$$\frac{1}{\sqrt{3}}\begin{bmatrix} 1 \\ 1 \\ -1 \end{bmatrix}, \quad \frac{1}{\sqrt{6}}\begin{bmatrix} 1 \\ 1 \\ 2 \end{bmatrix}, \quad \frac{1}{\sqrt{2}}\begin{bmatrix} 1 \\ -1 \\ 0 \end{bmatrix}$$

Let

$$Q=\frac{1}{\sqrt{6}}\begin{bmatrix} \sqrt{2} & 1 & \sqrt{3} \\ \sqrt{2} & 1 & -\sqrt{3} \\ -\sqrt{2} & 2 & 0 \end{bmatrix}$$

Q diagonalizes A and, since Q is orthogonal and $|Q|=1$, it follows that Q is a rotation.

$$Q^T A Q = D = \begin{bmatrix} 2 & 0 & 0 \\ 0 & -1 & 0 \\ 0 & 0 & 3 \end{bmatrix}$$

If we set $\mathbf{x}=Q\mathbf{x}'$, the equation of the surface becomes

$$(\mathbf{x}')^T D\mathbf{x}' = 1$$

or

$$2(x')^2-(y')^2+3(z')^2=1$$

The graph of this equation will be a hyperboloid of one sheet.

EXAMPLE 5. Given the quadric surface

$$2(x^2+y^2-xy-xz-yz)-z=0$$

find a change of coordinates so that the resulting equation is in standard form and classify the surface.

SOLUTION. This equation can be written in the form

$$\mathbf{x}^T A\mathbf{x} + B\mathbf{x} = 0$$

where A is the same as in Example 4 and $B = (0 \quad 0 \quad -1)$. Using the same matrix Q as before and substituting $\mathbf{x} = Q\mathbf{x}'$, we get

$$(\mathbf{x}')D\mathbf{x}' + BQ\mathbf{x}' = 0$$

or

$$2(x')^2 - (y')^2 + 3(z')^2 + \frac{1}{\sqrt{3}}x' - \frac{2}{\sqrt{6}}y' = 0$$

Completing the square gives

$$2\left(x' + \frac{1}{4\sqrt{3}}\right)^2 - \left(y' + \frac{1}{\sqrt{6}}\right)^2 + 3(z')^2 = -\frac{1}{8}$$

and if we let

$$x'' = x' + \frac{1}{4\sqrt{3}}, \qquad y'' = y' + \frac{1}{\sqrt{6}}, \qquad z'' = z'$$

the equation becomes

$$2(x'')^2 - (y'')^2 + 3(z'')^2 = -\tfrac{1}{8}$$

The graph of this equation will be a hyperboloid of two sheets.

EXERCISES

1. Find the matrix associated with each of the following quadratic forms.
(a) $3x^2 - 5xy + y^2$
(b) $2x^2 + 3y^2 + z^2 + xy - 2xz + 3yz$
(c) $x^2 + 2y^2 + z^2 + xy - 2xz + 3yz$

2. In each of the following find a suitable change of coordinates (i.e., a rotation and/or a translation) so that the resulting conic section is in standard form.
(a) $x^2 + xy + y^2 - 6 = 0$
(b) $3x^2 + 8xy + 3y^2 + 28 = 0$
(c) $-3x^2 + 6xy + 5y^2 - 24 = 0$
(d) $x^2 + 2xy + y^2 + 3x + y - 1 = 0$

3. Determine whether or not the following equations represent nondegenerate quadric surfaces. Classify each nondegenerate quadric.
(a) $\lambda_1 x^2 + \lambda_2 y^2 + \lambda_3 z^2 = 1$, $\lambda_1 > 0$, $\lambda_2 > 0$, $\lambda_3 > 0$
(b) $\lambda_1 x^2 + \lambda_2 y^2 + \lambda_3 z^2 = 0$, $\lambda_1 > 0$, $\lambda_2 > 0$, $\lambda_3 > 0$
(c) $\lambda_1 x^2 + \lambda_2 y^2 + z = 0$, $\lambda_1\lambda_2 > 0$
(d) $\lambda_1 x^2 + \lambda_2 y^2 - z = 0$, $\lambda_1\lambda_2 > 0$
(e) $\lambda_1 x^2 + \lambda_2 y^2 = 0$, $\lambda_1\lambda_2 > 0$
(f) $\lambda_1 x^2 + \lambda_2 y^2 + \lambda_3 z^2 = c^2$, $c \neq 0$, $\lambda_1 > 0$, $\lambda_2 > 0$, $\lambda_3 < 0$
(g) $\lambda_1 x^2 + \lambda_2 y^2 + \lambda_3 z^2 = 0$, $\lambda_1 > 0$, $\lambda_2\lambda_3 > 0$
(h) $\lambda_1 x^2 + \lambda_2 y^2 + \lambda_3 z^2 = 1$, $\lambda_1 > 0$, $\lambda_2 < 0$, $\lambda_3 < 0$

4. In each of the following, find a suitable change of coordinates so that the resulting quadric surface is in standard position.
(a) $2x^2 - y^2 - 12x - 4y - z + 16 = 0$
(b) $x^2 + y^2 + z^2 + 2xy + 2xz + 2yz - x + 2y - z = 0$
(c) $3x^2 + 2y^2 + 2z^2 + 2xz + 6x + y + z + 2 = 0$

5. Let A be a symmetric 2×2 matrix and let α be a nonzero scalar for which the equation $\mathbf{x}^T A \mathbf{x} = \alpha$ is consistent. Show that the corresponding conic section will be nondegenerate if and only if A is nonsingular.

6. Let A be a symmetric $n \times n$ matrix with eigenvalues $\lambda_1, \ldots, \lambda_n$. Show that there exists an orthonormal set of vectors $\{\mathbf{x}_1, \ldots, \mathbf{x}_n\}$ such that

$$\mathbf{x}^T A \mathbf{x} = \sum_{i=1}^{n} \lambda_i (\mathbf{x}^T \mathbf{x}_i)^2$$

for each $\mathbf{x} \in R^n$.

6. POSITIVE DEFINITE MATRICES

In this section we will consider the problem of maximizing and minimizing functions of several variables. In particular, we would like to determine the nature of the critical points of a real-valued vector function $w = F(\mathbf{x})$. If the function is a quadratic form, $w = \mathbf{x}^T A \mathbf{x}$, then $\mathbf{0}$ is a critical point. Whether it is a maximum, minimum, or saddle point depends upon the eigenvalues of A. More generally, if the function to be extremized is sufficiently differentiable, it behaves locally like a quadratic form. Thus each critical point can be tested by determining the signs of the eigenvalues of the matrix of an associated quadratic form.

Definition. Let $F(\mathbf{x})$ be a real-valued function on R^n. A point $\mathbf{x}_0$ in R^n is said to be a **stationary point** of F if all of the first partials of F at $\mathbf{x}_0$ are zero.

If $F(\mathbf{x})$ has either a local maximum or a local minimum at a point $\mathbf{x}_0$ and the first partials of F exist at $\mathbf{x}_0$, they will all be zero. Thus, if $F(\mathbf{x})$ has first partials everywhere, its local maxima and minima will occur at stationary points.

Consider the quadratic form

$$f(x, y) = ax^2 + 2bxy + cy^2$$

The first partials of f are

$$\begin{aligned} f_x &= 2ax + 2by \\ f_y &= 2bx + 2cy \end{aligned}$$

Setting these equal to zero, we see that (0, 0) is a stationary point. Moreover, if the matrix

$$A = \begin{pmatrix} a & b \\ b & c \end{pmatrix}$$

is nonsingular, this will be the only critical point. Thus, if A is nonsingular, f will have either a global minimum, a global maximum or a saddle point at (0, 0).

Suppose that f has a global minimum at (0, 0). Since $a = f(1, 0)$ and $c = f(0, 1)$, it follows that a and c must be positive. Let us rewrite f in the form

$$(1) \qquad f(x, y) = a\left(x + \frac{b}{a}y\right)^2 + \left(c - \frac{b^2}{a}\right)y^2$$

Let $(x_0, y_0) \neq (0, 0)$ be a point on the line $ax + by = 0$. Then

$$f(x_0, y_0) = \left(c - \frac{b^2}{a}\right)y_0^2 > 0$$

and hence $ac - b^2 > 0$. Thus, if f has a global minimum at (0, 0), then

$$a > 0, c > 0 \quad \text{and} \quad ac - b^2 > 0$$

It follows from (1) that these conditions are also sufficient to guarantee a global minimum.

The quadratic form f will have a global maximum at (0, 0) if and only if $-f$ has a global minimum at (0, 0). This will occur if and only if

$$a < 0, c < 0 \quad \text{and} \quad ac - b^2 > 0$$

(Note that the condition $ac - b^2 > 0$ implies that a and c have the same sign.)

The quadratic form f will have a saddle point at (0, 0) if and only if $ac - b^2 < 0$.

EXAMPLE 1. Let $f(x, y) = 4x^2 - 6xy + 3y^2$. Determine the nature of the stationary point (0, 0).

SOLUTION. The matrix A of the quadratic form is nonsingular and since

$$|A| = 4 \cdot 3 - (-3)^2 = 3 > 0$$

it follows that f assumes either a maximum or minimum at (0, 0). Since a and c are both positive, it must be a minimum.

Suppose now that we have a function $F(x, y)$ with a stationary point (x_0, y_0). If F has continuous third partials in a neighborhood of (x_0, y_0), it can be expanded in a Taylor series about that point.

$$\begin{aligned} F(x_0 + h, y_0 + k) &= F(x_0, y_0) + \left[hF_x(x_0, y_0) + kF_y(x_0 y_0)\right] \\ &\quad + \tfrac{1}{2}\left[h^2F_{xx}(x_0, y_0) + 2hkF_{xy}(x_0, y_0) + k^2F_{yy}(x_0, y_0)\right] + R \\ &= F(x_0, y_0) + \tfrac{1}{2}(ah^2 + 2bhk + ck^2) + R \end{aligned}$$

where

$$a = F_{xx}(x_0, y_0), \qquad b = F_{xy}(x_0, y_0), \qquad c = F_{yy}(x_0, y_0)$$

and the remainder R is given by

$$\begin{aligned} R &= \tfrac{1}{6}\left[h^3F_{xxx}(\mathbf{z}) + 3h^2kF_{xxy}(\mathbf{z}) + 3hk^2F_{xyy}(\mathbf{z}) + k^3F_{yyy}(\mathbf{z})\right] \\ \mathbf{z} &= (x_0 + \theta h, y_0 + \theta k); \quad 0 < \theta < 1 \end{aligned}$$

If h and k are sufficiently small, $|R|$ will be less than the modulus of $\frac{1}{2}\{ah^2 + 2bhk + ck^2\}$ and hence $[F(x_0 + h,\ y_0 + k) - F(x_0, y_0)]$ will have the same sign as $(ah^2 + 2bhk + ck^2)$. The expression

$$f(h, k) = ah^2 + 2bhk + ck^2$$

is a quadratic form in the variables h and k. Thus $F(x, y)$ will have a local minimum (maximum) at (x_0, y_0) if and only if $f(h, k)$ has a minimum (maximum) at (0, 0). Thus, if

$$F_{xx}(x_0, y_0)F_{yy}(x_0, y_0) - (F_{xy}(x_0, y_0))^2 > 0$$

then

(i) F has a minimum at (x_0, y_0) if $F_{xx}(x_0, y_0) > 0$.
(ii) F has a maximum at (x_0, y_0) if $F_{xx}(x_0, y_0) < 0$.

If

$$F_{xx}(x_0, y_0)F_{yy}(x_0, y_0) - (F_{xy}(x_0, y_0))^2 < 0$$

then (x_0, y_0) is a saddle point of F.

EXAMPLE 2. Find and describe all stationary points of the function

$$F(x, y) = \tfrac{1}{3}x^3 + xy^2 - 4xy + 1$$

SOLUTION. The first partials of F are

$$F_x = x^2 + y^2 - 4y$$
$$F_y = 2xy - 4x = 2x(y - 2)$$

Setting $F_y = 0$, we get $x = 0$ or $y = 2$. Setting $F_x = 0$, we see that if $x = 0$, then y must be either 0 or 4, and if $y = 2$, then $x = \pm 2$. Thus (0, 0), (0, 4), (2, 2)(−2, 2) are the stationary points of F. To classify the stationary points, we compute the second partials.

$$F_{xx} = 2x, \qquad F_{xy} = 2y - 4, \qquad F_{yy} = 2x$$

Stationary Point (x_0, y_0)	$F_{xx}(x_0, y_0)$	$F_{xx}F_{yy} - (F_{xy})^2$	*Description*
(0, 0)	0	−16	Saddle point
(0, 4)	0	−16	Saddle point
(2, 2)	4	16	Local minimum
(−2, 2)	−4	16	Local maximum

Definition. A quadratic form $f(\mathbf{x}) = \mathbf{x}^T A\mathbf{x}$ is said to be definite if it takes on only one sign as $\mathbf{x}$ varies over all nonzero vectors in R^n. The form is **positive definite** if $\mathbf{x}^T A\mathbf{x} > 0$ for all nonzero $\mathbf{x}$ in R^n and **negative definite** if $\mathbf{x}^T A\mathbf{x} < 0$ for all nonzero $\mathbf{x}$ in R^n. A quadratic form is said to be **indefinite** if it takes on values of different sign. If $f(\mathbf{x}) = \mathbf{x}^T A\mathbf{x} \geqslant 0$ and assumes the value 0 when $\mathbf{x} \neq \mathbf{0}$, then $f(\mathbf{x})$ is said to be **positive semidefinite**. If $f(\mathbf{x}) \leqslant 0$ and assumes the value 0 for $\mathbf{x} \neq \mathbf{0}$, then $f(\mathbf{x})$ is said to be **negative semidefinite**.

Thus, if

$$f(\mathbf{x}) = ax^2 + 2bxy + cy^2$$

then $f(\mathbf{x})$ is definite if and only if

$$ac - b^2 > 0$$

If $f(\mathbf{x})$ is definite, it will assume either a minimum or a maximum at (0, 0) depending on whether $a > 0$ or $a < 0$. $f(\mathbf{x})$ is indefinite if and only if

$$ac - b^2 < 0$$

and semidefinite if and only if

$$ac - b^2 = 0$$

Definition. A real symmetric matrix A is said to be

(i) **Positive definite** if $\mathbf{x}^T A\mathbf{x} > 0$ for all nonzero $\mathbf{x}$ in R^n
(ii) **Negative definite** if $\mathbf{x}^T A\mathbf{x} < 0$ for all nonzero $\mathbf{x}$ in R^n
(iii) **Positive semidefinite** if $\mathbf{x}^T A\mathbf{x} \geqslant 0$ for all nonzero $\mathbf{x}$ in R^n
(iv) **Negative semidefinite** if $\mathbf{x}^T A\mathbf{x} \leqslant 0$ for all nonzero $\mathbf{x}$ in R^n

We can now generalize our method of classifying stationary points to functions of more than two variables. Let $F(\mathbf{x}) = F(x_1, \ldots, x_n)$ be a real-valued function whose third partial derivatives are all continuous. Let $\mathbf{x}_0$ be a stationary point of F and define the matrix $H = H(\mathbf{x}_0)$ by

$$h_{ij} = F_{x_i x_j}(\mathbf{x}_0)$$

$H(\mathbf{x}_0)$ is called the *Hessian* of F at $\mathbf{x}_0$. By the same reasoning used in our discussion of the two-dimensional case, we conclude that:

(i) $\mathbf{x}_0$ is a local minimum of F if $H(\mathbf{x}_0)$ is positive definite.
(ii) $\mathbf{x}_0$ is a local maximum of F if $H(\mathbf{x}_0)$ is negative definite.
(iii) $\mathbf{x}_0$ is a saddle point of F if $H(\mathbf{x}_0)$ is indefinite.

Let us restrict ourselves to the problem of minimizing a function F. (The problem of maximizing F is the same as the problem of minimizing $-F$.) To determine whether or not F assumes a local minimum at a stationary point $\mathbf{x}_0$, one must determine whether or not the Hessian $H(\mathbf{x}_0)$ is positive definite. This can be done with the aid of the following theorem.

Theorem 6.6.1. *Let A be a real symmetric $n \times n$ matrix. Then A is positive definite if and only if all its eigenvalues are positive.*

PROOF. If A is positive definite and λ is an eigenvalue of A, then for any eigenvector $\mathbf{x}$ belonging to λ,

$$\mathbf{x}^T A\mathbf{x} = \lambda\mathbf{x}^T\mathbf{x} = \lambda\|\mathbf{x}\|^2$$

Hence

$$\lambda = \frac{\mathbf{x}^T A\mathbf{x}}{\|\mathbf{x}\|^2} > 0$$

Conversely, suppose that all the eigenvalues of A are positive. Let $\{\mathbf{x}_1, \ldots, \mathbf{x}_n\}$ be an orthonormal set of eigenvectors of A. If $\mathbf{x}$ is any nonzero vector in R^n, then $\mathbf{x}$ can be written in the form

$$\mathbf{x} = \alpha_1\mathbf{x}_1 + \alpha_2\mathbf{x}_2 + \cdots + \alpha_n\mathbf{x}_n$$

where

$$\alpha_i = \mathbf{x}^T\mathbf{x}_i \quad \text{for } i = 1, \ldots, n \qquad \text{and} \qquad \sum_{i=1}^{n} (\alpha_i)^2 = \|\mathbf{x}\|^2 > 0$$

It follows that

$$\begin{aligned}\mathbf{x}^T A\mathbf{x} &= (\alpha_1\mathbf{x}_1 + \cdots + \alpha_n\mathbf{x}_n)^T(\alpha_1\lambda_1\mathbf{x}_1 + \cdots + \alpha_n\lambda_n\mathbf{x}_n)\\ &= \sum_{i=1}^{n} (\alpha_i)^2\lambda_i\\ &\geqslant (\min \lambda_i)\|\mathbf{x}\|^2 > 0\end{aligned}$$

and hence A is positive definite.

EXAMPLE 3. Find the local minima of the function

$$F(x, y, z) = x^2 + xz - 3\cos y + z^2$$

SOLUTION. The first partials of F are

$$\begin{aligned}F_x &= 2x + z\\ F_y &= 3\sin y\\ F_z &= x + 2z\end{aligned}$$

It follows that (x, y, z) is a stationary point of F if and only if $x = z = 0$ and $y = n\pi$, where n is an integer. Let $\mathbf{x}_0 = (0, 2k\pi, 0)^T$. The Hessian of F at $\mathbf{x}_0$ is given by

$$H(\mathbf{x}_0) = \begin{bmatrix} 2 & 0 & 1\\ 0 & 3 & 0\\ 1 & 0 & 2\end{bmatrix}$$

The eigenvalues of $H(\mathbf{x}_0)$ are 3, 3, and 1. Since the eigenvalues are all positive, it follows that $H(\mathbf{x}_0)$ is positive definite and hence F has a local minimum at $\mathbf{x}_0$. On the other hand, at a stationary point of the form $\mathbf{x}_1 = (0, (2k-1)\pi, 0)^T$ the Hessian will be

$$H(\mathbf{x}_1) = \begin{bmatrix} 2 & 0 & 1\\ 0 & -3 & 0\\ 1 & 0 & 2\end{bmatrix}$$

The eigenvalues of $H(\mathbf{x}_1)$ are -3, 3, and 1. It follows that $H(\mathbf{x}_1)$ is indefinite and hence $\mathbf{x}_1$ is a saddle point of F.

EXERCISES

1. Which of the following matrices are positive definite? Negative definite? Indefinite?

(a) $\begin{pmatrix} 3 & 2\\ 2 & 2\end{pmatrix}$ (b) $\begin{pmatrix} 3 & 4\\ 4 & 1\end{pmatrix}$

(c) $\begin{bmatrix} -2 & 0 & 1\\ 0 & -1 & 0\\ 1 & 0 & -2\end{bmatrix}$ (d) $\begin{bmatrix} 1 & 2 & 1\\ 2 & 1 & 1\\ 1 & 1 & 2\end{bmatrix}$

2. For each of the following functions, determine whether the given stationary point corresponds to a local minimum, local maximum, or saddle point.
(a) $f(x, y) = 3x^2 - xy + y^2 \quad (0, 0)$
(b) $f(x, y) = \sin x + y^3 + 3xy + 2x - 3y \quad (0, -1)$
(c) $f(x, y, z) = x^3 + xyz + y^2 - 3x \quad (1, 0, 0)$
(d) $f(x, y, z) = -\frac{1}{4}(x^{-4} + y^{-4} + z^{-4}) + yz - x - 2y - 2z \quad (1, 1, 1)$

3. Prove that if B is a nonsingular $n \times n$ matrix, B^TB is positive definite.

4. Show that any positive definite symmetric matrix A can be factored into a product B^TB, where B is nonsingular.

5. Show that if A is a positive definite symmetric matrix, $|A| > 0$. Give an example of a 2×2 matrix with a positive determinant that is not positive definite.

6. Let A be a symmetric $n \times n$ matrix and let A_r be the matrix formed by deleting the last $n - r$ rows and columns of A. (A_r is called the *leading principal submatrix* of A of order r.)
(a) If $\mathbf{x}_r = (x_1, \ldots, x_r)^T$ and $\mathbf{x} = (x_1, \ldots, x_r, 0, \ldots, 0)^T$, show that

$$\mathbf{x}_r^T A_r \mathbf{x}_r = \mathbf{x}^T A \mathbf{x}$$

(b) Show that if A is positive definite, A_r is positive definite.

7. NONNEGATIVE MATRICES

In many of the types of linear systems that occur in applications, the entries of the coefficient matrix represent nonnegative quantities. This section deals with the study of such matrices and some of their properties.

Definition. An $n \times n$ matrix A with real entries is said to be **nonnegative** if $a_{ij} \geqslant 0$ for each i and j and **positive** if $a_{ij} > 0$ for each i and j.

Similarly, a vector $\mathbf{x} = (x_1, \ldots, x_n)^T$ is said to be **nonnegative** if each $x_i \geqslant 0$ and **positive** if each $x_i > 0$.

For an example of one of the applications of nonnegative matrices, we consider Leontief input–output models.

Application 1: The Open Model

Suppose that there are n industries producing n different products. Each industry requires input of the products from the other industries and possibly even of its own product. In the open model it is assumed that there is an additional demand for each of the products from an outside sector. The problem is to determine the output of each of the industries that is necessary to meet the total demand.

We will show that this problem can be represented by a linear system of equations and that the system has a unique nonnegative solution. Let a_{ij} denote the amount of input from the ith industry necessary to produce one unit of output in the jth industry. By a unit of input or output we mean \$1's worth of the product. Thus producing \$1's worth of the jth product will involve a cost of $\sum_{i=1}^{n} a_{ij}$ dollars. Clearly, production of the jth product will not be profitable unless $\sum_{i=1}^{n} a_{ij} < 1$. Let d_i denote the demand of the open sector for the ith product. Finally, let x_i represent the amount of output of the ith product necessary to meet the total demand. If the jth industry is to have an output of x_j, it will need an input of $a_{ij}x_j$ units from the ith industry. Thus the total demand for the ith product will be

$$a_{i1}x_1 + a_{i2}x_2 + \cdots + a_{in}x_n + d_i$$

and hence we require that

$$x_i = a_{i1}x_1 + a_{i2}x_2 + \cdots + a_{in}x_n + d_i$$

for $i = 1, \ldots, n$. This leads to the system

$$\begin{array}{lllll} (1 - a_{11})x_1 & - a_{12}x_2 - & \cdots & - a_{1n}x_n & = d_1 \\ - a_{21}x_1 + & (1 - a_{22})x_2 - & \cdots & - a_{2n}x_n & = d_2 \\ \vdots & & & & \\ - a_{n1}x_1 & - a_{n2}x_2 - & \cdots & -(1 - a_{nn})x_n & = d_n \end{array}$$

which may be written in the form

(1) $$(I - A)\mathbf{x} = \mathbf{d}$$

The entries of A have two important properties:

(i) $a_{ij} \geqslant 0$ for each i and j

(ii) $\sum_{i=1}^{n} a_{ij} < 1$

The vector $\mathbf{x}$ must not only be a solution to (1), it must be nonnegative. (It would not make any sense to have a negative output.) It follows from condition (ii) that $\|A\|_1 < 1$. Thus, if λ is any eigenvalue of A and $\mathbf{x}$ is an eigenvector belonging to λ, then

$$|\lambda| \, \|\mathbf{x}\|_1 = \|\lambda \mathbf{x}\|_1 = \|A\mathbf{x}\|_1 \leqslant \|A\|_1 \|\mathbf{x}\|_1$$

and hence

$$|\lambda| \leqslant \|A\|_1 < 1$$

Thus 1 is not an eigenvalue of A. It follows that $I - A$ is nonsingular and hence the system (1) has a unique solution.

$$\mathbf{x} = (I - A)^{-1}\mathbf{d}$$

We would like to show that this solution must be nonnegative. To do this, we will show that $(I - A)^{-1}$ is nonnegative. First note that

$$\|A^m\|_1 \leqslant \|A\|_1^m$$

Since $\|A\|_1 < 1$, it follows that

$$\|A^m\|_1 \to 0 \qquad \text{as} \quad m \to \infty$$

and hence A^m approaches the zero matrix as $m \to \infty$.

Since

$$(I - A)(I + A + \cdots + A^m) = I - A^{m+1}$$

it follows that

$$I + A + \cdots + A^m = (I - A)^{-1} - (I - A)^{-1}A^{m+1}$$

and hence $I + A + \cdots + A^m$ approaches $(I - A)^{-1}$ as $m \to \infty$. By condition (i), $I + A + \cdots + A^m$ is nonnegative for each m and therefore $(I - A)^{-1}$ must be nonnegative. Since $\mathbf{d}$ is nonnegative, it follows that the solution $\mathbf{x}$ must be nonnegative. We see, then, that conditions (i) and (ii) guarantee that the system (1) will have a unique nonnegative solution $\mathbf{x}$.

As you have probably guessed, there is also a closed version of the Leontief input–output model. In the closed version it is assumed that each industry must produce enough output to meet the input needs of only the other industries and itself. The open sector is ignored. Thus in place of the system (1), we have

$$(I - A)\mathbf{x} = \mathbf{0}$$

and we require that $\mathbf{x}$ be a positive solution. The existence of such an $\mathbf{x}$ in this case is a much deeper result than in the open version and requires some more advanced theorems.

Theorem 6.7.1 (Perron). *If A is a positive $n \times n$ matrix, A has a positive real eigenvalue r with the following properties*:

(i) *r is a simple root of the characteristic equation.*
(ii) *r has a positive eigenvector* $\mathbf{x}$.
(iii) *If λ is any other eigenvalue of A, then $|\lambda| < r$.*

The Perron theorem may be thought of as a special case of a more general theorem due to Frobenius. The Frobenius theorem applies to "irreducible" nonnegative matrices.

Definition. A nonnegative matrix A is said to be **reducible** if there exists a partition of the index set $\{1, 2, \ldots, n\}$ into nonempty disjoint sets I_1 and I_2 such that $a_{ij} = 0$ whenever $i \in I_1$ and $j \in I_2$. Otherwise, A is said to be **irreducible.**

EXAMPLE 1. Let A be a matrix of the form

$$\begin{pmatrix} \times & \times & 0 & 0 & \times \\ \times & \times & 0 & 0 & \times \\ \times & \times & \times & \times & \times \\ \times & \times & \times & \times & \times \\ \times & \times & 0 & 0 & \times \end{pmatrix}$$

Let $I_1 = \{1, 2, 5\}$ and $I_2 = \{3, 4\}$. Then $I_1 \cup I_2 = \{1, 2, 3, 4, 5\}$ and $a_{ij} = 0$ whenever $i \in I_1$ and $j \in I_2$. Therefore, A is reducible. Notice that if P is the permutation matrix formed by interchanging the third and fifth rows of the identity matrix I, then

$$PA = \begin{pmatrix} \times & \times & 0 & 0 & \times \\ \times & \times & 0 & 0 & \times \\ \times & \times & 0 & 0 & \times \\ \times & \times & \times & \times & \times \\ \times & \times & \times & \times & \times \end{pmatrix}$$

and

$$PAP = \left(\begin{array}{ccc|cc} \times & \times & \times & 0 & 0 \\ \times & \times & \times & 0 & 0 \\ \times & \times & \times & 0 & 0 \\ \hline \times & \times & \times & \times & \times \\ \times & \times & \times & \times & \times \end{array}\right)$$

In general, it can be shown that an $n \times n$ matrix A is reducible if and only if there exists a permutation matrix P such that PAP is a matrix of the form

$$\left(\begin{array}{c|c} B & O \\ \hline X & C \end{array}\right)$$

where B and C are square matrices.

Theorem 6.7.2 (Frobenius). *If A is an irreducible nonnegative matrix, then A has a positive real eigenvalue r with the following properties:*

(i) *r has a positive eigenvector* $\mathbf{x}$.

(ii) *If λ is any other eigenvalue of A, then $|\lambda| \leq r$. The eigenvalues of modulus r are all simple roots of the characteristic equation. Indeed, if there are m eigenvalues of modulus r, they must be of the form*

$$\lambda_k = r\left[\exp\left(\frac{2k\pi i}{m}\right)\right] \qquad k = 0, 1, \ldots, m-1$$

The proof of this theorem is beyond the scope of the text. We refer the reader to Gantmacher [2, Vol. 2]. Perron's theorem follows as a special case of the Frobenius theorem.

Application 2: The Closed Model

In the closed Leontief input–output model, we assume that there is no demand from the open sector, and we wish to find outputs to satisfy the demands of all n industries. Thus, defining the x_i's and a_{ij}'s as in the open model, we have

$$x_i = a_{i1}x_1 + a_{i2}x_2 + \cdots + a_{in}x_n$$

for $i = 1, \ldots, n$. The resulting system may be written in the form

$$(A - I)\mathbf{x} = \mathbf{0} \tag{2}$$

As before, we have the condition

$$a_{ij} \geqslant 0 \tag{i}$$

Since there is no open sector, the amount of output for the jth industry should be the same as the total input for that industry. Thus

$$x_j = \sum_{i=1}^{n} a_{ij}x_j$$

and hence we have as our second condition

$$\sum_{i=1}^{n} a_{ij} = 1 \qquad j = 1, \ldots, n \tag{ii}$$

Condition (ii) implies that $A - I$ is singular since the sum of its row vectors is $\mathbf{0}$. Therefore, 1 is an eigenvalue of A, and since $\|A\|_1 = 1$, it follows that all the eigenvalues of A have modules less than or equal to 1. Let us assume that enough of the coefficients of A are nonzero so that A is irreducible. Then by Theorem 6.7.2, $\lambda = 1$ has a positive eigenvector $\mathbf{x}$. Thus any positive multiple of $\mathbf{x}$ will be a positive solution to (2).

EXERCISES

1. Find the output vector $\mathbf{x}$ in the open version of the Leontief input–output model if

$$A = \begin{pmatrix} 0.2 & 0.4 & 0.4 \\ 0.4 & 0.2 & 0.2 \\ 0.0 & 0.2 & 0.2 \end{pmatrix} \qquad \text{and} \qquad \mathbf{d} = \begin{pmatrix} 16{,}000 \\ 8{,}000 \\ 24{,}000 \end{pmatrix}$$

2. Consider the closed version of the Leontief input–output model with input matrix

$$A = \begin{bmatrix} 0.5 & 0.4 & 0.1 \\ 0.5 & 0.0 & 0.5 \\ 0.0 & 0.6 & 0.4 \end{bmatrix}$$

If $\mathbf{x} = (x_1, x_2, x_3)^T$ is any output vector for this model, how are the coordinates x_1, x_2, and x_3 related?

3. Prove: If $A^m = O$ for some positive integer m, then $I - A$ is nonsingular.

4. Let

$$A = \begin{bmatrix} 0 & 1 & 1 \\ 0 & -1 & 1 \\ 0 & -1 & 1 \end{bmatrix}$$

(a) Compute $(I - A)^{-1}$.
(b) Compute A^2 and A^3. Verify that $(I - A)^{-1} = I + A + A^2$.

5. Which of the following matrices are reducible? For each reducible matrix, find a permutation matrix P such that PAP is of the form

$$\left(\begin{array}{c|c} B & O \\ \hline X & C \end{array}\right)$$

where B and C are square matrices.

(a) $\begin{bmatrix} 1 & 1 & 1 & 0 \\ 1 & 1 & 1 & 0 \\ 1 & 1 & 1 & 1 \\ 1 & 1 & 1 & 1 \end{bmatrix}$ (b) $\begin{bmatrix} 1 & 0 & 1 & 1 \\ 1 & 1 & 1 & 1 \\ 1 & 0 & 1 & 1 \\ 1 & 0 & 1 & 1 \end{bmatrix}$

(c) $\begin{bmatrix} 1 & 0 & 1 & 0 & 0 \\ 0 & 1 & 1 & 1 & 1 \\ 1 & 0 & 1 & 0 & 0 \\ 1 & 1 & 0 & 1 & 1 \\ 1 & 1 & 1 & 1 & 1 \end{bmatrix}$ (d) $\begin{bmatrix} 1 & 1 & 1 & 1 & 1 \\ 1 & 1 & 0 & 0 & 1 \\ 1 & 1 & 1 & 1 & 1 \\ 1 & 1 & 0 & 0 & 1 \\ 1 & 1 & 0 & 0 & 1 \end{bmatrix}$

6. Let A be a nonnegative irreducible 3×3 matrix whose eigenvalues satisfy $\lambda_1 = 2 = |\lambda_2| = |\lambda_3|$. Determine λ_2 and λ_3.

7. Let

$$A = \left(\begin{array}{c|c} B & O \\ \hline O & C \end{array}\right)$$

where B and C are square matrices.

(a) If λ is an eigenvalue of B with eigenvector $\mathbf{x} = (x_1, \ldots, x_k)^T$, show that λ is also an eigenvalue of A with eigenvector $\tilde{\mathbf{x}} = (x_1, \ldots, x_k, 0, \ldots, 0)^T$.

(b) If B and C are positive matrices, show that A has a positive real eigenvalue r with the property that $|\lambda| < r$ for any eigenvalue $\lambda \neq r$. Show also that the multiplicity of r is at most 2 and that r has a nonnegative eigenvector.

8. Prove that a 2×2 matrix A is reducible if and only if $a_{12}a_{21} = 0$.

9. Prove the Frobenius theorem in the case where A is a 2×2 matrix.

7

Numerical Linear Algebra

In this chapter we will consider computer methods for solving linear algebra problems. To understand these methods, one should be familiar with the type of number system used by the computer. When data are read into the computer, they are translated into the finite number system of the computer. This translation will usually involve some roundoff error. Further roundoff errors will occur when the algebraic operations of the algorithm are carried out. Because of this, we cannot expect to get the exact solution to the original problem. The best we can hope for is a good approximation to a slightly perturbed problem. Suppose, for example, that one wanted to solve $A\mathbf{x} = \mathbf{b}$. When the entries of A and $\mathbf{b}$ are read into the computer, roundoff errors will generally occur. Thus the program will actually be attempting to compute a good approximation to the solution of a system of the form $(A + E)\mathbf{x} = \hat{\mathbf{b}}$. An algorithm is said to be stable if it will produce a good approximation to a slightly perturbed problem. Algorithms that ordinarily would converge to the solution using exact arithmetic could very well fail to be stable, owing to the growth of error in the algebraic processes. Even using a stable algorithm, we may encounter problems that are very sensitive to perturbations. For example, if A is "nearly singular," the exact solutions of $A\mathbf{x} = \mathbf{b}$ and $(A + E)\mathbf{x} = \mathbf{b}$ may vary greatly even though all the entries of E are small. The major part of this chapter is devoted to numerical methods for solving linear systems. We shall pay particular attention to the growth of error and to the sensitivity of systems to small changes.

Another problem that is very important in numerical applications is the problem of finding the eigenvalues of a matrix. Two iterative methods for computing eigenvalues are presented in Section 8. The second of these methods is the powerful QR algorithm, which makes use of the special types of orthogonal transformations presented in Section 5.

In many matrix computations it is important to know whether or not a matrix is close to being singular. In Section 5 the concept of the singular values of a matrix is introduced. The smallest nonzero singular value can be used as a measure of how close the matrix is to a matrix of lower rank. In Section 9 we show how the singular values can be used in the solution of least squares problems. The Golub–Reinsch algorithm for computing the singular values is also presented in this section.

1. FLOATING-POINT NUMBERS

In solving a numerical problem on the computer, we do not usually expect to get the exact answer. Some amount of error is inevitable. Roundoff error may occur initially when the data are represented in the finite number system of the computer. Further roundoff error may occur whenever arithmetic operations are used. Like a cancer, these errors may grow to such an extent that the computed solution may be completely unreliable. To avoid this, one must understand how computational errors occur. To do this, one must be familiar with the type of numbers used by the computer.

Definition. A **floating-point number** in base b is a number of the form

$$\pm\left(\frac{d_1}{b} + \frac{d_2}{b^2} + \cdots + \frac{d_t}{b^t}\right) \times b^e$$

where $t, d_1, d_2, \ldots, d_t, b, e$ are all integers and

$$0 \leqslant d_i \leqslant b - 1 \qquad i = 1, \ldots, t$$

The integer t refers to the number of digits and this depends on the word length of the computer. The exponent e is restricted to be within certain bounds, $L \leqslant e \leqslant U$, which also depend on the particular computer. Most computers use base 2, although some use other bases such as 8 or 16. Hand calculators generally use base 10.

EXAMPLE 1. The following are five-digit decimal (base 10) floating-point numbers:

$$0.53216 \times 10^{-4}$$
$$-0.81724 \times 10^{21}$$
$$0.00112 \times 10^{8}$$
$$0.11200 \times 10^{6}$$

Note that the numbers 0.00112×10^8 and 0.11200×10^6 are equal. Thus the floating-point representation of a number need not be unique. Floating-point numbers that are written with no leading zeros are said to be normalized.

EXAMPLE 2. $(0.236)_8 \times 8^2$ and $(0.132)_8 \times 8^4$ are normalized three-digit base 8 floating-point numbers. Here $(0.236)_8$ represents

$$\frac{2}{8} + \frac{3}{8^2} + \frac{6}{8^3}$$

Thus $(0.236)_8 \times 8^2$ is the base 8 floating-point representation of the decimal number 19.75. Similarly,

$$(0.132)_8 \times 8^4 = \left(\frac{1}{8} + \frac{3}{8^2} + \frac{2}{8^3}\right) \times 8^4 = 720$$

To better understand the type of number system we are working with, it may help to look at a very simple example.

EXAMPLE 3. Suppose that $t = 1$, $L = -1$, $U = 1$, and $b = 10$. There are altogether 55 one-digit decimal floating-point numbers in this system. These are

$$0, \pm 0.1 \times 10^{-1}, \pm 0.2 \times 10^{-1}, \ldots, \pm 0.9 \times 10^{-1}$$
$$\pm 0.1 \times 10^{0}, \pm 0.2 \times 10^{0}, \ldots, \pm 0.9 \times 10^{0}$$
$$\pm 0.1 \times 10^{1}, \pm 0.2 \times 10^{1}, \ldots, \pm 0.9 \times 10^{1}$$

Although all these numbers lie in the interval $[-9, 9]$, over one third of the numbers have absolute value less than 0.1 and over two thirds of the numbers have absolute value less than 1. Figure 7.1.1 illustrates how the floating-point numbers in the interval $[0, 2]$ are distributed.

Most real numbers have to be rounded off in order to be represented as t-digit floating-point numbers. The difference between the floating-point number x' and the original number x is called the *roundoff error*. The size

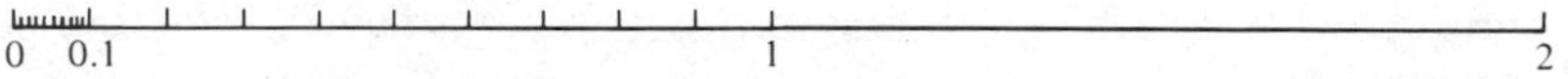

Figure 7.1.1

of the roundoff error is perhaps more meaningful when compared to the size of the original number.

Definition. If x is a real number and x' is its floating-point approximation, then the difference $x' - x$ is called the **absolute error**. Also, the quotient $(x' - x)/x$ is called the **relative error**.

Real number, x	*Four-digit decimal floating-point number,* x'	*Absolute error,* $x' - x$	*Relative error,* $(x' - x)/x$
62,133	0.6213×10^5	-3	$\dfrac{-3}{62{,}133} \approx -4.8 \times 10^{-5}$
0.12658	0.1266×10^0	2×10^{-5}	$\dfrac{1}{6329} \approx 1.6 \times 10^{-4}$
47.213	0.4721×10^2	-3.0×10^{-3}	$\dfrac{-0.003}{47.213} \approx -6.4 \times 10^{-5}$
π	0.3142×10^1	$3.142 - \pi \approx 4 \times 10^{-4}$	$\dfrac{3.142 - \pi}{\pi} \approx 1.3 \times 10^{-4}$

When arithmetic operations are applied to floating-point numbers, additional roundoff errors may occur.

EXAMPLE 4. Let $a' = 0.263 \times 10^4$ and $b' = 0.446 \times 10^1$ be three-digit decimal floating-point numbers. If these numbers are added, the exact sum will be

$$a' + b' = 0.263446 \times 10^4$$

However, the floating-point representation of this sum is 0.263×10^4. This then should be the computed sum. We will denote the floating-point sum by $fl(a' + b')$. The absolute error in the sum is

$$fl(a' + b') - (a' + b') = -4.46$$

and the relative error is

$$\frac{-4.46}{0.26344 \times 10^4} \approx -0.17 \times 10^{-2}$$

The actual value of $a'b'$ is 11,729.8; however $fl(a'b')$, is 0.117×10^5. The absolute error in the product is -29.8 and the relative error is approximately 0.25×10^{-2}. Floating-point subtraction and division can be done in a similar manner.

The relative error in approximating a number x by its floating-point representation x' is usually denoted by the symbol δ. Thus

$$(1) \qquad \delta = \frac{x' - x}{x} \qquad \text{or} \qquad x' = x(1 + \delta)$$

$|\delta|$ can be bounded by a positive constant ϵ, called the *machine precision* or the *machine epsilon*. The machine epsilon is defined to be the smallest floating-point number ϵ for which

$$fl(1 + \epsilon) > 1$$

For example, if the computer uses three-digit decimal floating-point numbers, then

$$fl(1 + 0.499 \times 10^{-2}) = 1$$

while

$$fl(1 + 0.500 \times 10^{-2}) = 1.01$$

In this case the machine epsilon would be 0.500×10^{-2}.

It follows from (1) that if a' and b' are two floating-point numbers, then

$$\begin{aligned} fl(a' + b') &= (a' + b')(1 + \delta_1) \\ fl(a'b') &= (a'b')(1 + \delta_2) \\ fl(a' - b') &= (a' - b')(1 + \delta_3) \\ fl(a' \div b') &= (a' \div b')(1 + \delta_4) \end{aligned}$$

The δ_i's are the relative errors and will all have absolute values less than ϵ. Note that in Example 4, $\delta_1 \approx -0.17 \times 10^{-2}$, $\delta_2 \approx 0.25 \times 10^{-2}$, and $\epsilon = 0.5 \times 10^{-2}$.

If the numbers you are working with involve some slight errors, arithmetic operations may compound these errors. If two numbers agree to k decimal places and one number is subtracted from the other, there will be a loss of significant digits in your answer. In this case the relative error in the difference will be many times as great as the relative error in either of the numbers.

EXAMPLE 5. Let $c = 3.4215298$ and $d = 3.4213851$. Calculate $c - d$ and $1/(c - d)$ using six-digit decimal floating-point arithmetic.

SOLUTION

(i) The first step is to represent c and d by six-digit decimal floating-point numbers.

$$\begin{aligned} c' &= 0.342153 \times 10^1 \\ d' &= 0.342139 \times 10^1 \end{aligned}$$

The relative errors in c and d are

$$\frac{c' - c}{c} \approx 0.6 \times 10^{-7} \qquad \text{and} \qquad \frac{d' - d}{d} \approx 1.4 \times 10^{-6}$$

(ii) $fl(c' - d') = c' - d' = 0.140000 \times 10^{-3}$.
The actual value of $c - d$ is 0.1447×10^{-3}. The absolute and relative errors in approximating $c - d$ by $fl(c' - d')$ are

$$fl(c' - d') - (c - d) = -0.47 \times 10^{-5}$$

and

$$\frac{fl(c' - d') - (c - d)}{c - d} \approx -3.2 \times 10^{-2}$$

Note that the magnitude of the relative error in the difference is more than 10^4 times the relative error in either c or d.

(iii) $fl[1/(c' - d')] = 0.714286 \times 10^4$, and the correct answer to six significant figures is

$$\frac{1}{c - d} \approx 0.691085 \times 10^4$$

The absolute and relative errors are approximately 232 and 0.03.

EXERCISES

1. Find the three-digit decimal floating-point representation of each of the following numbers.
(a) 2312 (b) 32.56
(c) 0.01277 (d) 82,431

2. Find the absolute error and the relative error when each of the real numbers in Exercise 1 is approximated by a three-digit decimal floating-point number.

3. Do each of the following using four-digit decimal floating-point arithmetic and calculate the absolute and relative errors in your answers.
(a) $10{,}420 + 0.0018$ (b) $10{,}424 - 10{,}416$
(c) $0.12347 - 0.12342$ (d) $(3626.6) \cdot (22.656)$

4. Let $x_1 = 94{,}210$, $x_2 = 8631$, $x_3 = 1440$, $x_4 = 133$, and $x_5 = 34$. Calculate each of the following using four-digit decimal floating-point arithmetic.
(a) $(((x_1 + x_2) + x_3) + x_4) + x_5$
(b) $x_1 + ((x_2 + x_3) + (x_4 + x_5))$
(c) $(((x_5 + x_4) + x_3) + x_2) + x_1$

5. How many floating-point numbers are there in the system if $t = 2$, $L = -2$, $U = 2$, and $b = 2$?

2. GAUSSIAN ELIMINATION

In this section we will discuss the problem of solving a system of n linear equations in n unknowns using Gaussian elimination. Gaussian elimination is generally considered to be the most efficient computational method, since it involves the least amount of arithmetic operations.

Gaussian Elimination Without Interchanges

Let $A = A^{(1)} = (a_{ij}^{(1)})$ be a nonsingular matrix. Then A can be reduced to triangular form using row operations I and III. For simplicity, let us assume that the reduction can be done using only row operation III.

$$A^{(1)} = \begin{bmatrix} \boxed{a_{11}^{(1)}} & a_{12}^{(1)} & \cdots & a_{1n}^{(1)} \\ a_{21}^{(1)} & a_{22}^{(1)} & \cdots & a_{2n}^{(1)} \\ \vdots & & & \\ a_{n1}^{(1)} & a_{n2}^{(1)} & \cdots & a_{nn}^{(1)} \end{bmatrix}$$

STEP 1. Let $m_{k1} = a_{k1}^{(1)}/a_{11}^{(1)}$ for $k = 2, \ldots, n$ [by our assumption, $a_{11}^{(1)} \neq 0$]. The first step of the elimination process is to apply row operation III $n - 1$ times to eliminate the entries below the diagonal in the first column of A. Note that m_{k1} is the multiple of the first row that is to be subtracted from the kth row. The new matrix obtained will be

$$A^{(2)} = \begin{bmatrix} a_{11}^{(1)} & a_{12}^{(1)} & \cdots & a_{1n}^{(1)} \\ 0 & a_{22}^{(2)} & \cdots & a_{2n}^{(2)} \\ \vdots & & & \\ 0 & a_{n2}^{(2)} & \cdots & a_{nn}^{(2)} \end{bmatrix}$$

where

$$a_{kj}^{(2)} = a_{kj}^{(1)} - m_{k1}a_{1j}^{(1)} \qquad (2 \leqslant k \leqslant n, 2 \leqslant j \leqslant n)$$

The first step of the elimination process requires $n - 1$ divisions, $(n - 1)^2$ multiplications, and $(n - 1)^2$ additions/subtractions.

STEP 2. If $a_{22}^{(2)} \neq 0$, then it can be used as a pivotal element to eliminate $a_{32}^{(2)}, a_{42}^{(2)}, \ldots, a_{n2}^{(2)}$. For $k = 3, \ldots, n$, set

$$m_{k2} = \frac{a_{k2}^{(2)}}{a_{22}^{(2)}}$$

and subtract m_{k2} times the second row of $A^{(2)}$ from the kth row. The new matrix obtained will be

$$A^{(3)} = \begin{bmatrix} a_{11}^{(1)} & a_{12}^{(1)} & a_{13}^{(1)} & \cdots & a_{1n}^{(1)} \\ 0 & a_{22}^{(2)} & a_{23}^{(2)} & \cdots & a_{2n}^{(2)} \\ 0 & 0 & a_{33}^{(3)} & \cdots & a_{3n}^{(3)} \\ \vdots & \vdots & \vdots & & \vdots \\ 0 & 0 & a_{n3}^{(3)} & \cdots & a_{nn}^{(3)} \end{bmatrix}$$

The second step requires $n-2$ divisions, $(n-2)^2$ multiplications, and $(n-2)^2$ additions/subtractions.

After $n-1$ steps we will end up with a triangular matrix $U = A^{(n)}$. The operation count for the entire process can be determined as follows:

Divisions: $$(n-1) + (n-2) + \cdots + 1 = \frac{n(n-1)}{2}$$

Multiplications: $$(n-1)^2 + (n-2)^2 + \cdots + 1^2 = \frac{n(2n-1)(n-1)}{6}$$

Additions and/or subtractions: $$(n-1)^2 + \cdots + 1^2 = \frac{n(2n-1)(n-1)}{6}$$

The elimination process is summarized in the following algorithm.

Algorithm 7.2.1 (Gaussian Elimination Without Interchanges)

```
For i = 1, 2, . . . , n − 1
    For k = i + 1, . . . , n
        Set m_ki = a_ki^(i) / a_ii^(i)      (provided that a_ii^(i) ≠ 0)
        For j = i + 1, . . . , n
            Set a_kj^(i+1) = a_kj^(i) − m_ki a_ij^(i)
```

To solve the system $A\mathbf{x} = \mathbf{b}$, we could augment A by $\mathbf{b}$. Thus $\mathbf{b}$ would be stored in an extra column of A. The reduction process could then be done using Algorithm 7.2.1 and letting j run from $i+1$ to $n+1$ instead of from $i+1$ to n. The triangular system could then be solved using back substitution.

Most of the work involved in solving a system $A\mathbf{x} = \mathbf{b}$ occurs in the reduction of A to triangular form. Suppose that after having solved $A\mathbf{x} = \mathbf{b}$, we want to solve another system, $A\mathbf{x} = \mathbf{b}_1$. We know the triangular form U from the first system, and consequently we would like to be able to solve the new system without having to go through the entire reduction process again. We can do this if we keep track of the numbers m_{ki} used in

Algorithm 7.2.1. These numbers are called *multipliers*. The multiplier m_{ki} is the multiple of the ith row that is subtracted from the kth row during the ith step of the reduction process. To see how the multipliers can be used to solve $A\mathbf{x} = \mathbf{b}_1$, it is helpful to view the reduction process in terms of matrix multiplications.

Triangular Factorization

The first step of the reduction process involves multiplying A by $n-1$ elementary matrices,

$$A^{(2)} = E_{n1} \cdots E_{31}E_{21}A^{(1)}$$

where

$$E_{k1} = \begin{pmatrix} 1 & & & & & \\ 0 & 1 & & & O & \\ \vdots & & \ddots & & & \\ -m_{k1} & & & 1 & & \\ \vdots & & O & & \ddots & \\ 0 & & & & & 1 \end{pmatrix}$$

Each E_{k1} is nonsingular with

$$E_{k1}^{-1} = \begin{pmatrix} 1 & & & & & \\ 0 & 1 & & & O & \\ \vdots & & \ddots & & & \\ m_{k1} & & & 1 & & \\ \vdots & & O & & \ddots & \\ 0 & & & & & 1 \end{pmatrix}$$

Let

$$M_1 = E_{n1} \cdots E_{31}E_{21} = \begin{pmatrix} 1 & & & & \\ -m_{21} & 1 & & O & \\ -m_{31} & 0 & 1 & & \\ \vdots & & & \ddots & \\ -m_{n1} & 0 & 0 & \cdots & 1 \end{pmatrix}$$

Thus $A^{(2)} = M_1A$. The matrix M_1 is nonsingular and

$$M_1^{-1} = E_{21}^{-1}E_{31}^{-1} \cdots E_{n1}^{-1} = \begin{pmatrix} 1 & & & & \\ m_{21} & 1 & & O & \\ m_{31} & 0 & 1 & & \\ \vdots & & & \ddots & \\ m_{n1} & 0 & 0 & \cdots & 1 \end{pmatrix}$$

Similarly,

$$\begin{aligned} A^{(3)} &= E_{n2} \cdots E_{42}E_{32}A^{(2)} \\ &= M_2A^{(2)} \\ &= M_2M_1A \end{aligned}$$

where

$$M_2 = E_{n2} \cdots E_{32} = \begin{bmatrix} 1 & & & & \\ 0 & 1 & & O & \\ 0 & -m_{32} & 1 & & \\ \vdots & \vdots & \vdots & \ddots & \\ 0 & -m_{n2} & 0 & \cdots & 1 \end{bmatrix}$$

and

$$M_2^{-1} = \begin{bmatrix} 1 & & & O & \\ 0 & 1 & & & \\ 0 & m_{32} & 1 & & \\ \vdots & \vdots & \vdots & \ddots & \\ 0 & m_{n2} & 0 & & 1 \end{bmatrix}$$

At the end of the reduction process, we have

$$U = A^{(n)} = M_{n-1} \cdots M_2M_1A$$

It follows that

$$A = M_1^{-1}M_2^{-1} \cdots M_{n-1}^{-1}U$$

The M_j^{-1}'s multiply out to give the following lower triangular matrix when they are taken in this order:

$$L = M_1^{-1}M_2^{-1} \cdots M_{n-1}^{-1} = \begin{bmatrix} 1 & 0 & 0 & \cdots & 0 \\ m_{21} & 1 & 0 & \cdots & 0 \\ m_{31} & m_{32} & 1 & \cdots & 0 \\ \vdots & & & & \\ m_{n1} & m_{n2} & m_{n3} & \cdots & 1 \end{bmatrix}$$

Thus $A = LU$, where L is lower triangular and U is upper triangular.

EXAMPLE 1. Let

$$A = \begin{bmatrix} 2 & 3 & 1 \\ 4 & 1 & 4 \\ 3 & 4 & 6 \end{bmatrix}$$

The first step is performed using the multipliers $m_{21} = 2$ and $m_{31} = \frac{3}{2}$.

$$M_1A^{(1)} = \begin{bmatrix} 1 & 0 & 0 \\ -2 & 1 & 0 \\ -\frac{3}{2} & 0 & 1 \end{bmatrix}\begin{bmatrix} 2 & 3 & 1 \\ 4 & 1 & 4 \\ 3 & 4 & 6 \end{bmatrix} = \begin{bmatrix} 2 & 3 & 1 \\ 0 & -5 & 2 \\ 0 & -\frac{1}{2} & \frac{9}{2} \end{bmatrix} = A^{(2)}$$

The multiplier in the second step is $m_{32} = \frac{1}{10}$.

$$M_2A^{(2)} = \begin{bmatrix} 1 & 0 & 0 \\ 0 & 1 & 0 \\ 0 & -\frac{1}{10} & 1 \end{bmatrix}\begin{bmatrix} 2 & 3 & 1 \\ 0 & -5 & 2 \\ 0 & -\frac{1}{2} & \frac{9}{2} \end{bmatrix} = \begin{bmatrix} 2 & 3 & 1 \\ 0 & -5 & 2 \\ 0 & 0 & 4.3 \end{bmatrix}$$

Let

$$L = \begin{bmatrix} 1 & 0 & 0 \\ m_{21} & 1 & 0 \\ m_{31} & m_{32} & 1 \end{bmatrix} = \begin{bmatrix} 1 & 0 & 0 \\ 2 & 1 & 0 \\ \frac{3}{2} & \frac{1}{10} & 1 \end{bmatrix}$$

and

$$U = \begin{bmatrix} 2 & 3 & 1 \\ 0 & -5 & 2 \\ 0 & 0 & 4.3 \end{bmatrix}$$

The reader may verify that $LU = A$.

Once A has been reduced to triangular form and the factorization LU has been determined, the system $A\mathbf{x} = \mathbf{b}$ can be solved in two steps.

STEP 1 FORWARD SUBSTITUTION. The system $A\mathbf{x} = \mathbf{b}$ can be written in the form

$$LU\mathbf{x} = \mathbf{b}$$

Let $\mathbf{y} = U\mathbf{x}$. It follows that

$$L\mathbf{y} = LU\mathbf{x} = \mathbf{b}$$

Thus one can find $\mathbf{y}$ by solving the lower triangular system

$$\begin{aligned}
&y_1 &&= b_1\\
&m_{21}y_1 + y_2 &&= b_2\\
&m_{31}y_1 + m_{32}y_2 + y_3 &&= b_3\\
&\vdots\\
&m_{n1}y_1 + m_{n2}y_2 + m_{n3}y_3 + \cdots + y_n &&= b_n
\end{aligned}$$

It follows from the first equation that $y_1 = b_1$. This value can be used in the second equation to solve for y_2. The values of y_1 and y_2 can be used in the third equation to solve for y_3, and so on. This method of solving a lower triangular system is called *forward substitution.*

STEP 2 BACK SUBSTITUTION. Once $\mathbf{y}$ has been determined, we need only solve the upper triangular system $U\mathbf{x} = \mathbf{y}$ to find the solution $\mathbf{x}$ to the system. The upper triangular system is solved by back substitution.

EXAMPLE 2. Solve the system

$$\begin{aligned}
2x_1 + 3x_2 + x_3 &= -4\\
4x_1 + x_2 + 4x_3 &= 9\\
3x_1 + 4x_2 + 6x_3 &= 0
\end{aligned}$$

SOLUTION. The coefficient matrix for this system is the matrix A in Example 1. Since L and U have been determined, the system can be solved using forward and back substitution.

$$\left[\begin{array}{ccc|c} 1 & 0 & 0 & -4\\ 2 & 1 & 0 & 9\\ \frac{3}{2} & \frac{1}{10} & 1 & 0 \end{array}\right] \qquad \begin{aligned} y_1 &= -4\\ y_2 &= 9 - 2y_1 = 17\\ y_3 &= 0 - \tfrac{3}{2}y_1 - \tfrac{1}{10}y_2 = 4.3 \end{aligned}$$

$$\left[\begin{array}{ccc|c} 2 & 3 & 1 & -4\\ 0 & -5 & 2 & 17\\ 0 & 0 & 4.3 & 4.3 \end{array}\right] \qquad \begin{aligned} 2x_1 + 3x_2 + x_3 &= -4\\ -5x_2 + 2x_3 &= 17\\ 4.3x_3 &= 4.3 \end{aligned} \qquad \begin{aligned} x_1 &= 2\\ x_2 &= -3\\ x_3 &= 1 \end{aligned}$$

The solution to the system is $\mathbf{x} = (2, -3, 1)^T$.

Algorithm 7.2.2 (Forward and Back Substitution)

For $k = 1, \ldots, n$
- *Set* $y_k = b_k - \sum_{i=1}^{k-1} m_{ki}y_i$

For $k = n, n-1, \ldots, 1$
- *Set* $x_k = \dfrac{y_k - \sum_{j=k+1}^{n} u_{kj}x_j}{u_{kk}}$

Operation Count

Algorithm 7.2.2 requires n divisions, $n(n-1)$ multiplications, and $n(n-1)$ additions/subtractions. The total operation count for solving a system $A\mathbf{x} = \mathbf{b}$ using Algorithms 7.2.1 and 7.2.2 is then

$$\text{Multiplications/divisions:} \quad \tfrac{1}{3}n^3 + n^2 - \tfrac{1}{3}n$$

$$\text{Additions/subtractions:} \quad \tfrac{1}{3}n^3 + n^2 - \tfrac{4}{3}n$$

In both cases the $\frac{1}{3}n^3$ is the dominant term. We will say that solving a system by Gaussian elimination involves roughly $\frac{1}{3}n^3$ multiplications/divisions and $\frac{1}{3}n^3$ additions/subtractions.

Storage

It is not necessary to store the multipliers in a separate matrix L. Each multiplier m_{ki} can be stored in the matrix A in place of the entry $a_{ki}^{(i)}$ eliminated. At the end of the reduction process, A is being used to store the m_{ki}'s and the u_{ij}'s.

$$\begin{bmatrix} u_{11} & u_{12} & \cdots & u_{1,n-1} & u_{1n} \\ m_{21} & u_{22} & \cdots & u_{2,n-1} & u_{2n} \\ \vdots & & & & \\ m_{n1} & m_{n2} & \cdots & m_{n,n-1} & u_{nn} \end{bmatrix}$$

Algorithm 7.2.1 breaks down if at any step $a_{kk}^{(k)}$ is 0. If this happens, it is necessary to perform row interchanges. In the next section we will see how to incorporate interchanges into our elimination algorithm.

EXERCISES

1. Let

$$A = \begin{bmatrix} 1 & 1 & 1 \\ 2 & 4 & 1 \\ -3 & 1 & -2 \end{bmatrix}$$

Factor A into a product LU, where L is lower triangular with 1's along the diagonal and U is upper triangular.

2. Let A be the matrix in Exercise 1. Solve $A\mathbf{x} = \mathbf{b}$ for each of the following choices of $\mathbf{b}$.
(a) $(4, 3, -13)^T$ (b) $(3, 1, -10)^T$
(c) $(7, 23, 0)^T$

3. Let A and B be $n \times n$ matrices and let $\mathbf{x} \in R^n$.
(a) How many scalar additions and multiplications are necessary to compute the product $A\mathbf{x}$?
(b) How many scalar additions and multiplications are necessary to compute the product AB?

4. Let E_{ki} be the elementary matrix formed by subtracting α times the ith row of the identity matrix from the kth row.
(a) Show that $E_{ki} = I - \alpha \mathbf{e}_k \mathbf{e}_i^T$.
(b) Let $E_{ji} = I - \beta \mathbf{e}_j \mathbf{e}_i^T$. Show that $E_{ji}E_{ki} = I - (\alpha \mathbf{e}_k + \beta \mathbf{e}_j)\mathbf{e}_i^T$.

5. Let A be a nonsingular $n \times n$ matrix that can be reduced to triangular form without row interchanges.
(a) Show that A can be factored into a product LDU, where L is lower triangular, D is diagonal, and U is upper triangular and the diagonal elements of L and U are all 1's.
(b) Show that if A has two such factorizations $L_1D_1U_1$ and $L_2D_2U_2$, then $L_1 = L_2$, $D_1 = D_2$, and $U_1 = U_2$. (*HINT:* L_2^{-1} is lower triangular and U_1^{-1} is upper triangular. Compare both sides of the equation $D_2^{-1}L_2^{-1}L_1D_1 = U_2U_1^{-1}$.)
(c) Show that if $A = LDU$ is symmetric, then $U = L^T$.

6. Write an algorithm for solving the tridiagonal system

$$\begin{bmatrix} a_1 & b_1 & & & \\ c_1 & a_2 & \cdot & & \\ & \cdot & \ddots & \cdot & \\ & & \cdot & a_{n-1} & b_{n-1} \\ & & & c_{n-1} & a_n \end{bmatrix} \begin{bmatrix} x_1 \\ x_2 \\ \vdots \\ x_{n-1} \\ x_n \end{bmatrix} = \begin{bmatrix} d_1 \\ d_2 \\ \vdots \\ d_{n-1} \\ d_n \end{bmatrix}$$

using Gaussian elimination with the diagonal elements as pivots. How many additions/subtractions and multiplications/divisions are necessary?

7. Let $A = LU$, where L is lower triangular with 1's on the diagonal and U is upper triangular.
(a) How many scalar additions and multiplications are necessary to solve $L\mathbf{y} = \mathbf{e}_j$ using forward substitution?
(b) How many additions/subtractions and multiplications/divisions are necessary to solve $A\mathbf{x} = \mathbf{e}_j$? The solution $\mathbf{x}_j$ to $A\mathbf{x} = \mathbf{e}_j$ will be the jth column of A^{-1}.

(c) Given the factorization $A = LU$, how many additional multiplications/divisions and additions/subtractions are needed to compute A^{-1}?

8. Suppose A^{-1} and the LU factorization of A have already been determined. How many scalar additions and multiplications are necessary to compute $A^{-1}\mathbf{b}$? Compare this with the number of operations required to solve $LU\mathbf{x} = \mathbf{b}$ using Algorithm 7.2.2. Suppose that we have a number of systems to solve with the same coefficient matrix A. Is it worthwhile to compute A^{-1}? Explain.

3. PIVOTING STRATEGIES

In this section we will present an algorithm for Gaussian elimination with row interchanges. At each step of the algorithm it will be necessary to choose a pivotal row. One can often avoid unnecessarily large error accumulations by choosing the pivotal rows in a reasonable manner.

Gaussian Elimination with Interchanges

Consider the following example.

EXAMPLE 1. Let

$$A = \begin{bmatrix} 6 & -4 & 2 \\ 4 & 2 & 1 \\ 2 & -1 & 1 \end{bmatrix}$$

We wish to reduce A to triangular form using row operations I and III. To keep track of the interchanges, we will use a row vector $\mathbf{p}$. The coordinates of $\mathbf{p}$ will be denoted by $p(1), p(2), p(3)$. Initially, we set $\mathbf{p} = (1, 2, 3)$. Suppose that at the first step of the reduction process, the third row is chosen as the pivotal row. Instead of interchanging the first and third rows, we will interchange the first and third entries of $\mathbf{p}$. Setting $p(1) = 3$ and $p(3) = 1$, the vector $\mathbf{p}$ becomes $(3, 2, 1)$. The vector $\mathbf{p}$ is used to keep track of the reordering of the rows. We can think of $\mathbf{p}$ as a renumbering of the rows. The actual physical reordering of the rows can be deferred until the end of the reduction process.

$$\begin{array}{c} \text{row} \\ p(3) \\ p(2) \\ p(1) \end{array} \quad \begin{bmatrix} \boxed{6} & -4 & 2 \\ \boxed{4} & 2 & 1 \\ \textcircled{2} & -1 & 1 \end{bmatrix} \rightarrow \begin{bmatrix} 0 & -1 & -1 \\ 0 & 4 & -1 \\ 2 & -1 & 1 \end{bmatrix}$$

If at the second step row $p(3)$ is chosen as the pivotal row, the entries of $p(3)$ and $p(2)$ are switched. The final step of the elimination process is then carried out.

$$\begin{matrix} p(2) \\ p(3) \\ p(1) \end{matrix} \begin{bmatrix} 0 & \textcircled{-1} & -1 \\ 0 & \boxed{4} & -1 \\ 2 & -1 & 1 \end{bmatrix} \to \begin{bmatrix} 0 & -1 & -1 \\ 0 & 0 & -5 \\ 2 & -1 & 1 \end{bmatrix}$$

If the rows of this last matrix are reordered in the order $(p(1), p(2), p(3)) = (3, 1, 2)$, then the resulting matrix will be in triangular form.

$$\begin{matrix} p(1) = 3 \\ p(2) = 1 \\ p(3) = 2 \end{matrix} \quad \begin{bmatrix} 2 & -1 & 1 \\ 0 & -1 & -1 \\ 0 & 0 & -5 \end{bmatrix}$$

Had the rows been written in the order (3, 1, 2) to begin with, then the reduction would have been exactly the same except there would have been no need for interchanges. Reordering the rows of A in the order (3, 1, 2) is the same as premultiplying A by the permutation matrix

$$P = \begin{bmatrix} 0 & 0 & 1 \\ 1 & 0 & 0 \\ 0 & 1 & 0 \end{bmatrix}$$

Let us perform the reduction on A and PA simultaneously and compare the results. The multipliers used in the reduction process were 3, 2, and -4. These will be stored in the place of the terms eliminated and underlined to distinguish them from the other entries of the matrix.

$$A = \begin{bmatrix} 6 & -4 & 2 \\ 4 & 2 & 1 \\ 2 & -1 & 1 \end{bmatrix} \qquad PA = \begin{bmatrix} 2 & -1 & 1 \\ 6 & -4 & 2 \\ 4 & 2 & 1 \end{bmatrix}$$

$$\downarrow \qquad\qquad\qquad\qquad \downarrow$$

$$\begin{bmatrix} \underline{3} & -1 & -1 \\ \underline{2} & 4 & -1 \\ 2 & -1 & 1 \end{bmatrix} \qquad \begin{bmatrix} 2 & -1 & 1 \\ \underline{3} & -1 & -1 \\ \underline{2} & 4 & -1 \end{bmatrix}$$

$$\downarrow \qquad\qquad\qquad\qquad \downarrow$$

$$\begin{bmatrix} \underline{3} & -1 & -1 \\ \underline{2} & -\underline{4} & -5 \\ 2 & -1 & 1 \end{bmatrix} \qquad \begin{bmatrix} 2 & -1 & 1 \\ \underline{3} & -1 & -1 \\ \underline{2} & -\underline{4} & -5 \end{bmatrix}$$

After reordering,

$$\begin{bmatrix} 2 & -1 & 1 \\ \underline{3} & -1 & -1 \\ \underline{2} & \underline{-4} & -5 \end{bmatrix}$$

the results are the same. The matrix PA can be factored into a product LU, where

$$L = \begin{bmatrix} 1 & 0 & 0 \\ 3 & 1 & 0 \\ 2 & -4 & 1 \end{bmatrix} \quad \text{and} \quad U = \begin{bmatrix} 2 & -1 & 1 \\ 0 & -1 & -1 \\ 0 & 0 & -5 \end{bmatrix}$$

On the computer it is not necessary to actually interchange the rows of A. We simply treat row $p(k)$ as the kth row and use $a_{p(k),j}$ in place of $a_{k,j}$.

Algorithm 7.3.1 (Gaussian Elimination with Interchanges)

For $i = 1, \ldots, n$
  Set $p(i) = i$

For $i = 1, \ldots, n$
  (1) *Choose a pivot element* $a_{p(j),i}$ *from the elements*

$$a_{p(i),i}, a_{p(i+1),i}, \ldots, a_{p(n),i}$$

  (*Strategies for doing this will be discussed later in the section.*)
  (2) *Switch the* ith *and* jth *entries of* p.
  (3) *For* $k = i + 1, \ldots, n$
    Set $m_{p(k),i} = a_{p(k),i}/a_{p(i),i}$
    For $j = i + 1, \ldots, n$
      Set $a_{p(k),j} = a_{p(k),j} - m_{p(k),i}a_{p(i),j}$

Remarks

1. The multiplier $m_{p(k),i}$ is stored in the position of the element $a_{p(k),i}$ being eliminated.
2. The vector **p** can be used to form a permutation matrix P whose ith row is the $p(i)$th row of the identity matrix.
3. The matrix PA can be factored into a product LU, where

$$l_{ki} = \begin{cases} m_{p(k),i} & \text{if } k > i \\ 1 & \text{if } k = i \\ 0 & \text{if } k < i \end{cases} \quad \text{and} \quad u_{ki} = \begin{cases} a_{p(k),i} & \text{if } k \leqslant i \\ 0 & \text{if } k > i \end{cases}$$

4. Since P is nonsingular, the system $A\mathbf{x} = \mathbf{b}$ is equivalent to the system $PA\mathbf{x} = P\mathbf{b}$. Let $\hat{\mathbf{b}} = P\mathbf{b}$. Since $PA = LU$, it follows that the system is equivalent to

$$LU\mathbf{x} = \hat{\mathbf{b}}$$

5. If $PA = LU$, then $A = P^{-1}LU = P^TLU$.

It follows from Remarks 4 and 5 that if $A = P^TLU$, the system $A\mathbf{x} = \mathbf{b}$ can be solved in three steps:

STEP 1 (REORDERING). Reorder the entries of $\mathbf{b}$ to form $\hat{\mathbf{b}} = P\mathbf{b}$.

STEP 2 (FORWARD SUBSTITUTION). Solve the system $L\mathbf{y} = \hat{\mathbf{b}}$ for $\mathbf{y}$.

STEP 3 (BACK SUBSTITUTION). Solve $U\mathbf{x} = \mathbf{y}$.

EXAMPLE 2. Solve the system

$$\begin{aligned} 6x_1 - 4x_2 + 2x_3 &= -2 \\ 4x_1 + 2x_2 + x_3 &= 4 \\ 2x_1 - x_2 + x_3 &= -1 \end{aligned}$$

SOLUTION. The coefficient matrix of this system is the matrix A from Example 1. P, L, and U have already been determined and they can be used to solve the system as follows:

STEP 1. $\hat{\mathbf{b}} = P\mathbf{b} = (-1, -2, 4)^T$

STEP 2.

$$\begin{aligned} y_1 \qquad\quad &= -1 & y_1 &= -1 \\ 3y_1 + y_2 \qquad &= -2 & y_2 &= -2 + 3 = 1 \\ 2y_1 - 4y_2 + y_3 &= 4 & y_3 &= 4 + 2 + 4 = 10 \end{aligned}$$

STEP 3.

$$\begin{aligned} 2x_1 - x_2 + x_3 &= -1 & x_1 &= 1 \\ -x_2 - x_3 &= 1 & x_2 &= 1 \\ -5x_3 &= 10 & x_3 &= -2 \end{aligned}$$

The solution to the system is $\mathbf{x} = (1, 1, -2)^T$.

In Algorithm 7.3.1 we did not specify how to choose the pivotal row at each step. It is important that the pivotal rows be chosen in a reasonable manner. If this is not done, one may end up with unreliable computational results. In the following example we will solve a system using three-digit decimal arithmetic. No row interchanges will be used. For comparison we will also solve the system using exact arithmetic.

EXAMPLE 3. Solve the following system using both three-digit decimal floating-point arithmetic and exact arithmetic. Compare the solutions.

$$\begin{aligned} x_1 - 2.01x_2 + \quad (0.5 \times 10^{-2})x_3 &= 0 \\ -x_1 + 2.02x_2 + (0.545 \times 10^{-2})x_3 &= 0 \\ 2x_1 + 1.98x_2 + \qquad\qquad 6.33x_3 &= 1 \end{aligned}$$

SOLUTION

$$\left[\begin{array}{ccc|c} 1.00 & -2.01 & 0.500 \times 10^{-2} & 0 \\ -1.00 & 2.02 & 0.545 \times 10^{-2} & 0 \\ 2.00 & 1.98 & 6.33 & 1 \end{array}\right]$$

three-digit floating point arithmetic

$$\left[\begin{array}{ccc|c} 1 & -2.01 & 0.005 & 0 \\ 0 & 0.01 & 0.0105 & 0 \\ 0 & 6.00 & 6.32 & 1 \end{array}\right] \rightarrow \left[\begin{array}{ccc|c} 1 & -2.01 & 0.005 & 0 \\ 0 & 0.01 & 0.0105 & 0 \\ 0 & 0 & 0.02 & 1 \end{array}\right]$$

exact arithmetic

$$\left[\begin{array}{ccc|c} 1 & -2.01 & 0.005 & 0 \\ 0 & 0.01 & 0.01045 & 0 \\ 0 & 6.00 & 6.32 & 1 \end{array}\right] \rightarrow \left[\begin{array}{ccc|c} 1 & -2.01 & 0.005 & 0 \\ 0 & 0.01 & 0.01045 & 0 \\ 0 & 0 & 0.05 & 1 \end{array}\right]$$

The multipliers used in the reduction process were $m_{21} = -1$, $m_{31} = 2$, and $m_{32} = 600$. The computer solution $\mathbf{x}'$ using three-digit decimal floating-point arithmetic is $(-106, -52.5, 50)^T$, while the exact solution is $\mathbf{x} = (-42.109, -20.9, 20)^T$. The relative error in each coordinate of the solution is greater than or equal to 1.5.

Let us reexamine Example 1 to see how the error was accumulated and if the accumulation could have been avoided. After the first step of the elimination process, the only error was in the (3, 2) entry of the matrix.

$$\begin{aligned} &\text{Calculated entry:} & (a_{32}^{(2)})' &= 0.0105 \\ &\text{Exact entry:} & a_{32}^{(2)} &= 0.01045 \\ &\text{Error:} & \epsilon &= (a_{32}^{(2)})' - a_{32}^{(2)} = 0.00005 \end{aligned}$$

In the next step of the algorithm, $(a_{32}^{(2)})'$ was multiplied by $m_{32} = 600$.

$$\begin{aligned} m_{32}(a_{32}^{(2)})' &= m_{32}(a_{32}^{(2)} + \epsilon) = 600a_{32}^{(2)} + 600\epsilon \\ (a_{33}^{(3)})' &= a_{33}^{(2)} - m_{32}(a_{32}^{(2)})' \\ &= \left[a_{33}^{(2)} - 600a_{32}^{(2)}\right] - (600\epsilon) \\ &= 0.05 - 0.03 \end{aligned}$$

The error in $(a_{33}^{(3)})'$ is a result of multiplying the error in $(a_{32}^{(2)})'$ by 600. Because of the error in $(a_{33}^{(3)})'$, the computed value of x_3 is inaccurate. This, in turn, causes the computed values of x_2 and x_1 to be inaccurate.

Suppose that at the second step of the elimination process in Example 1, the second and third rows had been interchanged.

$$\left[\begin{array}{ccc|c} 1 & -2.01 & 0.005 & 0 \\ 0 & 0.01 & 0.0105 & 0 \\ 0 & 6.00 & 6.32 & 1 \end{array}\right] \rightarrow \left[\begin{array}{ccc|c} 1 & -2.01 & 0.005 & 0 \\ 0 & 6.00 & 6.32 & 1 \\ 0 & 0.01 & 0.0105 & 0 \end{array}\right]$$

The new second row is used as a pivotal row with multiplier $m_{32} = 0.00167$. If there were an error in the (3, 2) entry of the matrix, it would be multiplied by 0.00167 in this step of the algorithm.

$$\left[\begin{array}{ccc|c} 1 & -2.01 & 0.005 & 0 \\ 0 & 6.00 & 6.32 & 1 \\ 0 & 0 & -0.0001 & -0.00167 \end{array}\right]$$

The computed solution is

$$\mathbf{x}' = (-35.3,\ -17.5,\ 16.7)$$

The relative error in each coordinate is about -0.16.

Multipliers that are of modulus greater than 1 generally contribute to the propagation of error, whereas multipliers whose magnitudes are less than 1 generally retard the growth of error. By careful selection of the pivotal elements, one can keep the multipliers less than 1 in modulus at each step. The most commonly used method for doing this is called *partial pivoting*.

Partial Pivoting

At the ith step of the reduction process, there are $n - i + 1$ candidates for the pivotal element:

$$a_{p(i),\,i},\ a_{p(i+1),\,i},\ \ldots,\ a_{p(n),\,i}$$

Choose the candidate $a_{p(j),\,i}$ with maximum modulus

$$|a_{p(j),\,i}| = \max_{i \leqslant k \leqslant n} |a_{p(k),\,i}|$$

and interchange the ith and jth entries of $\mathbf{p}$. The pivot element $a_{p(i),\,i}$ has the property

$$|a_{p(i),\,i}| \geqslant |a_{p(k),\,i}|$$

for $k = i + 1, \ldots, n$. Thus the multipliers will all satisfy

$$|m_{p(k),\,i}| = \left|\frac{a_{p(k),\,i}}{a_{p(i),\,i}}\right| \leqslant 1$$

One could always carry things one step further and do *complete pivoting*. In complete pivoting the pivotal element is chosen to be the element of maximum modulus among all the elements in the remaining rows and columns. In this case one must keep track of both the rows and the columns. At the ith step the element $a_{p(j)q(k)}$ is chosen so that

$$|a_{p(j)q(k)}| = \max_{\substack{i \leqslant s \leqslant n \\ i \leqslant t \leqslant n}} |a_{p(s)q(t)}|$$

The ith and jth entries of $\mathbf{p}$ are interchanged and the ith and kth entries of $\mathbf{q}$ are interchanged. The new pivot element is $a_{p(i)q(i)}$. The major drawback to complete pivoting is that at each step one must search for a pivotal element among $(n - i + 1)^2$ elements of A. This may be too costly in terms of computer time.

EXERCISES

1. Let

$$A = \begin{bmatrix} 0 & 3 & 1 \\ 1 & 2 & -2 \\ 2 & 5 & 4 \end{bmatrix} \quad \text{and} \quad \mathbf{b} = \begin{pmatrix} 1 \\ 7 \\ -1 \end{pmatrix}$$

(a) Reorder the rows of $(A|\mathbf{b})$ in the order (2, 3, 1) and then solve this reordered system.

(b) Factor A into a product P^TLU, where P is the permutation matrix corresponding to the reordering in part (a).

2. Let A be the matrix in Exercise 1. Use the factorization P^TLU to solve $A\mathbf{x} = \mathbf{c}$ for each of the following choices of $\mathbf{c}$.

(a) $(8, 1, 20)^T$
(b) $(-9, -2, -7)^T$
(c) $(4, 1, 11)^T$

3. The exact solution to the system

$$\begin{aligned} 0.6000x_1 + 2000x_2 &= 2003 \\ 0.3076x_1 - 0.4010x_2 &= 1.137 \end{aligned}$$

is $\mathbf{x} = (5, 1)^T$. Suppose that the calculated value of x_2 is $x_2' = 1 + \epsilon$. Use this value in the first equation and solve for x_1. What will the error be? Calculate the relative error in x_1 if $\epsilon = 0.001$.

4. Solve the system in Exercise 3 using four-digit decimal floating-point arithmetic and Gaussian elimination with partial pivoting.

5. Solve the system in Exercise 3 using four-digit decimal floating-point arithmetic and Gaussian elimination with complete pivoting.

6. Use four-digit decimal floating-point arithmetic and scale the system in Exercise 3 by multiplying the first equation through by $1/2000$ and the second equation through by $1/0.4010$. Solve the scaled system using partial pivoting.

4. THE CONDITION NUMBER OF A MATRIX

In this section we will be concerned with the accuracy of computed solutions to linear systems. How accurate can we expect the computed solutions to be, and how can one test their accuracy? The answer to these questions depends largely upon how sensitive the coefficient matrix of the system is to small changes. Consider the following example.

EXAMPLE 1. Solve the following system.

$$\begin{aligned} 2.0000x_1 + 2.0000x_2 &= 6.0000 \\ 2.0000x_1 + 2.0005x_2 &= 6.0010 \end{aligned} \tag{1}$$

If one uses five-digit decimal floating-point arithmetic, the computed solution will be the exact solution $\mathbf{x} = (1, 2)^T$. Suppose, however, that one is forced to use four-digit decimal floating-point numbers. Thus in place of (1), we have

$$\begin{aligned} 2.000x_1 + 2.000x_2 &= 6.000 \\ 2.000x_1 + 2.001x_2 &= 6.001 \end{aligned} \tag{2}$$

The computed solution to system (2) is the exact solution $\mathbf{x}' = (2, 1)^T$.

The systems (1) and (2) agree except for the coefficient a_{22}. The relative error in this coefficient is

$$\frac{a'_{22} - a_{22}}{a_{22}} \approx 0.00025$$

However, the relative errors in the coordinates of the solutions $\mathbf{x}$ and $\mathbf{x}'$ are

$$\frac{x'_1 - x_1}{x_1} = 1.0 \qquad \text{and} \qquad \frac{x'_2 - x_2}{x_2} = -0.5$$

Definition. A matrix A is said to be **ill conditioned** if a relatively small change in one or more entries of A results in a relatively large change in the solution to $A\mathbf{x} = \mathbf{b}$. A is said to be **well conditioned** if a relatively small change in one of the entries of A results in a relatively small change in the solution to $A\mathbf{x} = \mathbf{b}$.

If the matrix A is ill conditioned, the computed solution to $A\mathbf{x} = \mathbf{b}$ generally will not be very accurate. Even if the entries of A can be represented exactly as floating-point numbers, small roundoff errors occurring in the reduction process may have a drastic effect on the computed solution. On the other hand, if the matrix is well conditioned and the proper pivoting strategy is used, one should be able to compute solutions quite accurately. In general, the accuracy of the solution depends upon the conditioning of the matrix. If we could measure the conditioning of A, this measure could be used to derive a bound for the relative error in the computed solution.

Let A be an $n \times n$ nonsingular matrix and consider the system $A\mathbf{x} = \mathbf{b}$. If $\mathbf{x}$ is the exact solution to the system and $\mathbf{x}'$ is the calculated solution, then the error can be represented by the vector $\mathbf{e} = \mathbf{x} - \mathbf{x}'$. If $\| \; \|$ is a norm on R^n, then $\|\mathbf{e}\|$ is a measure of the absolute error and $\|\mathbf{e}\|/\|\mathbf{x}\|$ is a measure of the relative error. In general, we have no way of determining the exact values of $\|\mathbf{e}\|$ and $\|\mathbf{e}\|/\|\mathbf{x}\|$. One possible way of testing the accuracy of $\mathbf{x}'$ is to put it back into the original system and to see how close $\mathbf{b}' = A\mathbf{x}'$ comes to $\mathbf{b}$. The vector

$$\mathbf{r} = \mathbf{b} - \mathbf{b}' = \mathbf{b} - A\mathbf{x}'$$

is called the *residual* and can be easily calculated. The quantity

$$\frac{\|\mathbf{b} - A\mathbf{x}'\|}{\|\mathbf{b}\|} = \frac{\|\mathbf{r}\|}{\|\mathbf{b}\|}$$

is called the *relative residual*. Is the relative residual a good estimate of the relative error? The answer to this depends upon the conditioning of A. In Example 1 the residual for the computed solution $\mathbf{x}' = (2, 1)$ is

$$\mathbf{r} = \mathbf{b} - A\mathbf{x}' = (0, 0.0005)^T$$

The relative residual in terms of the ∞ norm is

$$\frac{\|\mathbf{r}\|_\infty}{\|\mathbf{b}\|_\infty} = \frac{0.0005}{6.0010} \approx 0.000083$$

and the relative error is given by

$$\frac{\|\mathbf{e}\|_\infty}{\|\mathbf{x}\|_\infty} = 0.5$$

The relative error is more than 6000 times the relative residual. In general, we will show that if A is ill conditioned, the relative residual may be much smaller than the relative error. On the other hand, for well-conditioned matrices, the relative residual and the relative error are quite close. To show this, we need to make use of matrix norms. Recall that if $\| \cdot \|$ is a compatible matrix norm on $M_{n,n}$, then for any $n \times n$ matrix C and any vector $\mathbf{y} \in R^n$, we have

$$\|C\mathbf{y}\| \leqslant \|C\| \, \|\mathbf{y}\| \tag{3}$$

Now

$$\mathbf{r} = \mathbf{b} - A\mathbf{x}' = A\mathbf{x} - A\mathbf{x}' = A\mathbf{e}$$

and consequently

$$\mathbf{e} = A^{-1}\mathbf{r}$$

It follows from property (3) that

$$\|\mathbf{e}\| \leqslant \|A^{-1}\| \, \|\mathbf{r}\|$$

and

$$\|\mathbf{r}\| = \|A\mathbf{e}\| \leqslant \|A\| \, \|\mathbf{e}\|$$

Therefore,

$$\frac{\|\mathbf{r}\|}{\|A\|} \leqslant \|\mathbf{e}\| \leqslant \|A^{-1}\| \, \|\mathbf{r}\| \tag{4}$$

Now $\mathbf{x}$ is the exact solution to $A\mathbf{x} = \mathbf{b}$, and hence $\mathbf{x} = A^{-1}\mathbf{b}$. By the same reasoning used to derive (4), we have

$$\frac{\|\mathbf{b}\|}{\|A\|} \leqslant \|\mathbf{x}\| \leqslant \|A^{-1}\| \, \|\mathbf{b}\| \tag{5}$$

It follows from (4) and (5) that

$$\frac{1}{\|A\| \, \|A^{-1}\|} \frac{\|\mathbf{r}\|}{\|\mathbf{b}\|} \leqslant \frac{\|\mathbf{e}\|}{\|\mathbf{x}\|} \leqslant \|A\| \, \|A^{-1}\| \frac{\|\mathbf{r}\|}{\|\mathbf{b}\|}$$

The number $\|A\| \, \|A^{-1}\|$ is called the *condition number* of A and will be denoted by cond (A). Thus

$$\frac{1}{\text{cond}\,(A)} \frac{\|\mathbf{r}\|}{\|\mathbf{b}\|} \leqslant \frac{\|\mathbf{e}\|}{\|\mathbf{x}\|} \leqslant \text{cond}\,(A) \frac{\|\mathbf{r}\|}{\|\mathbf{b}\|} \tag{6}$$

Inequality (6) relates the size of the relative error $\|\mathbf{e}\|/\|\mathbf{x}\|$ to the relative residual $\|\mathbf{r}\|/\|\mathbf{b}\|$. If the condition number is close to 1, the relative error and the relative residual will be close. If the condition number is large, the relative error could be many times as large as the relative residual.

EXAMPLE 2. Let

$$A = \begin{pmatrix} 3 & 3 \\ 4 & 5 \end{pmatrix}$$

Then

$$A^{-1} = \tfrac{1}{3}\begin{pmatrix} 5 & -3 \\ -4 & 3 \end{pmatrix}$$

$\|A\|_\infty = 9$ and $\|A^{-1}\|_\infty = \frac{8}{3}$. (We use $\|\cdot\|_\infty$, because it is easy to calculate.) Thus

$$\text{cond}_\infty (A) = 9 \cdot \tfrac{8}{3} = 24$$

Theoretically, the relative error in the calculated solution to a system $A\mathbf{x} = \mathbf{b}$ could be as much as 24 times the relative residual.

EXAMPLE 3. Suppose that $\mathbf{x}' = (2.0, 0.1)^T$ is the calculated solution to

$$\begin{aligned} 3x_1 + 3x_2 &= 6 \\ 4x_1 + 5x_2 &= 9 \end{aligned}$$

Determine the residual $\mathbf{r}$ and the relative residual $\|\mathbf{r}\|_\infty / \|\mathbf{b}\|_\infty$.

SOLUTION

$$\mathbf{r} = \begin{pmatrix} 6 \\ 9 \end{pmatrix} - \begin{pmatrix} 3 & 3 \\ 4 & 5 \end{pmatrix}\begin{pmatrix} 2.0 \\ 0.1 \end{pmatrix} = \begin{pmatrix} -0.3 \\ 0.5 \end{pmatrix}$$

$$\frac{\|\mathbf{r}\|_\infty}{\|\mathbf{b}\|_\infty} = \frac{0.5}{9} = \frac{1}{18}$$

We can see by inspection that the actual solution to the system in the example above is $\mathbf{x} = \begin{pmatrix} 1 \\ 1 \end{pmatrix}$. The error $\mathbf{e}$ is given by

$$\mathbf{e} = \mathbf{x} - \mathbf{x}' = \begin{pmatrix} -1.0 \\ 0.9 \end{pmatrix}$$

The relative error is given by

$$\frac{\|\mathbf{e}\|_\infty}{\|\mathbf{x}\|_\infty} = \frac{1.0}{1} = 1$$

The relative error is 18 times the relative residual. This is not surprising, since cond $(A) = 24$. The results are similar using $\|\cdot\|_1$. In this case

$$\frac{\|\mathbf{r}\|_1}{\|\mathbf{b}\|_1} = \frac{0.8}{15} = \frac{4}{75} \qquad \text{and} \qquad \frac{\|\mathbf{e}\|_1}{\|\mathbf{x}\|_1} = \frac{1.9}{2} = \frac{19}{20}$$

The condition number of a nonsingular matrix A actually gives us valuable information about the conditioning of A. Let $\hat{A}$ be a new matrix formed by altering the entries of A slightly. Let $E = \hat{A} - A$. Thus $\hat{A} = A + E$, where the entries of E are small relative to the entries of A. A will be ill conditioned if for some such E the solutions to $\hat{A}\mathbf{x} = \mathbf{b}$ and $A\mathbf{x} = \mathbf{b}$ vary greatly. Let $\hat{\mathbf{x}}$ be the exact solution to $\hat{A}\mathbf{x} = \mathbf{b}$ and $\mathbf{y}$ be the solution to $A\mathbf{x} = \mathbf{b}$. The condition number allows us to compare the change in the solution relative to $\hat{\mathbf{x}}$ and the relative change in the matrix A.

$$\mathbf{y} = A^{-1}\mathbf{b} = A^{-1}\hat{A}\hat{\mathbf{x}} = A^{-1}(A + E)\hat{\mathbf{x}} = \hat{\mathbf{x}} + A^{-1}E\hat{\mathbf{x}}$$

Hence

$$\mathbf{y} - \hat{\mathbf{x}} = A^{-1}E\hat{\mathbf{x}}$$

Using inequality (3), we see that

$$\|\mathbf{y} - \hat{\mathbf{x}}\| \leqslant \|A^{-1}\| \, \|E\| \, \|\hat{\mathbf{x}}\|$$

or

$$\frac{\|\mathbf{y} - \hat{\mathbf{x}}\|}{\|\hat{\mathbf{x}}\|} \leqslant \|A^{-1}\| \, \|E\| = \operatorname{cond}(A)\frac{\|E\|}{\|A\|}$$

EXERCISES

1. Let A be the coefficient matrix of the system (1) in Example 1. Find A^{-1} and $\operatorname{cond}_\infty A$.

2. Solve the two systems below and compare the two solutions. Are the coefficient matrices well conditioned? Ill conditioned? Explain.

$$\begin{aligned} 1.0x_1 + 2.0x_2 &= 1.12 \\ 2.0x_1 + 3.9x_2 &= 2.16 \end{aligned} \qquad \begin{aligned} 1.000x_1 + 2.011x_2 &= 1.120 \\ 2.000x_1 + 3.982x_2 &= 2.160 \end{aligned}$$

3. Let

$$A_n = \begin{pmatrix} 1 & 1 \\ 1 & 1 - \dfrac{1}{n} \end{pmatrix}$$

for each positive integer n. Calculate:
(i) A_n^{-1}
(ii) $\operatorname{cond}_\infty (A_n)$
(iii) $\lim_{n\to\infty} \operatorname{cond}_\infty (A_n)$

4. Let A be a nonsingular $n \times n$ matrix. Show that $\operatorname{cond}(A) \geqslant 1$.

5. Let

$$A = \begin{pmatrix} 1 & 0 & 1 \\ 2 & 2 & 3 \\ 1 & 1 & 2 \end{pmatrix}$$

Calculate $\operatorname{cond}_\infty (A) = \|A\|_\infty \|A^{-1}\|_\infty$.

6. Let A be the matrix in Exercise 5 and let $\mathbf{b} = (4, 7, 5)^T$. Suppose that $\mathbf{x}' = (0.8, -1.8, 3.3)^T$ is the calculated solution to $A\mathbf{x} = \mathbf{b}$.
(a) Calculate the residual $\mathbf{r}$ and $\|\mathbf{r}\|_\infty$.

(b) Without computing the exact solution $\mathbf{x}$, find a bound for the relative error $\|\mathbf{e}\|_\infty/\|\mathbf{x}\|_\infty$.

(c) Compute the exact solution $\mathbf{x}$ and the relative error $\|\mathbf{e}\|_\infty/\|\mathbf{x}\|_\infty$.

7. Let

$$A = \begin{bmatrix} -0.50 & 0.75 & -0.25 \\ -0.50 & 0.25 & 0.25 \\ 1.00 & -0.50 & 0.50 \end{bmatrix}$$

Calculate $\text{cond}_1(A) = \|A\|_1 \|A^{-1}\|_1$.

8. Let A be the matrix in Exercise 7 and let

$$\hat{A} = \begin{bmatrix} -0.5 & 0.8 & -0.3 \\ -0.5 & 0.3 & 0.3 \\ 1.0 & -0.5 & 0.5 \end{bmatrix}$$

Let $\mathbf{y}$ and $\hat{\mathbf{x}}$ be the solutions to $A\mathbf{x} = \mathbf{b}$ and $\hat{A}\mathbf{x} = \mathbf{b}$, respectively, for some $\mathbf{b} \in R^3$. Find a bound for the relative error $(\|\hat{\mathbf{x}} - \mathbf{y}\|_1)/\|\hat{\mathbf{x}}\|_1$.

9. Let A and B be nonsingular $n \times n$ matrices. Show that

$$\text{cond}(AB) \leqslant \text{cond}(A)\,\text{cond}(B)$$

5. ORTHOGONAL TRANSFORMATIONS

Orthogonal transformations are one of the most important tools in numerical linear algebra. The types of orthogonal transformations that will be introduced in this section are easy to work with and do not require much storage. Most important, processes that involve orthogonal transformations are inherently stable. For example, let $\mathbf{x} \in R^n$ and $\mathbf{x}' = \mathbf{x} + \mathbf{e}$ be an approximation to $\mathbf{x}$; if Q is an orthogonal matrix, then

$$Q\mathbf{x}' = Q\mathbf{x} + Q\mathbf{e}$$

The error in $Q\mathbf{x}'$ is $Q\mathbf{e}$. With respect to the 2-norm, the vector $Q\mathbf{e}$ is the same size as $\mathbf{e}$:

$$\|Q\mathbf{e}\|_2 = \|\mathbf{e}\|_2$$

Similarly, if $A' = A + E$, then

$$QA = QA + QE$$

and

$$\|QE\|_2 = \|E\|_2$$

When an orthogonal transformation is applied to a vector or a matrix, the error will not grow with respect to the 2-norm.

Elementary Orthogonal Transformations

By an *elementary orthogonal matrix*, we mean a matrix of the form

$$Q = I - 2\mathbf{u}\mathbf{u}^T$$

where $\mathbf{u} \in R^n$ and $\|\mathbf{u}\|_2 = 1$. To see that Q is orthogonal, note that

$$Q^T = (I - 2\mathbf{u}\mathbf{u}^T)^T = I - 2\mathbf{u}\mathbf{u}^T = Q$$

and

$$\begin{aligned} Q^TQ &= Q^2 = (I - 2\mathbf{u}\mathbf{u}^T)(I - 2\mathbf{u}\mathbf{u}^T) \\ &= I - 4\mathbf{u}\mathbf{u}^T + 4\mathbf{u}(\mathbf{u}^T\mathbf{u})\mathbf{u}^T \\ &= I \end{aligned}$$

Thus, if Q is an elementary orthogonal matrix, then

$$Q^T = Q^{-1} = Q$$

The matrix $Q = I - 2\mathbf{u}\mathbf{u}^T$ is completely determined by the unit vector $\mathbf{u}$. Rather than store all n^2 entries of Q, we need store only the vector $\mathbf{u}$. To compute $Q\mathbf{x}$, note that

$$\begin{aligned} Q\mathbf{x} &= (I - 2\mathbf{u}\mathbf{u}^T)\mathbf{x} \\ &= \mathbf{x} - 2\alpha\mathbf{u} \qquad \text{where} \quad \alpha = \mathbf{u}^T\mathbf{x} \end{aligned}$$

The matrix product QA is computed as follows:

$$QA = (Q\mathbf{a}_1, Q\mathbf{a}_2, \ldots, Q\mathbf{a}_n)$$

where

$$Q\mathbf{a}_i = \mathbf{a}_i - 2\alpha_i\mathbf{u} \qquad \alpha_i = \mathbf{u}^T\mathbf{a}_i$$

Elementary orthogonal transformations can be used to obtain a QR factorization of A, and this in turn can be used to solve a linear system $A\mathbf{x} = \mathbf{b}$. As with Gaussian elimination, the elementary matrices are chosen so as to produce zeros in the coefficient matrix. To see how this is done, let us consider the problem of finding a unit vector $\mathbf{u}$ such that

$$(I - 2\mathbf{u}\mathbf{u}^T)\mathbf{x} = (\alpha, 0, \ldots, 0)^T = \alpha\mathbf{e}_1$$

for a given vector $\mathbf{x} \in R^n$.

Householder Transformations

Let $H = I - 2\mathbf{u}\mathbf{u}^T$. If $H\mathbf{x} = \alpha\mathbf{e}_1$, then, since H is orthogonal, it follows that

$$|\alpha| = \|\alpha\mathbf{e}_1\|_2 = \|H\mathbf{x}\|_2 = \|\mathbf{x}\|_2$$

If we take $\alpha = \|\mathbf{x}\|_2$ and $H\mathbf{x} = \alpha\mathbf{e}_1$, then, since H is its own inverse, it follows that

$$\text{(1)} \qquad \mathbf{x} = H(\alpha\mathbf{e}_1) = \alpha(\mathbf{e}_1 - (2u_1)\mathbf{u})$$

Thus

$$\begin{aligned} x_1 &= \alpha(1 - 2u_1^2) \\ x_2 &= -2\alpha u_1 u_2 \\ &\vdots \\ x_n &= -2\alpha u_1 u_n \end{aligned}$$

Solving for the u_i's, we get

$$u_1 = \pm\left(\frac{\alpha - x_1}{2\alpha}\right)^{1/2}$$

$$u_i = \frac{-x_i}{2\alpha u_1} \qquad \text{for} \quad i = 2, \ldots, n$$

If we let

$$u_1 = -\left(\frac{\alpha - x_1}{2\alpha}\right)^{1/2} \qquad \text{and set} \qquad \beta = \alpha(\alpha - x_1)$$

then

$$-2\alpha u_1 = [2\alpha(\alpha - x_1)]^{1/2} = (2\beta)^{1/2}$$

It follows that

$$\begin{aligned} \mathbf{u} &= \left(-\frac{1}{2\alpha u_1}\right)(-2\alpha u_1^2, x_2, \ldots, x_n)^T \\ &= \frac{1}{\sqrt{2\beta}}(x_1 - \alpha, x_2, \ldots, x_n)^T \end{aligned}$$

If we set $\mathbf{v} = (x_1 - \alpha, x_2, \ldots, x_n)^T$, then

$$\|\mathbf{v}\|_2^2 = (x_1 - \alpha)^2 + \sum_{i=2}^{n} x_i^2 = 2\alpha(\alpha - x_1)$$

and hence

$$\|\mathbf{v}\|_2 = \sqrt{2\beta}$$

Thus

$$\mathbf{u} = \frac{1}{\sqrt{2\beta}}\mathbf{v} = \frac{1}{\|\mathbf{v}\|_2}\mathbf{v}$$

In summation, given a vector $\mathbf{x} \in R^n$, if we set

$$\begin{aligned} \alpha &= \|\mathbf{x}\|_2, \qquad \beta = \alpha(\alpha - x_1) \\ \mathbf{v} &= (x_1 - \alpha, x_2, \ldots, x_n) \\ \mathbf{u} &= \frac{1}{\|\mathbf{v}\|_2}\mathbf{v} = \frac{1}{\sqrt{2\beta}}\mathbf{v} \end{aligned}$$

and

$$H = I - 2\mathbf{u}\mathbf{u}^T = I - \frac{1}{\beta}\mathbf{v}\mathbf{v}^T$$

then

$$H\mathbf{x} = \alpha\mathbf{e}_1$$

The matrix H formed in this way is called a *Householder transformation*. The matrix H is determined by the vector $\mathbf{v}$ and the scalar β. For any vector $\mathbf{y} \in R^n$,

$$H\mathbf{y} = \left(I - \frac{1}{\beta}\mathbf{v}\mathbf{v}^T\right)\mathbf{y} = \mathbf{y} - \left(\frac{1}{\beta}\mathbf{v}^T\mathbf{y}\right)\mathbf{v}$$

Rather than store all n^2 entries of H, we need store only $\mathbf{v}$ and β.

Suppose now that we wish to zero out only the last $n - k$ components of a vector $\mathbf{x} = (x_1, \ldots, x_k, x_{k+1}, \ldots, x_n)^T$. To do this, we let $\mathbf{x}^{(1)} = (x_1, \ldots, x_{k-1})^T$ and $\mathbf{x}^{(2)} = (x_k, x_{k+1}, \ldots, x_n)^T$. Let $I^{(1)}$ and $I^{(2)}$ denote the $(k-1) \times (k-1)$ and $(n-k+1) \times (n-k+1)$ identity matrices, respectively. By the methods just described, we can construct a Householder matrix $H_k^{(2)} = I^{(2)} - (1/\beta_k)\mathbf{v}_k\mathbf{v}_k^T$ such that

$$H_k^{(2)}\mathbf{x}^{(2)} = \|\mathbf{x}^{(2)}\|_2\mathbf{e}_1$$

Let

$$H_k = \begin{pmatrix} I^{(1)} & O \\ O & H_k^{(2)} \end{pmatrix}$$

It follows that

$$\begin{aligned} H_k\mathbf{x} &= \begin{pmatrix} I^{(1)} & O \\ O & H_k^{(2)} \end{pmatrix}\begin{pmatrix} \mathbf{x}^{(1)} \\ x^{(2)} \end{pmatrix} \\ &= \begin{pmatrix} I^{(1)}\mathbf{x}^{(1)} \\ H_k^{(2)}\mathbf{x}^{(2)} \end{pmatrix} \\ &= \left(x_1, \ldots, x_{k-1}, \left(\sum_{i=k}^{n} x_i^2\right)^{1/2}, 0, \ldots, 0\right)^T \end{aligned}$$

Remarks

1. The Householder matrix H_k defined above is an elementary orthogonal matrix. If we let $\mathbf{v} = \begin{pmatrix} \mathbf{0} \\ \mathbf{v}_k \end{pmatrix}$ and $\mathbf{u} = (1/\|\mathbf{v}\|)\mathbf{v}$, then

$$H_k = I - \frac{1}{\beta_k}\mathbf{v}\mathbf{v}^T = I - 2\mathbf{u}\mathbf{u}^T$$

2. H_k acts like the identity on the first $k-1$ coordinates of any vector $\mathbf{y} \in R^n$. If $\mathbf{y} = (y_1, \ldots, y_{k-1}, y_k, \ldots, y_n)^T$, $\mathbf{y}^{(1)} = (y_1, \ldots, y_{k-1})^T$ and $\mathbf{y}^{(2)} = (y_k, \ldots, y_n)^T$, then

$$H_k\mathbf{y} = \begin{pmatrix} I^{(1)} & O \\ O & H_k^{(2)} \end{pmatrix} \begin{pmatrix} \mathbf{y}^{(1)} \\ \mathbf{y}^{(2)} \end{pmatrix} = \begin{pmatrix} \mathbf{y}^{(1)} \\ H_k^{(2)}\mathbf{y}^{(2)} \end{pmatrix}$$

In particular, if $\mathbf{y}^{(2)} = \mathbf{0}$, then $H_k\mathbf{y} = \mathbf{y}$.
3. It is generally not necessary to store the entire matrix H_k. It suffices to store the $n-k+1$ vector $\mathbf{v}_k$ and the scalar β_k.

We are now ready to apply Householder transformations to solve linear systems. If A is a nonsingular $n \times n$ matrix, we can use Householder transformations to reduce A to triangular form. To begin with, we can find a Householder transformation $H_1 = I - (1/\beta_1)\mathbf{v}_1\mathbf{v}_1^T$ which when applied to the first column of A will give a multiple of $\mathbf{e}_1$. Thus H_1A will be of the form

$$\begin{bmatrix} \times & \times & \cdots & \times \\ 0 & \times & \cdots & \times \\ 0 & \times & \cdots & \times \\ \vdots & & & \\ 0 & \times & \cdots & \times \end{bmatrix}$$

We can then find a Householder transformation H_2 that will zero out the last $n-2$ elements in the second column of H_1A while leaving the first element in that column unchanged. It follows from remark (2) that H_2 will have no effect on the first column of H_1A.

$$H_2H_1A = \begin{bmatrix} \times & \times & \times & \cdots & \times \\ 0 & \times & \times & \cdots & \times \\ 0 & 0 & \times & \cdots & \times \\ \vdots & & & & \\ 0 & 0 & \times & \cdots & \times \end{bmatrix}$$

We can continue applying Householder transformations in this fashion until we end up with an upper triangular matrix, which we will denote by R. Thus

$$H_{n-1} \cdots H_2H_1A = R$$

It follows that

$$\begin{aligned} A &= H_1^{-1}H_2^{-1} \cdots H_{n-1}^{-1}R \\ &= H_1H_2 \cdots H_{n-1}R \end{aligned}$$

Let $Q = H_1 H_2 \cdots H_{n-1}$. The matrix Q is orthogonal and A can be factored into an orthogonal matrix times an upper triangular matrix:

$$A = QR$$

Once A has been factored into a product QR, the system $A\mathbf{x} = \mathbf{b}$ is easily solved.

$$\begin{aligned} A\mathbf{x} &= \mathbf{b} \\ QR\mathbf{x} &= \mathbf{b} \\ R\mathbf{x} &= Q^T\mathbf{b} \end{aligned}$$

$$R\mathbf{x} = H_{n-1} \cdots H_2 H_1 \mathbf{b} \tag{2}$$

Once $H_{n-1} \cdots H_2 H_1 \mathbf{b}$ has been calculated, the system (2) can be solved using back substitution.

Storage

The vector $\mathbf{v}_k$ can be stored in the kth column of A. Since $\mathbf{v}_k$ has $n - k + 1$ nonzero entries and there are only $n - k$ zeros in the kth column of the reduced matrix, it is necessary to store r_{kk} elsewhere. The diagonal elements of R can either be stored in an n vector or in an additional row added to A. The β_k's can also be stored in an additional row of A.

$$\begin{pmatrix} v_{11} & r_{12} & r_{13} & r_{14} \\ v_{12} & v_{22} & r_{23} & r_{24} \\ v_{13} & v_{23} & v_{33} & r_{34} \\ v_{14} & v_{24} & v_{34} & 0 \\ r_{11} & r_{22} & r_{33} & r_{44} \\ \beta_1 & \beta_2 & \beta_3 & 0 \end{pmatrix}$$

Operation Count

In solving an $n \times n$ system using Householder transformations, most of the work is done in reducing A to triangular form. The number of operations required is approximately $\frac{2}{3}n^3$ multiplications, $\frac{2}{3}n^3$ additions, and $n - 1$ square roots.

Often it will be desirable to have a transformation that annihilates only a single entry of a vector. In this case it is convenient to use either a rotation or a reflection. Let us consider first the two-dimensional case.

Let

$$R = \begin{pmatrix} \cos\theta & -\sin\theta \\ \sin\theta & \cos\theta \end{pmatrix} \quad \text{and} \quad G = \begin{pmatrix} \cos\theta & \sin\theta \\ \sin\theta & -\cos\theta \end{pmatrix}$$

and let

$$\mathbf{x} = \begin{pmatrix} x_1 \\ x_2 \end{pmatrix} = \begin{pmatrix} r\cos\alpha \\ r\sin\alpha \end{pmatrix}$$

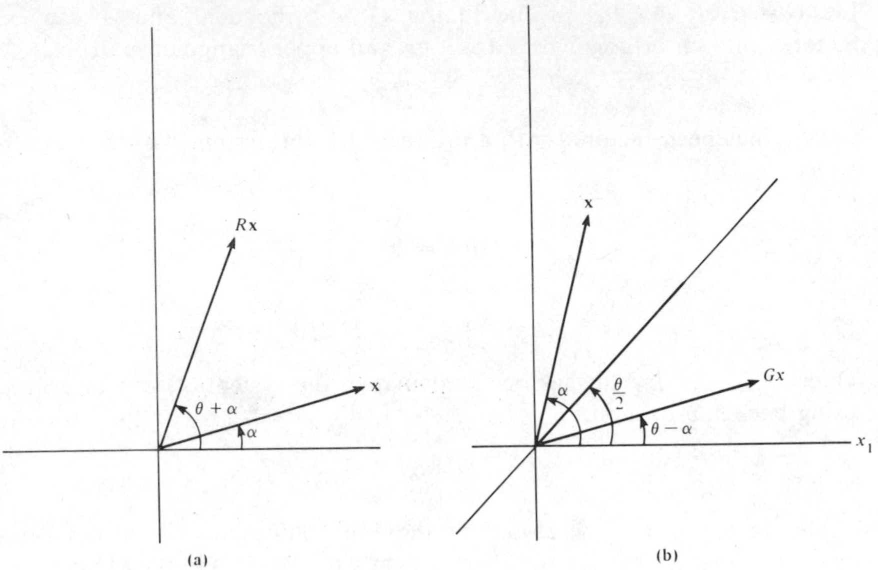

Figure 7.5.1

be a vector in R^2.

$$R\mathbf{x} = \begin{pmatrix} r\cos(\theta+\alpha) \\ r\sin(\theta+\alpha) \end{pmatrix} \quad \text{and} \quad G\mathbf{x} = \begin{pmatrix} r\cos(\theta-\alpha) \\ r\sin(\theta-\alpha) \end{pmatrix}$$

R represents a rotation in the plane by an angle θ. The matrix G has the effect of reflecting $\mathbf{x}$ about the line $x_2 = [\tan(\theta/2)]x_1$ (see Figure 7.5.1).

If we set $\cos\theta = x_1/r$ and $\sin\theta = -x_2/r$, then

$$R\mathbf{x} = \begin{pmatrix} x_1\cos\theta - x_2\sin\theta \\ x_1\sin\theta + x_2\cos\theta \end{pmatrix} = \begin{pmatrix} r \\ 0 \end{pmatrix}$$

If we set $\cos\theta = x_1/r$ and $\sin\theta = x_2/r$, then

$$G\mathbf{x} = \begin{pmatrix} x_1\cos\theta + x_2\sin\theta \\ x_1\sin\theta - x_2\cos\theta \end{pmatrix} = \begin{pmatrix} r \\ 0 \end{pmatrix}$$

Both R and G are orthogonal. The matrix G is also symmetric. Indeed, G is an elementary orthogonal matrix. If we let $u = (\sin\theta/2, -\cos\theta/2)^T$, then $G = I - 2\mathbf{u}\mathbf{u}^T$.

Let us now consider the n-dimensional case. Let R and G be $n \times n$ matrices with

$$\begin{aligned} r_{ii} &= r_{jj} = \cos\theta & \qquad g_{ii} &= \cos\theta, \quad g_{jj} = -\cos\theta \\ r_{ji} &= \sin\theta, \quad r_{ij} = -\sin\theta & \qquad g_{ji} &= g_{ij} = \sin\theta \end{aligned}$$

and $r_{st} = g_{st} = \delta_{st}$ for all other entries of R and G. Thus R and G resemble the identity matrix except for the ii, ij, jj, and ji positions. Let $c = \cos\theta$ and $s = \sin\theta$. If $\mathbf{x} \in R^n$, then

$$R\mathbf{x} = (x_1, \ldots, x_{i-1}, x_i c - x_j s, x_{i+1}, \ldots, x_{j-1}, x_i s + x_j c, x_{j+1}, \ldots, x_n)^T$$

and

$$G\mathbf{x} = (x_1, \ldots, x_{i-1}, x_i c + x_j s, x_{i+1}, \ldots, x_{j-1}, x_i s - x_j c, x_{j+1}, \ldots, x_n)^T$$

The transformations R and G only alter the ith and jth components of a vector. They have no effect on the other coordinates. We will refer to R as a *plane rotation* and to G as a *Givens transformation* or a *Givens reflection*. If we set $c = x_i/r$ and $s = -x_j/r$, $(r = \sqrt{x_i^2 + x_j^2})$, the jth component of $R\mathbf{x}$ will be 0. If we set $c = x_i/r$ and $s = x_j/r$, the jth component of $G\mathbf{x}$ will be 0.

Given a nonsingular $n \times n$ matrix A, we can use either plane rotations or Givens transformations to obtain a QR factorization of A. Let G_{21} be the Givens transformation acting on the first and second coordinates, which when applied to A results in a zero in the (2, 1) position. We can apply another Givens transformation, G_{31}, to $G_{21}A$ to obtain a zero in the (3, 1) position. This process can be continued until the last $n - 1$ entries in the first column have been eliminated.

$$G_{n1} \cdots G_{31}G_{21}A = \begin{bmatrix} \times & \times & \cdots & \times \\ 0 & \times & \cdots & \times \\ 0 & \times & \cdots & \times \\ \vdots & & & \\ 0 & \times & \cdots & \times \end{bmatrix}$$

At the next step, Givens transformations $G_{32}, G_{42}, \ldots, G_{n2}$ are used to eliminate the last $n - 2$ entries in the second column. This process is continued until all the elements below the diagonal have been eliminated.

$$(G_{n,n-1}) \cdots (G_{n2} \cdots G_{32})(G_{n1} \cdots G_{21})A = R \qquad (R \text{ upper triangular})$$

If we let $Q^T = (G_{n,n-1}) \cdots (G_{n2} \cdots G_{32})(G_{n1} \cdots G_{21})$, then $A = QR$ and the system $A\mathbf{x} = \mathbf{b}$ is equivalent to the system

$$R\mathbf{x} = Q^T\mathbf{b}$$

This system can be solved by back substitution.

Operation Count

The QR factorization of A using Givens transformations or plane rotations requires roughly $\frac{4}{3}n^3$ multiplications, $\frac{2}{3}n^3$ additions, and $\frac{1}{2}n^2$ square roots.

EXERCISES

1. Suppose that you wish to eliminate the last coordinate of a vector $\mathbf{x}$ and leave the first $n-2$ coordinates unchanged. How many operations are necessary if this is to be done by a Givens transformation G? A Householder transformation H? If A is an $n \times n$ matrix, how many operations are required to compute GA and HA?

2. Let $H_k = I - 2\mathbf{u}\mathbf{u}^T$ be a Householder transformation with $\mathbf{u} = (0, \ldots, 0, u_k, u_{k+1}, \ldots, u_n)^T$. Let $\mathbf{b} \in R^n$ and let A be an $n \times n$ matrix. How many additions and multiplications are necessary to compute $H_k\mathbf{b}$? H_kA?

3. Let $Q^T = G_{n-k} \cdots G_2G_1$, where each G_i is a Givens transformation. Let $\mathbf{b} \in R^n$ and let A be an $n \times n$ matrix. How many additions and multiplications are necessary to compute $Q^T\mathbf{b}$? Q^TA?

4. Let R_1 and R_2 be two 2×2 rotation matrices and let G_1 and G_2 be two 2×2 Givens transformations. What type of transformations are each of the following?
(a) R_1R_2 (b) G_1G_2 (c) R_1G_1 (d) G_1R_1

5. Let $\mathbf{x}$ and $\mathbf{y}$ be distinct vectors in R^n with $\|\mathbf{x}\|_2 = \|\mathbf{y}\|_2$. Define

$$\mathbf{u} = \frac{1}{\|\mathbf{x}-\mathbf{y}\|_2}(\mathbf{x}-\mathbf{y}) \qquad \text{and} \qquad Q = I - 2\mathbf{u}\mathbf{u}^T$$

Show that:
(a) $\|\mathbf{x}-\mathbf{y}\|_2^2 = 2(\mathbf{x}-\mathbf{y})^T\mathbf{x}$ (b) $Q\mathbf{x} = \mathbf{y}$

6. Let $A = Q_1R_1 = Q_2R_2$, where Q_1 and Q_2 are orthogonal and R_1 and R_2 are upper triangular.
(a) Show that $Q_1^TQ_2$ is diagonal.
(b) How do R_1 and R_2 compare? Explain.

6. ITERATIVE METHODS FOR SOLVING LINEAR SYSTEMS

In this section we will study iterative methods for solving a linear system $A\mathbf{x} = \mathbf{b}$. Iterative methods start out with an initial approximation $\mathbf{x}^{(0)}$ to the solution and go through a fixed procedure to obtain a better approximation, $\mathbf{x}^{(1)}$. The same procedure is then repeated on $\mathbf{x}^{(1)}$ to obtain an improved approximation, $\mathbf{x}^{(2)}$; and so on. The iterations terminate when a desired accuracy has been achieved.

Iterative methods are most useful in solving large sparse systems. Such systems occur, for example, in the solution of boundary value problems for partial differential equations. The amount of storage required for a sparse coefficient matrix A is proportional to n. Gaussian elimination and the other direct methods studied in Section 5 usually tend to fill in the zeros of A and hence require an amount of storage proportional to n^2. This can present quite a problem when n is very large, say $n \geqslant 500$. The iterative methods we will describe will not require any extra storage and hence are more practical. One disadvantage they do have is that after solving $A\mathbf{x} = \mathbf{b}_1$, one must start all over again from the beginning in order to solve $A\mathbf{x} = \mathbf{b}_2$.

Given a system $A\mathbf{x} = \mathbf{b}$, we write the coefficient matrix A in the form $A = C - M$, where C is a nonsingular matrix which is in some form that is easily invertible (e.g., diagonal or triangular). The system can then be rewritten in the form

$$\begin{aligned} C\mathbf{x} &= M\mathbf{x} + \mathbf{b} \\ \mathbf{x} &= C^{-1}M\mathbf{x} + C^{-1}\mathbf{b} \end{aligned}$$

If we set

$$B = C^{-1}M = I - C^{-1}A \qquad \text{and} \qquad \mathbf{c} = C^{-1}\mathbf{b}$$

then

$$\mathbf{x} = B\mathbf{x} + \mathbf{c} \tag{1}$$

To solve (1), we start out with an initial guess $\mathbf{x}^{(0)}$, which may be any vector in R^n. We then set

$$\begin{aligned} \mathbf{x}^{(1)} &= B\mathbf{x}^{(0)} + \mathbf{c} \\ \mathbf{x}^{(2)} &= B\mathbf{x}^{(1)} + \mathbf{c} \end{aligned}$$

and in general

$$\mathbf{x}^{(k+1)} = B\mathbf{x}^{(k)} + \mathbf{c}$$

Let $\mathbf{x}$ be a solution to (1). If $\|\cdot\|$ denotes some vector norm on R^n and the corresponding matrix norm of B is less than 1, we claim that $\|\mathbf{x}^{(k)} - \mathbf{x}\| \to 0$ as $k \to \infty$. Indeed,

$$\begin{aligned} \mathbf{x}^{(1)} - \mathbf{x} &= (B\mathbf{x}^{(0)} + \mathbf{c}) - (B\mathbf{x} + \mathbf{c}) = B(\mathbf{x}^{(0)} - \mathbf{x}) \\ \mathbf{x}^{(2)} - \mathbf{x} &= (B\mathbf{x}^{(1)} + \mathbf{c}) - (B\mathbf{x} + \mathbf{c}) = B(\mathbf{x}^{(1)} - \mathbf{x}) = B^2(\mathbf{x}^{(0)} - \mathbf{x}) \end{aligned}$$

and so on. In general,

$$\mathbf{x}^{(k)} - \mathbf{x} = B^k(\mathbf{x}^{(0)} - \mathbf{x})$$

and hence

$$\begin{aligned}\|\mathbf{x}^{(k)} - \mathbf{x}\| &= \|B^k(\mathbf{x}^{(0)} - \mathbf{x})\| \\ &\leqslant \|B^k\| \, \|\mathbf{x}^{(0)} - \mathbf{x}\| \\ &\leqslant \|B\|^k \|\mathbf{x}^{(0)} - \mathbf{x}\|\end{aligned}$$

Thus, if $\|B\| < 1$, then $\|\mathbf{x}^{(k)} - \mathbf{x}\| \to 0$ as $k \to \infty$.

The foregoing result holds for any norm on R^n, although in practice it is simplest to use the $\|\cdot\|_\infty$ or the $\|\cdot\|_1$. Essentially, then, we require that the matrix C be easily invertible and that C^{-1} be a good enough approximation to A^{-1} so that

$$\|I - C^{-1}A\| = \|B\| < 1$$

This last condition implies that all the eigenvalues of B are less than 1 in modulus.

Definition. Let $\lambda_1, \ldots, \lambda_n$ be the eigenvalues of B and let $\rho(B) = \max_{1 \leqslant i \leqslant n} |\lambda_i|$. The constant $\rho(B)$ is called the *spectral radius* of B.

Theorem 7.6.1. *Let $\mathbf{x}^{(0)}$ be an arbitrary vector in R^n and define $\mathbf{x}^{(i+1)} = B\mathbf{x}^{(i)} + \mathbf{c}$ for $i = 0, 1, \ldots$. If $\mathbf{x}$ is the solution to* (1), *then a necessary and sufficient condition for $\mathbf{x}^{(k)} \to \mathbf{x}$ is that $\rho(B) < 1$.*

PROOF. We will prove the theorem only in the case where B has n linearly independent eigenvectors. The case where B is not diagonalizable is beyond the scope of this text. If $\mathbf{x}_1, \ldots, \mathbf{x}_n$ are n linearly independent eigenvectors of B, we can write

$$\mathbf{x}^{(0)} - \mathbf{x} = \alpha_1 \mathbf{x}_1 + \cdots + \alpha_n \mathbf{x}_n$$

and

$$\begin{aligned}\mathbf{x}^{(k)} - \mathbf{x} &= B^k(\alpha_1 \mathbf{x}_1 + \cdots + \alpha_n \mathbf{x}_n) \\ &= \alpha_1 \lambda_1^k \mathbf{x}_1 + \cdots + \alpha_n \lambda_n^k \mathbf{x}_n\end{aligned}$$

It follows that

$$(\mathbf{x}^{(k)} - \mathbf{x}) \to \mathbf{0}$$

if and only if $|\lambda_i| < 1$ for $i = 1, \ldots, n$. Thus $\mathbf{x}^{(k)} \to \mathbf{x}$ if and only if $\rho(B) < 1$.

The simplest choice of C is to let C be a diagonal matrix whose diagonal elements are the diagonal elements of A. The iteration scheme with this choice of C is called *Jacobi iteration*.

Jacobi Iteration

Let

$$C = \begin{bmatrix} a_{11} & 0 & \cdots & 0 \\ 0 & a_{22} & & \\ \vdots & & \ddots & \\ 0 & 0 & & a_{nn} \end{bmatrix} \quad \text{and } M = -\begin{bmatrix} 0 & a_{12} & \cdots & a_{1n} \\ a_{21} & 0 & & a_{2n} \\ \vdots & & \ddots & \\ a_{n1} & a_{n2} & & 0 \end{bmatrix}$$

and set $B = C^{-1}M$ and $\mathbf{c} = C^{-1}\mathbf{b}$. Thus

$$B = \begin{bmatrix} 0 & \dfrac{-a_{12}}{a_{11}} & \cdots & \dfrac{-a_{1n}}{a_{11}} \\ \dfrac{-a_{21}}{a_{22}} & 0 & \cdots & \dfrac{-a_{2n}}{a_{22}} \\ \vdots & & & \\ \dfrac{-a_{n1}}{a_{nn}} & \dfrac{-a_{n2}}{a_{nn}} & \cdots & 0 \end{bmatrix} \quad \text{and } \mathbf{c} = \begin{bmatrix} \dfrac{b_1}{a_{11}} \\ \dfrac{b_2}{a_{22}} \\ \vdots \\ \dfrac{b_n}{a_{nn}} \end{bmatrix}$$

At the $(i+1)$st iteration, the vector $\mathbf{x}^{(i+1)}$ is calculated by

$$(2) \qquad x_j^{(i+1)} = \frac{1}{a_{jj}}\left[-\sum_{\substack{k=1\\k\neq j}}^{n} a_{jk}x_k^{(i)} + b_j\right] \qquad j = 1, \ldots, n$$

The vector $\mathbf{x}^{(i)}$ is used in calculating $\mathbf{x}^{(i+1)}$. Consequently, these two vectors must be stored separately.

If the diagonal elements of A are much larger than the off-diagonal elements, the entries of B should all be small and the Jacobi iteration should converge. We say that A is *diagonally dominant* if

$$|a_{ii}| > \sum_{\substack{j=1\\j\neq i}}^{n} |a_{ij}| \qquad \text{for } i = 1, \ldots, n$$

If A is diagonally dominant, the matrix B of the Jacobi iteration will have the property

$$\sum_{j=1}^{n} |b_{ij}| = \sum_{\substack{j=1\\j\neq i}}^{n} \frac{|a_{ij}|}{|a_{ii}|} < 1 \qquad \text{for } i = 1, \ldots, n$$

Thus

$$\|B\|_\infty = \max_{1<i<n}\left(\sum_{j=1}^{n} |b_{ij}|\right) < 1$$

It follows, then, that if A is diagonally dominant, the Jacobi iteration will converge to the solution of $A\mathbf{x} = \mathbf{b}$.

An alternative to the Jacobi iteration is to take C to be the lower triangular part of A (i.e., $c_{ij} = a_{ij}$ if $i \geqslant j$ and $c_{ij} = 0$ if $i < j$). Since C is a better approximation to A than the diagonal matrix in the Jacobi iteration, we would expect that C^{-1} is a better approximation to A^{-1}, and hopefully B will have a smaller norm. The iteration scheme with this choice of C is called *Gauss–Seidel iteration*. It usually converges faster than Jacobi iteration.

Gauss–Seidel Iteration

Let

$$L = -\begin{bmatrix} 0 & 0 & \cdots & 0 & 0 \\ a_{21} & 0 & \cdots & 0 & 0 \\ \vdots & & & & \\ a_{n-1,1} & a_{n-1,2} & & 0 & 0 \\ a_{n1} & a_{n2} & \cdots & a_{n,n-1} & 0 \end{bmatrix}$$

$$D = \begin{bmatrix} a_{11} & 0 & \cdots & 0 \\ 0 & a_{22} & & 0 \\ \vdots & & \ddots & \\ 0 & 0 & \cdots & a_{nn} \end{bmatrix}$$

and

$$U = -\begin{bmatrix} 0 & a_{12} & \cdots & a_{1,n-1} & a_{1n} \\ 0 & 0 & & a_{2,n-1} & a_{2n} \\ \vdots & & & & \\ 0 & 0 & & 0 & a_{n-1,n} \\ 0 & 0 & & 0 & 0 \end{bmatrix}$$

Set $C = D - L$ and $M = U$. Let $\mathbf{x}^{(0)}$ be an arbitrary nonzero vector in R^n. We have

$$\begin{aligned} C\mathbf{x}^{(i+1)} &= M\mathbf{x}^{(i)} + \mathbf{b} \\ (D - L)\mathbf{x}^{(i+1)} &= U\mathbf{x}^{(i)} + \mathbf{b} \\ D\mathbf{x}^{(i+1)} &= L\mathbf{x}^{(i+1)} + U\mathbf{x}^{(i)} + \mathbf{b} \end{aligned}$$

We can solve this last equation for $\mathbf{x}^{(i+1)}$ one coordinate at a time. The first coordinate of $\mathbf{x}^{(i+1)}$ is given by

$$x_1^{(i+1)} = \frac{1}{a_{11}}\left(-\sum_{k=2}^{n} a_{1k}x_k^{(i)} + b_1\right)$$

The second coordinate of $\mathbf{x}^{(i+1)}$ can be solved for in terms of the first coordinate and the last $n-2$ coordinates of $x^{(i)}$.

$$x_2^{(i+1)} = \frac{1}{a_{22}}\left(-a_{21}x_1^{(i+1)} - \sum_{k=3}^{n} a_{2k}x_k^{(i)} + b_2\right)$$

In general,

$$x_j^{(i+1)} = \frac{1}{a_{jj}}\left(-\sum_{k=1}^{j-1} a_{jk}x_k^{(i+1)} - \sum_{k=j+1}^{n} a_{jk}x_k^{(i)} + b_j\right) \tag{3}$$

It is interesting to compare (2) and (3). The difference between the Jacobi and Gauss–Seidel iterations is that in the latter case, one is using the coordinates of $\mathbf{x}^{(i+1)}$ as soon as they are calculated rather than in the next iteration. The program for the Gauss–Seidel iteration is actually simpler than the program for the Jacobi iteration. The vectors $\mathbf{x}^{(i)}$ and $\mathbf{x}^{(i+1)}$ are both stored in the same vector, $\mathbf{x}$. As a coordinate of $\mathbf{x}^{(i+1)}$ is calculated, it replaces the corresponding coordinate of $\mathbf{x}^{(i)}$.

***Theorem* 7.6.2.** *If A is diagonally dominant, then the Gauss–Seidel iteration converges to a solution of $A\mathbf{x} = \mathbf{b}$.*

PROOF. For $j = 1, \ldots, n$, let

$$\alpha_j = \sum_{i=1}^{j-1} |a_{ji}|, \qquad \beta_j = \sum_{i=j+1}^{n} |a_{ji}|, \qquad \text{and} \qquad M_j = \frac{\beta_j}{(|a_{jj}| - \alpha_j)}$$

Since A is diagonally dominant, it follows that

$$|a_{jj}| > \alpha_j + \beta_j$$

and consequently $M_j < 1$ for $j = 1, \ldots, n$. Thus

$$M = \max_{1 \leqslant j \leqslant n} M_j < 1$$

We will show that

$$\|B\|_\infty = \max_{\mathbf{x} \neq \mathbf{0}} \frac{\|B\mathbf{x}\|_\infty}{\|\mathbf{x}\|_\infty} \leqslant M < 1$$

Let $\mathbf{x}$ be a nonzero vector in R^n and let $\mathbf{y} = B\mathbf{x}$. Choose k so that

$$\|\mathbf{y}\|_\infty = \max_{1 \leqslant i \leqslant n} |y_i| = |y_k|$$

It follows from the definition of B that

$$\mathbf{y} = B\mathbf{x} = (D - L)^{-1}U\mathbf{x}$$

and hence

$$\mathbf{y} = D^{-1}(L\mathbf{y} + U\mathbf{x})$$

Comparing the kth coordinates of each side, we see that

$$y_k = \frac{1}{a_{kk}}\left(-\sum_{i=1}^{k-1} a_{ki}y_i - \sum_{i=k+1}^{n} a_{ki}x_i\right)$$

and hence

$$(4) \qquad \|\mathbf{y}\|_\infty = |y_k| \leqslant \frac{1}{|a_{kk}|}(\alpha_k\|\mathbf{y}\|_\infty + \beta_k\|\mathbf{x}\|_\infty)$$

It follows from (4) that

$$\frac{\|B\mathbf{x}\|_\infty}{\|\mathbf{x}\|_\infty} = \frac{\|\mathbf{y}\|_\infty}{\|\mathbf{x}\|_\infty} \leqslant M_k \leqslant M$$

Thus

$$\|B\|_\infty = \max_{\mathbf{x}\neq\mathbf{0}} \frac{\|B\mathbf{x}\|_\infty}{\|\mathbf{x}\|_\infty} \leqslant M < 1$$

and hence the iteration will converge to the solution of $A\mathbf{x} = \mathbf{b}$.

EXERCISES

1. Let

$$A = \begin{bmatrix} 1 & 1 & 1 \\ 0 & 1 & 1 \\ 0 & 0 & 1 \end{bmatrix}, \qquad \mathbf{b} = \begin{bmatrix} 3 \\ 2 \\ 1 \end{bmatrix}, \qquad \text{and} \qquad \mathbf{x}^{(0)} = \begin{bmatrix} 1 \\ 0 \\ 0 \end{bmatrix}$$

Use Jacobi iteration to compute $\mathbf{x}^{(1)}$, $\mathbf{x}^{(2)}$, $\mathbf{x}^{(3)}$, and $\mathbf{x}^{(4)}$.

2. Let

$$A = \begin{bmatrix} 10 & 1 & 1 \\ 1 & 10 & 1 \\ 1 & 1 & 10 \end{bmatrix}, \qquad \mathbf{b} = \begin{bmatrix} 12 \\ 12 \\ 12 \end{bmatrix}, \qquad \text{and} \qquad \mathbf{x}^{(0)} = \begin{bmatrix} 1 \\ 0 \\ 0 \end{bmatrix}$$

(a) Calculate $\mathbf{x}^{(1)}$ using Jacobi iteration.
(b) Calculate $\mathbf{x}^{(1)}$ using Gauss–Seidel iteration.
(c) Compare your answers to (a) and (b) with the correct solution $\mathbf{x} = (1, 1, 1)^T$. Which is closer?

3. For which of the following matrices, will the iteration scheme

$$\mathbf{x}^{(k+1)} = B\mathbf{x}^{(k)} + \mathbf{c}$$

converge to a solution of $\mathbf{x} = B\mathbf{x} + \mathbf{c}$? Explain.

(a) $B = \begin{pmatrix} 1 & 1 & 1 \\ 0 & 1 & 1 \\ 0 & 0 & 1 \end{pmatrix}$ (b) $B = \begin{pmatrix} 0.9 & 1 & 1 \\ 0 & 0.9 & 1 \\ 0 & 0 & 0.9 \end{pmatrix}$

(c) $B = \begin{pmatrix} \frac{1}{2} & 10 & 100 \\ 0 & \frac{1}{2} & 10 \\ 0 & 0 & \frac{1}{2} \end{pmatrix}$ (d) $B = \begin{pmatrix} \frac{1}{4} & \frac{1}{4} & \frac{1}{4} \\ \frac{1}{4} & \frac{1}{2} & \frac{1}{8} \\ \frac{1}{2} & \frac{1}{4} & \frac{1}{8} \end{pmatrix}$

(e) $B = \begin{pmatrix} \frac{1}{3} & \frac{1}{3} & \frac{1}{3} \\ \frac{1}{2} & \frac{1}{3} & \frac{1}{6} \\ 0 & \frac{1}{6} & \frac{1}{3} \end{pmatrix}$

4. Let $\mathbf{x}$ be the solution to $\mathbf{x} = B\mathbf{x} + \mathbf{c}$. Let $\mathbf{x}^{(0)}$ be an arbitrary vector in R^n and define

$$\mathbf{x}^{(k+1)} = B\mathbf{x}^{(k)} + \mathbf{c}$$

for $k = 0, 1, \ldots$. Prove that if B^m is the zero matrix, then $\mathbf{x}^{(m)} = \mathbf{x}$.

5. Let A be a nonsingular upper triangular matrix. Show that the Jacobi iteration will give the exact solution (assuming no roundoff errors) to $A\mathbf{x} = \mathbf{b}$ after n iterations.

6. Let $A = D - L - U$, where D, L, and U are defined as in Gauss–Seidel iteration, and let α be a nonzero scalar. Determine B and $\mathbf{c}$ if
(a) $C = \alpha D$ (b) $C = \alpha D - L$

7. Let $\mathbf{x}$ be the solution to $\mathbf{x} = B\mathbf{x} + \mathbf{c}$. Let $\mathbf{x}^{(0)}$ be an arbitrary vector in R^n and define

$$\mathbf{x}^{(i+1)} = B\mathbf{x}^{(i)} + \mathbf{c}$$

for $i = 0, 1, \ldots$. If $\|B\| = \alpha < 1$, show that

$$\|\mathbf{x}^{(k)} - \mathbf{x}\| \leqslant \frac{\alpha}{1-\alpha}\|\mathbf{x}^{(k)} - \mathbf{x}^{(k-1)}\|$$

7. THE SINGULAR VALUE DECOMPOSITION

In many applications it is necessary to either determine the rank of a matrix or to determine whether or not the matrix is deficient in rank. Theoretically, one can use Gaussian elimination to reduce the matrix to row echelon form and then count the number of nonzero rows. However, this approach is not practical when working in finite precision arithmetic.

If A is rank deficient and U is the computed echelon form, then because of roundoff errors in the elimination process, it is unlikely that U will have the proper number of nonzero rows. In practice, the coefficient matrix A usually involves some error. This may be due to errors in the data or to the finite number system. Thus it is generally more practical to ask whether A is "close" to a rank deficient matrix. However, it may well turn out that A is close to being rank deficient and the computed echelon form U is not.

In this section we will assume throughout that A is an $m \times n$ matrix with $m \geqslant n$. (This assumption is made for convenience only; all the results will also hold if $m < n$.) We will present a method for determining how close A is to a matrix of smaller rank. The method involves factoring A into a product $U\Sigma V^T$, where U is an $m \times m$ orthogonal matrix, V is an $n \times n$ orthogonal matrix, and Σ is an $m \times n$ matrix whose off diagonal entries are all 0's and whose diagonal elements satisfy

$$\sigma_1 \geqslant \sigma_2 \geqslant \cdots \geqslant \sigma_n \geqslant 0$$

$$\Sigma = \begin{bmatrix} \sigma_1 & & & \\ & \sigma_2 & O & \\ & & \ddots & \\ & O & & \sigma_n \\ & & & \end{bmatrix}$$

The σ_i's determined by this factorization are unique and are called the *singular values* of A. The factorization $U\Sigma V^T$ is called the *singular value decomposition* of A. We will show that the rank of A equals the number of nonzero singular values and that the magnitudes of the nonzero singular values provide a measure of how close A is to a matrix of lower rank.

We begin by showing that such a decomposition is always possible.

Theorem 7.7.1. *If A is an $m \times n$ matrix, then A has a singular value decomposition.*

PROOF. The matrix A^TA is positive semidefinite, since

$$\mathbf{x}^T A^T A \mathbf{x} = \|A\mathbf{x}\|_2^2 \geqslant 0$$

for all $\mathbf{x} \in R^n$. Thus A^TA has nonnegative eigenvalues. The eigenvalues may be ordered so that

$$\lambda_1 \geqslant \lambda_2 \geqslant \cdots \geqslant \lambda_n \geqslant 0$$

Define

$$\sigma_i = \sqrt{\lambda_i} \qquad i = 1, \ldots, n$$

Without loss of generality, one may assume that exactly r of the σ_i's are nonzero, so that

$$\sigma_1 \geqslant \sigma_2 \geqslant \cdots \geqslant \sigma_r > 0 \quad \text{and} \quad \sigma_{r+1} = \sigma_{r+2} = \cdots = \sigma_n = 0$$

Let

$$\Sigma = \begin{bmatrix} \sigma_1 & & & & & \\ & \sigma_2 & & & & \\ & & \ddots & & O & \\ & & & \sigma_r & & \\ & & & & 0 & \\ & & & & & \ddots \\ & & O & & & & 0 \end{bmatrix} = \begin{pmatrix} \Sigma_r & O \\ O & O \end{pmatrix}$$

where Σ_r is an $r \times r$ diagonal matrix whose diagonal elements are $\sigma_1, \ldots, \sigma_r$. Since A^TA is symmetric, there is an orthogonal matrix V that diagonalizes A^TA.

$$V^TA^TAV = \Sigma^T\Sigma$$

Let $\mathbf{v}_1, \mathbf{v}_2, \ldots, \mathbf{v}_n$ be the column vectors of V. Thus $\mathbf{v}_i$ is an eigenvector belonging to σ_i^2 for $i = 1, \ldots, r$ and $\mathbf{v}_{r+1}, \ldots, \mathbf{v}_n$ are eigenvectors belonging to 0. Let $V_1 = (\mathbf{v}_1, \mathbf{v}_2, \ldots, \mathbf{v}_r)$ and let $V_2 = (\mathbf{v}_{r+1}, \ldots, \mathbf{v}_n)$. Since

$$A^TA\mathbf{v}_i = \mathbf{0}$$

for $i = r + 1, \ldots, n$, it follows that

$$(AV_2)^TAV_2 = V_2^TA^TAV_2 = O$$

and hence

$$AV_2 = O$$

Since

$$A^TAV_1 = V_1\Sigma_r^2$$

it follows that

$$\Sigma_r^{-1}V_1^TA^TAV_1\Sigma_r^{-1} = I_r$$

where I_r denotes the $r \times r$ identity matrix. If we set

$$U_1 = AV_1\Sigma_r^{-1}$$

then

$$U_1^TU_1 = I_r$$

and hence U_1 is an $m \times r$ matrix with orthonormal column vectors $\mathbf{u}_1, \ldots, \mathbf{u}_r$. The set $\{\mathbf{u}_1, \ldots, \mathbf{u}_r\}$ can be extended to form an orthonormal basis for R^m. Let $U_2 = (\mathbf{u}_{r+1}, \ldots, \mathbf{u}_m)$ and let $U = (U_1 \quad U_2)$. It follows that

$$U^TAV = \begin{pmatrix} U_1^T \\ U_2^T \end{pmatrix} A(V_1 \quad V_2)$$

$$= \begin{pmatrix} U_1^T \\ U_2^T \end{pmatrix} (AV_1 \quad O)$$

$$= \begin{pmatrix} U_1^TAV_1 & O \\ U_2^TAV_1 & O \end{pmatrix}$$

But

$$U_1^TAV_1 = \Sigma_r^{-1}V_1^TA^TAV_1 = \Sigma_r$$

and

$$U_2^TAV_1 = U_2^TU_1\Sigma_r = O$$

Therefore, $U^TAV = \Sigma$ and hence $A = U\Sigma V^T$.

Observations

Let A be an $m \times n$ matrix with a singular value decomposition $U\Sigma V^T$.

1. Since

$$A^TA = V\Sigma^T\Sigma V^T$$

and

$$\Sigma^T\Sigma = \begin{bmatrix} \sigma_1^2 & & & O \\ & \sigma_2^2 & & \\ & & \ddots & \\ O & & & \sigma_n^2 \end{bmatrix}$$

it follows that the eigenvalues of A^TA are $\sigma_1^2, \sigma_2^2, \ldots, \sigma_n^2$. Since the σ_i's are the nonnegative square roots of the eigenvalues of A^TA, they are unique. However, the orthogonal matrices U and V are not unique.

2. Since V diagonalizes A^TA, it follows that the $\mathbf{v}_j$'s are eigenvectors of A^TA.

3. Since $AA^T = U\Sigma\Sigma^TU^T$, it follows that U diagonalizes AA^T and that the $\mathbf{u}_j$'s are eigenvectors of AA^T.

4. Comparing the jth columns of each side of the equation

$$AV = U\Sigma$$

we get

$$A\mathbf{v}_j = \sigma_j\mathbf{u}_j \qquad j = 1, \ldots, n$$

Similarly,

$$A^T U = V\Sigma^T$$

and hence

$$A^T \mathbf{u}_j = \sigma_j \mathbf{v}_j \qquad \text{for } j = 1, \ldots, n$$

$$A^T \mathbf{u}_j = \mathbf{0} \qquad \text{for } j = n+1, \ldots, m$$

The $\mathbf{v}_j$'s are called the *right singular vectors* of A and the $\mathbf{u}_j$'s are called the *left singular vectors* of A.

5. Since multiplying a matrix on either side by a nonsingular matrix does not alter its rank, it follows that A and Σ have the same rank. Thus, if $\sigma_1 \geqslant \sigma_2 \geqslant \cdots \geqslant \sigma_r > 0$ and $\sigma_{r+1} = \cdots = \sigma_n = 0$, then A has rank r. The rank of a matrix is the number of nonzero singular values (where singular values are counted according to multiplicity). The reader should be careful not to make a similar assumption about eigenvalues. The matrix

$$M = \begin{bmatrix} 0 & 1 & 0 & 0 \\ 0 & 0 & 1 & 0 \\ 0 & 0 & 0 & 1 \\ 0 & 0 & 0 & 0 \end{bmatrix}$$

for example, has rank 3 even though all its eigenvalues are 0.

EXAMPLE 1. Let

$$A = \begin{bmatrix} 1 & 1 \\ 1 & 1 \\ 0 & 0 \end{bmatrix}$$

Compute the singular values and the singular value decomposition of A.

SOLUTION. The matrix

$$A^T A = \begin{pmatrix} 2 & 2 \\ 2 & 2 \end{pmatrix}$$

has eigenvalues $\lambda_1 = 4$ and $\lambda_2 = 0$. Consequently, the singular values of A are $\sigma_1 = \sqrt{4} = 2$ and $\sigma_2 = 0$. The eigenvalue λ_1 has eigenvectors of the form $\alpha(1, 1)^T$ and λ_2 has eigenvectors $\beta(1, -1)^T$. Therefore, the orthogonal matrix

$$V = \frac{1}{\sqrt{2}} \begin{pmatrix} 1 & 1 \\ 1 & -1 \end{pmatrix}$$

diagonalizes $A^T A$. Let

$$\mathbf{u}_1 = AV_1\Sigma_1^{-1} = \begin{bmatrix} 1 & 1 \\ 1 & 1 \\ 0 & 0 \end{bmatrix} \begin{bmatrix} \dfrac{1}{\sqrt{2}} \\ \dfrac{1}{\sqrt{2}} \end{bmatrix} \left(\tfrac{1}{2}\right) = \begin{bmatrix} \dfrac{1}{\sqrt{2}} \\ \dfrac{1}{\sqrt{2}} \\ 0 \end{bmatrix}$$

The remaining column vectors of U must be eigenvectors of AA^T belonging to the eigenvalue 0. Thus we can determine $\mathbf{u}_2$ and $\mathbf{u}_3$ by finding an orthonormal basis for $N(AA^T)$. Since

$$AA^T = \begin{bmatrix} 2 & 2 & 0 \\ 2 & 2 & 0 \\ 0 & 0 & 0 \end{bmatrix}$$

we can choose $\mathbf{u}_2 = (\frac{1}{\sqrt{2}}, -\frac{1}{\sqrt{2}}, 0)^T$ and $\mathbf{u}_3 = (0, 0, 1)^T$. It follows then that

$$A = U\Sigma V^T = \begin{bmatrix} \frac{1}{\sqrt{2}} & \frac{1}{\sqrt{2}} & 0 \\ \frac{1}{\sqrt{2}} & -\frac{1}{\sqrt{2}} & 0 \\ 0 & 0 & 1 \end{bmatrix} \begin{bmatrix} 2 & 0 \\ 0 & 0 \\ 0 & 0 \end{bmatrix} \begin{bmatrix} \frac{1}{\sqrt{2}} & \frac{1}{\sqrt{2}} \\ \frac{1}{\sqrt{2}} & \frac{-1}{\sqrt{2}} \end{bmatrix}$$

If A is an $m \times n$ matrix of rank r and $0 < k < r$, we can use the singular value decomposition to find a matrix in $M_{m,n}$ of rank k that is closest to A with respect to the Frobenius norm. Let $\mathbb{S}$ be the set of all $m \times n$ matrices of rank k or less. It can be shown that there is a matrix X in $\mathbb{S}$ such that

$$(1) \qquad \|A - X\|_F = \min_{S \in \mathbb{S}} \|A - S\|_F$$

We will not prove this, since the proof is beyond the scope of this text. Assuming that the minimum is achieved, we will show how such a matrix X can be derived from the singular value decomposition of A. The following lemma will be useful.

***Lemma* 7.7.2.** *If A is an $m \times n$ matrix and Q is an $m \times m$ orthogonal matrix, then*

$$\|QA\|_F = \|A\|_F$$

PROOF

$$\begin{aligned} \|QA\|_F^2 &= \|(Q\mathbf{a}_1, Q\mathbf{a}_2, \ldots, Q\mathbf{a}_n)\|_F^2 \\ &= \sum_{i=1}^{n} \|Q\mathbf{a}_i\|_2^2 \\ &= \sum_{i=1}^{n} \|\mathbf{a}_i\|_2^2 \\ &= \|A\|_F^2 \end{aligned}$$

It follows from the lemma that if A has singular value decomposition $U\Sigma V^T$, then

$$\|A\|_F = \|\Sigma V^T\|_F$$

Since

$$\|\Sigma V^T\|_F = \|(\Sigma V^T)^T\|_F = \|V\Sigma^T\|_F = \|\Sigma^T\|_F$$

it follows that

$$\|A\|_F = (\sigma_1^2 + \sigma_2^2 + \cdots + \sigma_n^2)^{1/2}$$

Theorem 7.7.3. *Let $A = U\Sigma V^T$ be an $m \times n$ matrix and let $\mathbb{S}$ denote the set of all $m \times n$ matrices of rank k or less, where $0 < k < \text{rank}\,(A)$. If X is a matrix in $\mathbb{S}$ satisfying* (1), *then*

$$\|A - X\|_F = (\sigma_{k+1}^2 + \sigma_{k+2}^2 + \cdots + \sigma_n^2)^{1/2}$$

In particular, if $A' = U\Sigma' V^T$, where

$$\Sigma' = \left[\begin{array}{ccc|c} \sigma_1 & & O & \\ & \ddots & & O \\ O & & \sigma_k & \\ \hline & O & & O \end{array}\right] = \begin{pmatrix} \Sigma_k & O \\ O & O \end{pmatrix}$$

then

$$\|A - A'\|_F = (\sigma_{k+1}^2 + \cdots + \sigma_n^2)^{1/2} = \min_{S \in \mathbb{S}} \|A - S\|_F$$

PROOF. Let X be a matrix in $\mathbb{S}$ satisfying (1). Since $A' \in \mathbb{S}$, it follows that

$$(2) \qquad \|A - X\|_F \leqslant \|A - A'\|_F = (\sigma_{k+1}^2 + \cdots + \sigma_n^2)^{1/2}$$

We will show that

$$\|A - X\|_F \geqslant (\sigma_{k+1}^2 + \cdots + \sigma_n^2)^{1/2}$$

and hence that equality holds in (2). Let $Q\Omega P^T$ be the singular value decomposition of X.

$$\Omega = \left[\begin{array}{cccc|c} \omega_1 & & & & \\ & \omega_2 & & & \\ & & \ddots & & O \\ & & & \omega_k & \\ \hline & & O & & O \end{array}\right] = \begin{pmatrix} \Omega_k & O \\ O & O \end{pmatrix}$$

If we set $B = Q^TAP$, then $A = QBP^T$ and it follows that

$$\|A - X\|_F = \|Q(B - \Omega)P^T\|_F = \|B - \Omega\|_F$$

Let us partition B in the same manner as Ω.

$$B = \left[\begin{array}{c|c} \overbrace{B_{11}}^{k \times k} & \overbrace{B_{12}}^{k \times (n-k)} \\ \hline \underbrace{B_{21}}_{(m-k) \times k} & \underbrace{B_{22}}_{(m-k) \times (n-k)} \end{array}\right]$$

It follows that

$$\|A - X\|_F^2 = \|B_{11} - \Omega_k\|_F^2 + \|B_{12}\|_F^2 + \|B_{21}\|_F^2 + \|B_{22}\|_F^2$$

We claim $B_{12} = \mathrm{O}$. If not, then define

$$Y = Q\begin{pmatrix} B_{11} & B_{12} \\ \mathrm{O} & \mathrm{O} \end{pmatrix} P^T$$

$Y \in \mathbb{S}$ and

$$\|A - Y\|_F^2 = \|B_{21}\|_F^2 + \|B_{22}\|_F^2 < \|A - X\|_F^2$$

This contradicts the definition of X. Therefore, $B_{12} = \mathrm{O}$. In a similar manner it can be shown that B_{21} must equal O. If we set

$$Z = Q\begin{pmatrix} B_{11} & \mathrm{O} \\ \mathrm{O} & \mathrm{O} \end{pmatrix} P^T$$

then $Z \in \mathbb{S}$ and

$$\|A - Z\|_F^2 = \|B_{22}\|_F^2 \leqslant \|B_{11} - \Omega_k\|_F^2 + \|B_{22}\|_F^2 = \|A - X\|_F^2$$

It follows from the definition of X that B_{11} must equal Ω_k. If B_{22} has singular value decomposition $U_1 \Lambda V_1^T$, then

$$\|A - X\|_F = \|B_{22}\|_F = \|\Lambda\|_F$$

Let

$$U_2 = \begin{pmatrix} I_k & \mathrm{O} \\ \mathrm{O} & U_1 \end{pmatrix} \quad \text{and} \quad V_2 = \begin{pmatrix} I_k & \mathrm{O} \\ \mathrm{O} & V_1 \end{pmatrix}$$

Now

$$U_2^T Q^T A P V_2 = \begin{pmatrix} \Omega_k & \mathrm{O} \\ \mathrm{O} & \Lambda \end{pmatrix}$$

$$A = (QU_2)\begin{pmatrix} \Omega_k & \mathrm{O} \\ \mathrm{O} & \Lambda \end{pmatrix}(PV_2)^T$$

and hence it follows that the diagonal elements of Λ are singular values of A. Thus

$$\|A - X\|_F = \|\Lambda\|_F \geqslant (\sigma_{k+1}^2 + \cdots + \sigma_n^2)^{1/2}$$

It follows from (2) that

$$\|A - X\|_F = \left(\sigma_{k+1}^2 + \cdots + \sigma_n^2\right)^{1/2} = \|A - A'\|_F$$

If $A = U\Sigma V^T$ and we define $E_j = \mathbf{u}_j\mathbf{v}_j^T$ for $j = 1, \ldots, n$, then each E_j is of rank 1 and

$$A = \sigma_1 E_1 + \sigma_2 E_2 + \cdots + \sigma_n E_n \tag{3}$$

To prove (3), note that

$$\left(\mathbf{v}_1\mathbf{v}_1^T + \cdots + \mathbf{v}_n\mathbf{v}_n^T\right)\mathbf{x} = \left(\mathbf{v}_1^T\mathbf{x}\right)\mathbf{v}_1 + \cdots + \left(\mathbf{v}_n^T\mathbf{x}\right)\mathbf{v}_n = \mathbf{x}$$

for every $\mathbf{x} \in R^n$, since the column vectors of V form an orthonormal basis for R^n. Thus

$$N\left(\mathbf{v}_1\mathbf{v}_1^T + \cdots + \mathbf{v}_n\mathbf{v}_n^T - I\right) = R^n$$

and hence

$$\mathbf{v}_1\mathbf{v}_1^T + \cdots + \mathbf{v}_n\mathbf{v}_n^T = I$$

It follows that

$$\begin{aligned} A &= A\left(\mathbf{v}_1\mathbf{v}_1^T + \cdots + \mathbf{v}_n\mathbf{v}_n^T\right) \\ &= \sigma_1\mathbf{u}_1\mathbf{v}_1^T + \cdots + \sigma_n\mathbf{u}_n\mathbf{v}_n^T \\ &= \sigma_1 E_1 + \cdots + \sigma_n E_n \end{aligned}$$

If A is of rank n, then

$$A' = U\begin{bmatrix} \sigma_1 & & & & \\ & \sigma_2 & & O & \\ & & \ddots & & \\ & & & \sigma_{n-1} & \\ & O & & & \\ & & & & \end{bmatrix}V^T = \sigma_1 E_1 + \cdots + \sigma_{n-1}E_{n-1}$$

will be a matrix of rank $n-1$ which is nearest to A with respect to the Frobenius norm. Similarly, $A'' = \sigma_1 E_1 + \cdots + \sigma_{n-2}E_{n-2}$ will be the nearest matrix of rank $n - 2$, and so on. In particular, if A is a nonsingular $n \times n$ matrix, then A' is singular and $\|A - A'\|_F = \sigma_n$. Thus σ_n may be taken as a measure of how close a matrix is to being singular.

The reader should be careful not to use the value of $|A|$ as a measure of how close A is to being singular. If, for example, A is the 100×100 diagonal matrix whose diagonal entries are all $\frac{1}{2}$, then $|A| = 2^{-100}$; however, $\sigma_{100} = \frac{1}{2}$. On the other hand, the matrix in the following example is very close to being singular even though its determinant is 1 and all of its eigenvalues are equal to 1.

EXAMPLE 2. Let A be an $n \times n$ upper triangular matrix whose diagonal elements are all 1 and whose entries above the main diagonal are all -1.

$$A = \begin{bmatrix} 1 & -1 & -1 & \cdots & -1 & -1 \\ 0 & 1 & -1 & \cdots & -1 & -1 \\ 0 & 0 & 1 & \cdots & -1 & -1 \\ \vdots & & & & & \\ 0 & 0 & 0 & \cdots & 1 & -1 \\ 0 & 0 & 0 & \cdots & 0 & 1 \end{bmatrix}$$

Notice that $|A| = |A^{-1}| = 1$ and all the eigenvalues of A are 1. However, if n is large, then A is close to being singular. To see this, let

$$B = \begin{bmatrix} 1 & -1 & -1 & \cdots & -1 & -1 \\ 0 & 1 & -1 & \cdots & -1 & -1 \\ 0 & 0 & 1 & \cdots & -1 & -1 \\ \vdots & & & & & \\ 0 & 0 & 0 & \cdots & 1 & -1 \\ \dfrac{-1}{2^{n-2}} & 0 & 0 & \cdots & 0 & 1 \end{bmatrix}$$

B is singular, since the system $B\mathbf{x} = \mathbf{0}$ has a nontrivial solution $\mathbf{x} = (2^{n-2}, 2^{n-3}, \ldots, 2^0, 1)^T$. The matrices A and B only differ in the $(n, 1)$ position.

$$\|A - B\|_F = \frac{1}{2^{n-2}}$$

It follows from Theorem 7.7.3 that

$$\sigma_n \leqslant \min_{X \text{ singular}} \|A - X\|_F \leqslant \|A - B\|_F = \frac{1}{2^{n-2}}$$

Thus, if $n = 100$, then $\sigma_n \leqslant 1/2^{98}$, and consequently A is very close to singular.

The inverse of the matrix A in Example 2 is given by

$$A^{-1} = \begin{bmatrix} 1 & 1 & 2 & 4 & \cdots & 2^{n-2} \\ 0 & 1 & 1 & 2 & \cdots & 2^{n-3} \\ \vdots & & & & & \\ 0 & 0 & 0 & 0 & \cdots & 2^0 \\ 0 & 0 & 0 & 0 & \cdots & 1 \end{bmatrix}$$

Thus

$$\text{cond}_\infty A = \|A\|_\infty \|A^{-1}\|_\infty = n2^{n-1}$$

If n is large, A will be extremely ill conditioned. This is not surprising. We have already seen that A is very close to a singular matrix. Consequently, we would expect that a small change in the entries of A would have a drastic effect. It seems reasonable that there should be some relation between the condition number and the singular values of a matrix. This is indeed the case when the condition number is defined in terms of the 2-norm of the matrix. Recall that

$$\|A\|_2 = \max_{\mathbf{x}\neq\mathbf{0}} \frac{\|A\mathbf{x}\|_2}{\|\mathbf{x}\|_2}$$

$$\text{cond}_2 (A) = \|A\|_2 \|A^{-1}\|_2$$

***Theorem* 7.7.4.** *If A is an $m \times n$ matrix with singular decomposition $U\Sigma V^T$, then*

$$\|A\|_2 = \sigma_1 \qquad (\textit{the largest singular value})$$

PROOF. Since U and V are orthogonal,

$$\|A\|_2 = \|U\Sigma V^T\|_2 = \|\Sigma\|_2$$

Now

$$\|\Sigma\|_2 = \max_{\mathbf{x}\neq\mathbf{0}} \frac{\|\Sigma\mathbf{x}\|_2}{\|\mathbf{x}\|_2}$$

$$= \max_{\mathbf{x}\neq\mathbf{0}} \frac{\left(\sum_{i=1}^{n} (\sigma_i x_i)^2\right)^{1/2}}{\left(\sum_{i=1}^{n} x_i^2\right)^{1/2}}$$

$$\leqslant \sigma_1$$

However, if we choose $\mathbf{x} = \mathbf{e}_1$, then

$$\frac{\|\Sigma\mathbf{x}\|_2}{\|\mathbf{x}\|_2} = \sigma_1$$

and hence it follows that

$$\|A\|_2 = \|\Sigma\|_2 = \sigma_1$$

***Corollary* 7.7.5.** *If $A = U\Sigma V^T$ is nonsingular, then*

$$\text{cond}_2 (A) = \frac{\sigma_1}{\sigma_n}$$

PROOF. The singular values of $A^{-1} = V\Sigma^{-1}U^T$ arranged in decreasing order are

$$\frac{1}{\sigma_n} \geqslant \frac{1}{\sigma_{n-1}} \geqslant \cdots \geqslant \frac{1}{\sigma_1}$$

Therefore,

$$\|A^{-1}\|_2 = \frac{1}{\sigma_n} \quad \text{and} \quad \text{cond}_2\,(A) = \frac{\sigma_1}{\sigma_n}$$

We close this section by remarking that if two matrices A and B are close, their singular values must also be close. More precisely, if A has the singular values $\sigma_1 \geqslant \sigma_2 \geqslant \cdots \geqslant \sigma_n$ and B has the singular values $\omega_1 \geqslant \omega_2 \geqslant \cdots \geqslant \omega_n$, then

$$|\sigma_i - \omega_i| \leqslant \|A - B\|_2 \qquad i = 1, \ldots, n$$

(see Stewart [16], p. 321). Thus in computing the singular values of a matrix A, we need not worry that small changes in the entries of A will cause drastic changes in the computed singular values. In Section 9 we will learn an algorithm for computing the singular values of a matrix. We will also see how the singular value decomposition can be used to solve least squares problems.

EXERCISES

1. Show that A and A^T have the same nonzero singular values. How are their singular value decompositions related?

2. Use the method of Example 1 to find the singular value decomposition of each of the following matrices.

(a) $\begin{pmatrix} 1 & 1 \\ 0 & 1 \end{pmatrix}$ (b) $\begin{pmatrix} 2 & -2 \\ 1 & 2 \end{pmatrix}$

(c) $\begin{bmatrix} 1 & 3 \\ 3 & 1 \\ 0 & 0 \\ 0 & 0 \end{bmatrix}$ (d) $\begin{bmatrix} 2 & 0 & 0 \\ 0 & 2 & 1 \\ 0 & 1 & 2 \\ 0 & 0 & 0 \end{bmatrix}$

3. For each of the matrices in Exercise 2:

(i) Determine the rank.
(ii) Determine the 2-norm.
(iii) Find the closest (with respect to the Frobenius norm) matrix of rank 1.

4. Prove that if A is a symmetric matrix with eigenvalues $\lambda_1, \lambda_2, \ldots, \lambda_n$, then the singular values of A are $|\lambda_1|, |\lambda_2|, \ldots, |\lambda_n|$.

5. Let A be an $m \times n$ matrix with singular value decomposition $U\Sigma V^T$. Show that

$$\min_{\mathbf{x} \neq \mathbf{0}} \frac{\|A\mathbf{x}\|_2}{\|\mathbf{x}\|_2} = \sigma_n$$

6. Let A be an $m \times n$ matrix of rank n with singular value decomposition $U\Sigma V^T$. Let Σ^+ denote the $n \times m$ matrix

$$\begin{bmatrix} \frac{1}{\sigma_1} & & & & O \\ & \frac{1}{\sigma_2} & & & \\ & & \ddots & & O \\ & O & & \frac{1}{\sigma_n} & \end{bmatrix}$$

and define $A^+ = V\Sigma^+U^T$. Show that $\hat{\mathbf{x}} = A^+\mathbf{b}$ satisfies the normal equations $A^TA\mathbf{x} = A^T\mathbf{b}$.

7. Let A^+ be defined as in Exercise 6 and let $P = AA^+$. Show that $P^2 = P$ and $P^T = P$.

8. THE EIGENVALUE PROBLEM

In this section we will be concerned with numerical methods for computing the eigenvalues and eigenvectors of an $n \times n$ matrix A. The first method we will study is called the power method. The power method is an iterative method for finding the dominant eigenvalue of a matrix and a corresponding eigenvector. By the dominant eigenvalue we mean an eigenvalue λ_1 satisfying $|\lambda_1| > |\lambda_i|$ for $i = 2, \ldots, n$. If the eigenvalues of A

$$|\lambda_1| > |\lambda_2| > \cdots > |\lambda_n|$$

then the power method can be used to compute the eigenvalues one at a time. The second method is called the QR algorithm. The QR algorithm is an iterative method involving orthogonal similarity transformations. It has many advantages over the power method. It will converge whether or not A has a dominant eigenvalue and it calculates all of the eigenvalues at the same time.

In the examples in Chapter 6 the eigenvalues were determined by forming the characteristic polynomial and finding its roots. However, this procedure is generally not recommended for numerical computations. The difficulty with this procedure is that often a small change in one or more of the coefficients of the characteristic polynomial can result in a relatively large change in the computed zeros of the polynomial. For example, consider the polynomial $p(x) = x^{10}$. The lead coefficient is 1 and the remaining coefficients are all 0. If the constant term is altered by adding -10^{-10}, we obtain the polynomial $q(x) = x^{10} - 10^{-10}$. Although the coefficients of $p(x)$ and $q(x)$ only differ by 10^{-10}, the roots of $q(x)$ all have absolute value $\frac{1}{10}$, whereas the roots of $p(x)$ are all 0. Thus even when the coefficients of the characteristic polynomial have been determined quite accurately, the computed eigenvalues may involve significant error. For this reason, the methods presented in this section will not involve the characteristic polynomial. To see that there is some advantage to working directly with the matrix A, we must determine the effect that small changes in the entries of A have upon the eigenvalues. This is done in the following theorem.

Theorem 7.8.1. *Let A be an $n \times n$ matrix with a complete set of eigenvectors and let X be a matrix that diagonalizes A.*

$$X^{-1}AX = D = \begin{bmatrix} \lambda_1 & & & O \\ & \lambda_2 & & \\ & & \ddots & \\ O & & & \lambda_n \end{bmatrix}$$

If $A' = A + E$ and λ' is an eigenvalue of A', then

$$\min_{1 \leqslant i \leqslant n} |\lambda' - \lambda_i| \leqslant \operatorname{cond}_2(X)\|E\|_2 \tag{1}$$

PROOF. We may assume that λ' is unequal to any of the λ_i's (otherwise there is nothing to prove). Thus, if we set $D_1 = D - \lambda' I$, then D_1 is a nonsingular diagonal matrix. Since λ' is an eigenvalue of A', it is also an eigenvalue of $X^{-1}A'X$. Therefore, $X^{-1}A'X - \lambda' I$ is singular and hence $D_1^{-1}(X^{-1}A'X - \lambda' I)$ is also singular. But

$$\begin{aligned} D_1^{-1}(X^{-1}A'X - \lambda' I) &= D_1^{-1}X^{-1}(A + E - \lambda' I)X \\ &= D_1^{-1}X^{-1}EX + I \end{aligned}$$

Therefore, -1 is an eigenvalue of $D_1^{-1}X^{-1}EX$. It follows that

$$|-1| \leqslant \|D_1^{-1}X^{-1}EX\|_2 \leqslant \|D_1^{-1}\|_2 \operatorname{cond}_2(X)\|E\|_2$$

The 2-norm of D_1^{-1} is given by

$$\|D_1^{-1}\|_2 = \max_{1 \leqslant i \leqslant n} |\lambda' - \lambda_i|^{-1}$$

The index i that maximizes $|\lambda' - \lambda_i|^{-1}$ is the same index that minimizes $|\lambda' - \lambda_i|$. Thus

$$\min_{1 \leq i \leq n} |\lambda' - \lambda_i| \leq \text{cond}_2 (X) \|E\|_2$$

If the matrix A is symmetric, we can choose an orthogonal diagonalizing matrix. In general, if Q is any orthogonal matrix, then

$$\text{cond}_2 (Q) = \|Q\|_2 \|Q^{-1}\|_2 = 1$$

Hence (1) simplifies to

$$\min_{1 \leq i \leq n} |\lambda' - \lambda_i| \leq \|E\|_2$$

Thus, if A is symmetric and $\|E\|_2$ is small, the eigenvalues of A' will be close to the eigenvalues of A.

We are now ready to talk about some methods for calculating the eigenvalues and eigenvectors of an $n \times n$ matrix A. The first method we will present computes an eigenvector $\mathbf{x}$ of A by successively applying A to a given vector in R^n. To see the idea behind the method, let us assume that A has n linearly independent eigenvectors $\mathbf{x}_1, \ldots, \mathbf{x}_n$ and that the corresponding eigenvalues satisfy

$$(2) \qquad |\lambda_1| > |\lambda_2| \geq \cdots \geq |\lambda_n|$$

Given an arbitrary vector $\mathbf{v}_0$ in R^n, we can write

$$\begin{aligned} \mathbf{v}_0 &= \alpha_1 \mathbf{x}_1 + \alpha_2 \mathbf{x}_2 + \cdots + \alpha_n \mathbf{x}_n \\ A\mathbf{v}_0 &= \alpha_1 \lambda_1 \mathbf{x}_1 + \alpha_2 \lambda_2 \mathbf{x}_2 + \cdots + \alpha_n \lambda_n \mathbf{x}_n \\ A^2\mathbf{v}_0 &= \alpha_1 \lambda_1^2 \mathbf{x}_1 + \alpha_2 \lambda_2^2 \mathbf{x}_2 + \cdots + \alpha_n \lambda_n^2 \mathbf{x}_n \end{aligned}$$

and in general

$$A^k\mathbf{v}_0 = \alpha_1 \lambda_1^k \mathbf{x}_1 + \alpha_2 \lambda_2^k \mathbf{x}_2 + \cdots + \alpha_n \lambda_n^k \mathbf{x}_n$$

If we define

$$\mathbf{v}_k = A^k \mathbf{v}_0 \qquad k = 1, 2, \ldots$$

then

$$(3) \qquad \frac{1}{\lambda_1^k} \mathbf{v}_k = \alpha_1 \mathbf{x}_1 + \alpha_2 \left(\frac{\lambda_2}{\lambda_1}\right)^k \mathbf{x}_2 + \cdots + \alpha_n \left(\frac{\lambda_n}{\lambda_1}\right)^k \mathbf{x}_n$$

Since

$$\left|\frac{\lambda_i}{\lambda_1}\right| < 1 \qquad \text{for} \quad i = 2, 3, \ldots, n$$

it follows that

$$\frac{1}{\lambda_1^k}\mathbf{v}_k \to \alpha_1\mathbf{x}_1 \qquad \text{as} \quad k \to \infty$$

Thus, if $\alpha_1 \neq 0$, the sequence $\{(1/\lambda_1^k)\mathbf{v}_k\}$ converges to an eigenvector $\alpha_1\mathbf{x}_1$ of A. There are some obvious difficulties with the method as it has been presented so far. We cannot compute $(1/\lambda_1^k)\mathbf{v}_k$, since λ_1 is unknown. Even if λ_1 were known, there would be difficulties because of λ_1^k approaching 0 or $\pm\infty$. Fortunately, however, we do not have to scale the sequence $\{\mathbf{v}_k\}$ using $1/\lambda_1^k$. If the $\mathbf{v}_k$'s are scaled so that one obtains unit vectors at each step, the sequence will converge to a unit vector in the direction of $\mathbf{x}_1$. The eigenvalue λ_1 can be computed at the same time. This method of computing the eigenvalue of largest magnitude and a corresponding eigenvector is called the *power method.* It may be summarized as follows.

The Power Method

Two sequences $\{\mathbf{v}_k\}$ and $\{\mathbf{u}_k\}$ are defined recursively. To start, $\mathbf{u}_0$ can be any nonzero vector in R^n. Once $\mathbf{u}_k$ has been determined, the vectors $\mathbf{v}_{k+1}$ and $\mathbf{u}_{k+1}$ are calculated as follows:

(i) Set $\mathbf{v}_{k+1} = A\mathbf{u}_k$.
(ii) Find the coordinate j_{k+1} of $\mathbf{v}_{k+1}$ of maximum modulus.
(iii) Set $\mathbf{u}_{k+1} = (1/v_{j_{k+1}})\mathbf{v}_{k+1}$.

The sequence $\{\mathbf{u}_k\}$ has the property that for $k \geqslant 1$, $\|\mathbf{u}_k\|_\infty = u_{j_k} = 1$. If the eigenvalues of A satisfy (2) and $\mathbf{u}_0$ can be written as a linear combination of eigenvectors $\alpha_1\mathbf{x}_1 + \cdots + \alpha_n\mathbf{x}_n$ with $\alpha_1 \neq 0$, the sequence $\{\mathbf{u}_k\}$ will converge to an eigenvector $\mathbf{y}$ of λ_1. If k is large, then $\mathbf{u}_k$ will be a good approximation to $\mathbf{y}$ and $\mathbf{v}_{k+1} = A\mathbf{u}_k$ will be a good approximation to $\lambda_1\mathbf{y}$. Since the j_kth coordinate of $\mathbf{u}_k$ is 1, it follows that the j_kth coordinate of $\mathbf{v}_{k+1}$ will be a good approximation to λ_1.

In view of (3), we can expect that the $\mathbf{u}_k$'s will converge to $\mathbf{y}$ at the same rate at which $(\lambda_2/\lambda_1)^k$ is converging to 0. Thus, if $|\lambda_2|$ is nearly as large as $|\lambda_1|$, the convergence will be slow.

EXAMPLE 1. Let

$$A = \begin{pmatrix} 2 & 1 \\ 1 & 2 \end{pmatrix}$$

It is an easy matter to determine the exact eigenvalues of A. These turn out to be $\lambda_1 = 3$ and $\lambda_2 = 1$, with corresponding eigenvectors $\mathbf{x}_1 = (1, 1)^T$ and

$\mathbf{x}_2 = (1, -1)^T$. To illustrate how the vectors generated by the power method converge, we will apply the method with $\mathbf{u}_0 = (2, 1)^T$.

$$\mathbf{v}_1 = A\mathbf{u}_0 = \begin{pmatrix} 5 \\ 4 \end{pmatrix}, \qquad \mathbf{u}_1 = \frac{1}{5}\mathbf{v}_1 = \begin{pmatrix} 1.0 \\ 0.8 \end{pmatrix}$$

$$\mathbf{v}_2 = A\mathbf{u}_1 = \begin{pmatrix} 2.8 \\ 2.6 \end{pmatrix}, \qquad \mathbf{u}_2 = \frac{1}{2.8}\mathbf{v}_2 = \begin{bmatrix} 1 \\ \frac{13}{14} \end{bmatrix} \approx \begin{pmatrix} 1.00 \\ 0.93 \end{pmatrix}$$

$$\mathbf{v}_3 = A\mathbf{u}_2 = \frac{1}{14}\begin{pmatrix} 41 \\ 40 \end{pmatrix}, \qquad \mathbf{u}_3 = \frac{14}{41}\mathbf{v}_3 = \begin{bmatrix} 1 \\ \frac{40}{41} \end{bmatrix} \approx \begin{pmatrix} 1.00 \\ 0.98 \end{pmatrix}$$

$$\mathbf{v}_4 = A\mathbf{u}_3 \approx \begin{pmatrix} 2.98 \\ 2.95 \end{pmatrix}$$

If $\mathbf{u}_3 = (1.00, 0.98)^T$ is taken as an approximate eigenvector, then 2.98 is the approximate value of λ_1. Thus with only a few iterations, the approximation for λ_1 involves an error of only 0.02.

The power method can be used to compute the eigenvalue λ_1 of largest magnitude and a corresponding eigenvector $\mathbf{y}_1$. What about remaining eigenvalues and eigenvectors? If we could reduce the problem of finding the remaining eigenvalues of A to that of finding the eigenvalues of some $(n-1) \times (n-1)$ matrix A_1, then the power method could be applied to A_1. This can actually be done by a process called *deflation*.

Deflation

The idea behind deflation is to find a nonsingular matrix H such that HAH^{-1} is a matrix of the form

$$\begin{bmatrix} \lambda_1 & \times & \cdots & \times \\ \hline 0 & & & \\ \vdots & & A_1 & \\ 0 & & & \end{bmatrix} \tag{4}$$

Since A and HAH^{-1} are similar, they have the same characteristic polynomials. Thus, if HAH^{-1} is of the form (4), then

$$|A - \lambda I| = |HAH^{-1} - \lambda I| = (\lambda_1 - \lambda)|A_1 - \lambda I|$$

and it follows that the remaining $n-1$ eigenvalues of A are the eigenvalues of A_1. The question remains: How do we find such a matrix H? Note that the form (4) requires that the first column of HAH^{-1} be $\lambda_1\mathbf{e}_1$. The first column of HAH^{-1} is $HAH^{-1}\mathbf{e}_1$. Thus

$$HAH^{-1}\mathbf{e}_1 = \lambda_1\mathbf{e}_1$$

or, equivalently,

$$A(H^{-1}\mathbf{e}_1) = \lambda_1(H^{-1}\mathbf{e}_1)$$

So $H^{-1}\mathbf{e}_1$ is in the eigenspace corresponding to λ_1. Thus, for some eigenvector $\mathbf{x}_1$ belonging to λ_1,

$$H^{-1}\mathbf{e}_1 = \mathbf{x}_1 \qquad \text{or} \qquad H\mathbf{x}_1 = \mathbf{e}_1$$

We must find a matrix H such that $H\mathbf{x}_1 = \mathbf{e}_1$ for some eigenvector $\mathbf{x}_1$ belonging to λ_1. This can be done by means of a Householder transformation. If $\mathbf{y}_1$ is the computed eigenvector belonging to λ_1, set

$$\mathbf{x}_1 = \frac{1}{\|\mathbf{y}_1\|_2}\mathbf{y}_1$$

Since $\|\mathbf{x}_1\|_2 = 1$, one can find a Householder transformation H such that

$$H\mathbf{x}_1 = \mathbf{e}_1$$

Because H is a Householder transformation, it follows that $H^{-1} = H$ and hence HAH is the desired similarity transformation.

Reduction to Hessenberg Form

The standard methods for finding eigenvalues are all iterative. The amount of work required in each iteration is often prohibitively high unless initially A is in some special form that is easier to work with. If this is not the case, the standard procedure is to reduce A to a simpler form by means of similarity transformations. Generally, Householder matrices are used to transform A into a matrix of the form

$$\begin{bmatrix} \times & \times & \cdots & \times & \times & \times \\ \times & \times & \cdots & \times & \times & \times \\ 0 & \times & \cdots & \times & \times & \times \\ 0 & 0 & \cdots & \times & \times & \times \\ \vdots & & & & & \\ 0 & 0 & \cdots & \times & \times & \times \\ 0 & 0 & \cdots & 0 & \times & \times \end{bmatrix}$$

A matrix in this form is said to be in *upper Hessenberg form*. Thus B is an upper Hessenberg form if and only if $b_{ij} = 0$ whenever $i \geqslant j + 2$.

A matrix A can be transformed into upper Hessenberg form in the following manner. First, choose a Householder matrix H_1 so that H_1A is of the form

$$\begin{bmatrix} a_{11} & a_{12} & \cdots & a_{1n} \\ \times & \times & \cdots & \times \\ 0 & \times & \cdots & \times \\ \vdots & & & \\ 0 & \times & \cdots & \times \end{bmatrix}$$

The matrix H_1 will be of the form

$$\begin{bmatrix} 1 & 0 & \cdots & 0 \\ 0 & \times & \cdots & \times \\ \vdots & & & \\ 0 & \times & \cdots & \times \end{bmatrix}$$

and hence postmultiplication of H_1A_1 by H_1 will leave the first column unchanged. If $A^{(1)} = H_1AH_1$, then $A^{(1)}$ is a matrix of the form

$$\begin{bmatrix} a_{11}^{(1)} & a_{12}^{(1)} & \cdots & a_{1n}^{(1)} \\ a_{21}^{(1)} & a_{22}^{(1)} & \cdots & a_{2n}^{(1)} \\ 0 & a_{32}^{(1)} & \cdots & a_{3n}^{(1)} \\ \vdots & & & \\ 0 & a_{n2}^{(1)} & \cdots & a_{nn}^{(1)} \end{bmatrix}$$

Since H_1 is a Householder matrix, it follows that $H_1^{-1} = H_1$, and hence $A^{(1)}$ is similar to A. Next a Householder matrix H_2 is chosen so that

$$H_2(a_{12}^{(1)}, a_{22}^{(1)}, \ldots, a_{n2}^{(1)})^T = (a_{12}^{(1)}, a_{22}^{(1)}, \times, 0, \ldots, 0)^T$$

The matrix H_2 will be of the form

$$\begin{bmatrix} 1 & 0 & 0 & \cdots & 0 \\ 0 & 1 & 0 & \cdots & 0 \\ 0 & 0 & \times & \cdots & \times \\ \vdots & & & & \\ 0 & 0 & \times & \cdots & \times \end{bmatrix} = \left(\begin{array}{c|c} I_2 & O \\ \hline O & X \end{array}\right)$$

Multiplication of $A^{(1)}$ on the left by H_2 will leave the first two rows and the first column unchanged.

$$H_2A^{(1)} = \begin{bmatrix} a_{11}^{(1)} & a_{12}^{(1)} & a_{13}^{(1)} & \cdots & a_{1n}^{(1)} \\ a_{21}^{(1)} & a_{22}^{(1)} & a_{23}^{(1)} & \cdots & a_{2n}^{(1)} \\ 0 & \times & \times & \cdots & \times \\ 0 & 0 & \times & \cdots & \times \\ \vdots & & & & \\ 0 & 0 & \times & \cdots & \times \end{bmatrix}$$

Postmultiplication of $H_2A^{(1)}$ by H_2 will leave the first two columns unchanged. Thus $A^{(2)} = H_2A^{(1)}H_2$ is of the form

$$\begin{bmatrix} \times & \times & \times & \cdots & \times \\ \times & \times & \times & \cdots & \times \\ 0 & \times & \times & \cdots & \times \\ 0 & 0 & \times & \cdots & \times \\ \vdots & & & & \\ 0 & 0 & \times & \cdots & \times \end{bmatrix}$$

This process may be continued until one ends up with an upper Hessenberg matrix

$$H = A^{(n-2)} = H_{n-2} \cdots H_2H_1AH_1H_2 \cdots H_{n-2}$$

which is similar to A.

If in particular, A is symmetric, then since

$$\begin{aligned} H^T &= H_{n-2}^T \cdots H_2^T H_1^T A^T H_1^T H_2^T \cdots H_{n-2}^T \\ &= H_{n-2} \cdots H_2H_1AH_1H_2 \cdots H_{n-2} \\ &= H \end{aligned}$$

it follows that H is tridiagonal. Thus any $n \times n$ matrix A can be reduced by similarity transformations to upper Hessenberg form. If A is symmetric, the reduction will yield a tridiagonal matrix.

We close this section by outlining one of the best methods available for computing the eigenvalues of a matrix. The method is called the *QR* algorithm and was presented by K. G. F. Francis in 1961.

QR Algorithm

Given an $n \times n$ matrix A, factor it into a product Q_1R_1 where Q_1 is orthogonal and R_1 is upper triangular. Define

$$A_1 = A = Q_1R_1$$

and

$$A_2 = Q_1^T A Q_1 = R_1 Q_1$$

Factor A_2 into a product $Q_2 R_2$ where Q_2 is orthogonal and R_2 is upper triangular. Define

$$A_3 = Q_2^T A_2 Q_2 = R_2 Q_2$$

Note that $A_2 = Q_1^T A Q_1$ and $A_3 = (Q_1 Q_2)^T A (Q_1 Q_2)$ are both similar to A. We can continue in this manner to obtain a sequence of similar matrices. In general, if

$$A_k = Q_k R_k$$

then A_{k+1} is defined to be $R_k Q_k$. It can be shown that under very general conditions the sequence of matrices defined in this way converges to a matrix of the form

$$\begin{bmatrix} B_1 & & & \times \\ & B_2 & & \\ & & \ddots & \\ & O & & B_s \end{bmatrix}$$

where the B_i's are either 1×1 or 2×2 diagonal blocks. Each 2×2 block will correspond to a pair of complex conjugate eigenvalues of A. The eigenvalues of A will be the eigenvalues of the B_i's. In the case where A is symmetric, each of the A_k's will also be symmetric and the sequence will converge to a diagonal matrix.

EXAMPLE 2. Let A_1 be the matrix from Example 1. The QR factorization of A_1 requires only a single Givens transformation,

$$G_1 = \frac{1}{\sqrt{5}}\begin{pmatrix} 2 & 1 \\ 1 & -2 \end{pmatrix}$$

Thus

$$A_2 = G_1 A G_1 = \tfrac{1}{5}\begin{pmatrix} 2 & 1 \\ 1 & -2 \end{pmatrix}\begin{pmatrix} 2 & 1 \\ 1 & 2 \end{pmatrix}\begin{pmatrix} 2 & 1 \\ 1 & -2 \end{pmatrix} = \begin{pmatrix} 2.8 & -0.6 \\ -0.6 & 1.2 \end{pmatrix}$$

The QR factorization of A_2 can be accomplished using the Givens transformation

$$G_2 = \frac{1}{\sqrt{8.2}}\begin{pmatrix} 2.8 & -0.6 \\ -0.6 & -2.8 \end{pmatrix}$$

It follows that

$$A_3 = G_2 A_2 G_2 \approx \begin{pmatrix} 2.98 & 0.22 \\ 0.22 & 1.02 \end{pmatrix}$$

The off-diagonal elements are getting closer to 0 after each iteration, and the diagonal elements are approaching the eigenvalues $\lambda_1 = 3$ and $\lambda_2 = 1$.

Remarks

1. Because of the amount of work required at each iteration of the QR algorithm, it is important that the starting matrix A be in either Hessenberg or symmetric tridiagonal form. If this is not the case, we should perform similarity transformations on A to obtain a matrix A_1 that is in one of these forms.

2. If A_k is in upper Hessenberg form, the QR factorization can be carried out using $n - 1$ Givens transformations.

$$G_{n,\,n-1} \cdots G_{32}G_{21}A_k = R_k$$

Setting

$$Q_k^T = G_{n,\,n-1} \cdots G_{32}G_{21}$$

we have

$$A_k = Q_k R_k$$

and

$$A_{k+1} = Q_k^T A_k Q_k$$

To compute A_{k+1}, it is not necessary to determine Q_k explicitly. One need only keep track of the $n - 1$ Givens transformations. When R_k is postmultiplied by G_{21}, the resulting matrix will have the (2, 1) entry filled in. The other entries below the diagonals will all still be zero. Postmultiplying $R_k G_{21}$ by G_{32} will have the effect of filling in the (3, 2) position. Postmultiplication of $R_k G_{21} G_{32}$ by G_{43} will fill in the (4, 3) position, and so on. Thus the resulting matrix $A_{k+1} = R_k G_{21} G_{32} \cdots G_{n,\,n-1}$ will be in upper Hessenberg form. If A_1 is a symmetric tridiagonal matrix, each of the succeeding A_i's will be upper Hessenberg and symmetric. Thus $A_2, A_3, \ldots$ will all be tridiagonal.

3. As in the power method, convergence may be slow when some of the eigenvalues are close together. To speed up convergence, it is customary to introduce *origin shifts*. At the kth step a scalar α_k is chosen and $A_k - \alpha_k I$ (rather than A_k) is decomposed into a product $Q_k R_k$. The matrix A_{k+1} is defined by

$$A_{k+1} = R_k Q_k + \alpha_k I$$

Note that

$$Q_k^T A_k Q_k = Q_k^T (Q_k R_k + \alpha_k I) Q_k = R_k Q_k + \alpha_k I = A_{k+1}$$

so that A_k and A_{k+1} are similar. With the proper choice of shifts α_k, the convergence can be greatly accelerated.

4. In our brief discussion we have presented only an outline of the method. Many of the details, such as how to choose the origin shifts, have been omitted. For a more thorough discussion and a proof of convergence, see Wilkinson ([18], Chap. 8).

EXERCISES

1. Let

$$A = \begin{pmatrix} 1 & 1 \\ 1 & 1 \end{pmatrix}$$

(a) Apply one iteration of the power method with any nonzero starting vector.
(b) Apply one iteration of the QR algorithm to A.
(c) Determine the exact eigenvalues of A by solving the characteristic equation and determine the eigenspace corresponding to the largest eigenvalue. Compare your answers with those to parts (a) and (b).

2. Let

$$A = \begin{bmatrix} 2 & 1 & 0 \\ 1 & 3 & 1 \\ 0 & 1 & 2 \end{bmatrix} \quad \text{and} \quad \mathbf{u}_0 = \begin{bmatrix} 1 \\ 1 \\ 1 \end{bmatrix}$$

(a) Apply the power method to compute $\mathbf{v}_1$, $\mathbf{u}_1$, $\mathbf{v}_2$, $\mathbf{u}_2$, and $\mathbf{v}_3$. (Round off to two decimal places.)
(b) Determine an approximation λ_1' to the largest eigenvalue of A from the coordinates of $\mathbf{v}_3$. Determine the exact value of λ_1 and compare it to λ_1'. What is the relative error?

3. Let

$$A = \begin{pmatrix} 1 & 2 \\ -1 & -1 \end{pmatrix} \quad \text{and} \quad \mathbf{u}_0 = \begin{pmatrix} 1 \\ 1 \end{pmatrix}$$

(a) Compute $\mathbf{u}_1$, $\mathbf{u}_2$, $\mathbf{u}_3$, and $\mathbf{u}_4$ using the power method.
(b) Explain why the power method will fail to converge in this case.

4. Let

$$A = A_1 = \begin{pmatrix} 1 & 1 \\ 1 & 3 \end{pmatrix}$$

Compute A_2 and A_3 using the QR algorithm. Compute the exact eigenvalues of A and compare them to the diagonal elements of A_3. To how many decimal places do they agree?

5. Let A be an $n \times n$ matrix with distinct real eigenvalues $\lambda_1, \lambda_2, \ldots, \lambda_n$. Let λ be a scalar that is not an eigenvalue of A and let $B = (A - \lambda I)^{-1}$.
   (a) Show that the scalars $\mu_j = 1/(\lambda_j - \lambda)$, $j = 1, \ldots, n$ are the eigenvalues of B.
   (b) Show that if $\mathbf{x}_j$ is an eigenvector of B belonging to μ_j, then $\mathbf{x}_j$ is an eigenvector of A belonging to λ_j.
   (c) Show that if the power method is applied to B, then the sequence of vectors will converge to an eigenvector of A belonging to the eigenvalue that is closest to λ. [The convergence will be quite rapid if λ is much closer to one λ_i than to any of the others. This method of computing eigenvectors using powers of $(A - \lambda I)^{-1}$ is called the *inverse power method*.]

6. Let $\mathbf{x} = (x_1, \ldots, x_n)^T$ be an eigenvector of A belonging to λ. If $|x_i| = \|\mathbf{x}\|_\infty$, show that:
   (a)
   $$\sum_{j=1}^{n} a_{ij} x_j = \lambda x_i$$
   (b) $|\lambda - a_{ii}| \leqslant \sum_{\substack{j=1 \\ j \neq i}}^{n} |a_{ij}|$ (Gerschgorin's theorem)

7. Let A be a matrix with eigenvalues $\lambda_1, \ldots, \lambda_n$, and let λ be an eigenvalue of $A + E$. Let X be a matrix that diagonalizes A, and let $C = X^{-1}EX$. Prove:
   (a) For some i
   $$|\lambda - \lambda_i| \leqslant \sum_{j=1}^{n} |c_{ij}|$$
   [*HINT:* λ is an eigenvalue of $X^{-1}(A + E)X$. Apply Gerschgorin's theorem from Exercise 6.]
   (b) $\min_{1 \leqslant j \leqslant n} |\lambda - \lambda_j| \leqslant \text{cond}_\infty (X) \|E\|_\infty$

8. Let $A_k = Q_k R_k$, $k = 1, 2, \ldots$ be the sequence of matrices derived from $A = A_1$ by applying the QR algorithm. For each positive integer k, define
   $$P_k = Q_1 Q_2 \cdots Q_k \qquad \text{and} \qquad U_k = R_k \cdots R_2 R_1$$
   Show that
   $$P_k A_{k+1} = A P_k$$
   for all $k \geqslant 1$.

9. Let P_k and U_k be defined as in Exercise 7.
(a) Show that

$$P_{k+1}U_{k+1} = P_kA_{k+1}U_k = AP_kU_k$$

(b) Show that $P_kU_k = A^k$ and hence that $(Q_1Q_2 \cdots Q_k)(R_k \cdots R_2R_1)$ is the QR factorization of A^k.

9. LEAST SQUARES PROBLEMS

In this section we will study computational methods for finding least squares solutions to overdetermined systems. Let A be an $m \times n$ matrix with $m \geqslant n$ and let $\mathbf{b} \in R^m$. We will consider some methods for computing a vector $\hat{\mathbf{x}}$ that minimizes $\|\mathbf{b} - A\mathbf{x}\|_2^2$.

The Normal Equations

We saw in Chapter 4 that if $\hat{\mathbf{x}}$ satisfies the normal equations

$$A^TA\mathbf{x} = A^T\mathbf{b}$$

then $\hat{\mathbf{x}}$ is a solution to the least squares problem. If A is of full rank (rank n), then A^TA is nonsingular and hence the system will have a unique solution. Thus, if A^TA is invertible, one possible method for solving the least squares problem is to form the normal equations and then solve them using Gaussian elimination. An algorithm for doing this would have two main parts.

1. Compute $B = A^TA$ and $\mathbf{c} = A^T\mathbf{b}$.
2. Solve $B\mathbf{x} = \mathbf{c}$.

Note that forming the normal equations requires roughly $mn^2/2$ multiplications. Since A^TA is nonsingular, the matrix B is positive definite. For positive definite matrices there are reduction algorithms that require only half the usual number of multiplications. Thus the solution of $B\mathbf{x} = \mathbf{c}$ requires roughly $n^3/6$ multiplications. Most of the work then occurs in forming the normal equations rather than solving them. However, the main difficulty with this method is that in forming the normal equations, we may well end up transforming the problem into an ill-conditioned one. Recall from Section 4 that if $\mathbf{x}'$ is the computed solution to $B\mathbf{x} = \mathbf{c}$ and $\mathbf{x}$ is the exact solution, then the inequality

$$\frac{1}{\text{cond}(B)}\frac{\|\mathbf{r}\|}{\|\mathbf{c}\|} \leqslant \frac{\|\mathbf{x} - \mathbf{x}'\|}{\|\mathbf{x}\|} \leqslant \text{cond}(B)\frac{\|\mathbf{r}\|}{\|\mathbf{x}\|}$$

shows how the relative error compares to the relative residual. If A has singular values $\sigma_1 \geqslant \sigma_2 \geqslant \cdots \geqslant \sigma_n > 0$, then $\text{cond}_2(A) = \sigma_1/\sigma_n$. The singular values of B are $\sigma_1^2, \sigma_2^2, \ldots, \sigma_n^2$. Thus

$$\text{cond}_2(B) = \frac{\sigma_1^2}{\sigma_n^2} = [\text{cond}_2(A)]^2$$

If, for example, $\text{cond}_2(A) = 100$, the relative error in the computed solution to the normal equations could be 10^4 times as large as the relative residual. For this reason one should be very careful about using the normal equations to solve least squares problems.

In Chapter 4 we saw how to use the Gram–Schmidt process to obtain a QR factorization of a matrix A with full rank. In that case the matrix Q was an $m \times n$ matrix with orthonormal columns and R was an $n \times n$ upper triangular matrix. In the following numerical method, we will use Householder transformations to obtain a QR factorization of A. In this case Q will be an $m \times m$ orthogonal matrix and R will be an $m \times n$ matrix whose subdiagonal entries are all 0.

The QR Factorization

Given an $m \times n$ matrix A of full rank, we can apply $n - 1$ Householder transformations to zero out all the elements below the diagonal. Thus

$$H_{n-1}H_{n-2}\cdots H_1 A = R$$

where R is of the form

$$\begin{pmatrix} R_1 \\ O \end{pmatrix} = \begin{bmatrix} \times & \times & \times & \cdots & \times \\ & \times & \times & \cdots & \times \\ & & \times & \cdots & \times \\ & & & \ddots & \vdots \\ & & & & \times \\ & & O & & \end{bmatrix}$$

with nonzero diagonal. Let

$$Q^T = H_{n-1}\cdots H_1 = \begin{pmatrix} Q_1^T \\ Q_2^T \end{pmatrix}$$

where Q_1^T is an $n \times m$ matrix consisting of the first n rows of Q^T. Since

$$Q^T A = \begin{pmatrix} Q_1^T A \\ Q_2^T A \end{pmatrix} = \begin{pmatrix} R_1 \\ O \end{pmatrix}$$

it follows that $A = Q_1R_1$. Let

$$\mathbf{c} = Q^T\mathbf{b} = \begin{pmatrix} Q_1^T\mathbf{b} \\ Q_2^T\mathbf{b} \end{pmatrix} = \begin{pmatrix} \mathbf{c}_1 \\ \mathbf{c}_2 \end{pmatrix}$$

The normal equations can be written in the form

$$R_1^TQ_1^TQ_1R_1\mathbf{x} = R_1^TQ_1^T\mathbf{b}$$

Since $Q_1^TQ_1 = I$ and R_1^T is nonsingular, this simplifies to

$$R_1\mathbf{x} = \mathbf{c}_1$$

This system can be solved by back substitution. The solution $\mathbf{x} = R_1^{-1}\mathbf{c}_1$ will be the unique solution to the least squares problem. To compute the residual, note that

$$Q^T\mathbf{r} = \begin{pmatrix} \mathbf{c}_1 \\ \mathbf{c}_2 \end{pmatrix} - \begin{pmatrix} R_1 \\ O \end{pmatrix}\mathbf{x} = \begin{pmatrix} \mathbf{0} \\ \mathbf{c}_2 \end{pmatrix}$$

so that

$$\mathbf{r} = Q\begin{pmatrix} \mathbf{0} \\ \mathbf{c}_2 \end{pmatrix} \qquad \text{and} \qquad \|\mathbf{r}\|_2 = \|\mathbf{c}_2\|_2$$

In summation, when A has full rank, the least squares problem can be solved as follows:

1. Use Householder transformations to compute

$$R = H_{n-1} \cdots H_2H_1A \qquad \text{and} \qquad \mathbf{c} = H_{n-1} \cdots H_2H_1\mathbf{b}.$$

2. Use back substitution to solve $R_1\mathbf{x} = \mathbf{c}_1$.

Now consider the case where the matrix A has rank $r < n$. The singular value decomposition provides the key to solving the least squares problem in this case. It can be used to construct a generalized inverse of A. In the case where A is a nonsingular $n \times n$ matrix with singular value decomposition $U\Sigma V^T$, the inverse is given by

$$A^{-1} = V\Sigma^{-1}U^T$$

More generally, if $A = U\Sigma V^T$ is an $m \times n$ matrix of rank r, the matrix Σ will be an $m \times n$ matrix of the form

$$\Sigma = \left(\begin{array}{c|c} \Sigma_1 & O \\ \hline O & O \end{array}\right) = \left[\begin{array}{cccc|c} \sigma_1 & & & O & \\ & \sigma_2 & & & O \\ O & & \ddots & & \\ & & & \sigma_r & \\ \hline & & O & & O \end{array}\right]$$

and one can define

$$A^+ = V\Sigma^+ U^T \tag{1}$$

where Σ^+ is the $n \times m$ matrix

$$\Sigma^+ = \left(\begin{array}{c|c} \Sigma_1^{-1} & O \\ \hline O & O \end{array}\right) = \left[\begin{array}{ccc|c} \frac{1}{\sigma_1} & & O & \\ & \ddots & & O \\ O & & \frac{1}{\sigma_r} & \\ \hline & O & & O \end{array}\right]$$

Equation (1) gives a natural generalization of the inverse of a matrix. The matrix A^+ defined by (1) is called the *pseudoinverse* of A.

It is also possible to define A^+ by its algebraic properties. These properties are given in the following four conditions.

The Penrose Conditions (Penrose, 1955)

1. $AXA = A$
2. $XAX = X$
3. $(AX)^T = AX$
4. $(XA)^T = XA$

If A is an $m \times n$ matrix, we claim that there is a unique $n \times m$ matrix X that satisfies these conditions. Indeed, if we choose $X = A^+ = V\Sigma^+U^T$, then it is easily verified that X satisfies all four conditions. We leave this as an exercise for the reader. To show uniqueness, suppose that Y also satisfies the Penrose conditions. By successively applying these conditions, one can argue as follows:

$$\begin{aligned} X &= XAX & (2) \qquad Y &= YAY & (2) \\ &= A^TX^TX & (4) \qquad &= YY^TA^T & (3) \\ &= (AYA)^TX^TX & (1) \qquad &= YY^T(AXA)^T & (1) \\ &= (A^TY^T)(A^TX^T)X & \qquad &= Y(Y^TA^T)(X^TA^T) & \\ &= YAXAX & (4) \qquad &= YAYAX & (3) \\ &= YAX & (1) \qquad &= YAX & (1) \end{aligned}$$

Therefore, $X = Y$. Thus A^+ is the unique matrix satisfying the four Penrose conditions. These conditions are often used to define the pseudoinverse and A^+ is often referred to as the *Moore–Penrose pseudoinverse*.

To see how the pseudoinverse can be used in solving least squares problems, let us first consider the case where A is an $m \times n$ matrix of rank n. Thus Σ is of the form

$$\Sigma = \begin{pmatrix} \Sigma_1 \\ O \end{pmatrix}$$

where Σ_1 is a nonsingular $n \times n$ diagonal matrix. The matrix A^TA is nonsingular and

$$(A^TA)^{-1} = V(\Sigma^T\Sigma)^{-1}V^T$$

The solution to the normal equations is given by

$$\begin{aligned} \mathbf{x} &= (A^TA)^{-1}A^T\mathbf{b} \\ &= V(\Sigma^T\Sigma)^{-1}V^TV\Sigma^TU^T\mathbf{b} \\ &= V(\Sigma^T\Sigma)^{-1}\Sigma^TU^T\mathbf{b} \\ &= V\Sigma^+U^T\mathbf{b} \\ &= A^+\mathbf{b} \end{aligned}$$

Thus if A has full rank, $A^+\mathbf{b}$ is the solution to the least squares problem. What about the case where A has rank $r < n$? In this case there are infinitely many solutions to the least squares problem. The following theorem shows that not only is $A^+\mathbf{b}$ a solution, but it is the minimal solution with respect to the 2-norm.

Theorem 7.9.1. *If A is an $m \times n$ matrix of rank $r < n$ with singular value decomposition $U\Sigma V^T$, then the vector*

$$\mathbf{x} = A^+\mathbf{b} = V\Sigma^+U^T\mathbf{b}$$

minimizes $\|\mathbf{b} - A\mathbf{x}\|_2^2$. Moreover, if $\mathbf{z}$ is any other vector that minimizes $\|\mathbf{b} - A\mathbf{x}\|_2^2$, then $\|\mathbf{z}\|_2 > \|\mathbf{x}\|_2$.

PROOF. Let $\mathbf{x}$ be a vector in R^n and define

$$\mathbf{c} = U^T\mathbf{b} = \begin{pmatrix} \mathbf{c}_1 \\ \mathbf{c}_2 \end{pmatrix} \quad \text{and} \quad \mathbf{y} = V^T\mathbf{x} = \begin{pmatrix} \mathbf{y}_1 \\ \mathbf{y}_2 \end{pmatrix}$$

where $\mathbf{c}_1$ and $\mathbf{y}_1$ are vectors in R^r. Since U^T is orthogonal, it follows that

$$\begin{aligned} \|\mathbf{b} - A\mathbf{x}\|_2^2 &= \|U^T\mathbf{b} - \Sigma(V^T\mathbf{x})\|_2^2 \\ &= \|\mathbf{c} - \Sigma\mathbf{y}\|_2^2 \\ &= \left\| \begin{pmatrix} \mathbf{c}_1 \\ \mathbf{c}_2 \end{pmatrix} - \begin{pmatrix} \Sigma_1 & O \\ O & O \end{pmatrix} \begin{pmatrix} \mathbf{y}_1 \\ \mathbf{y}_2 \end{pmatrix} \right\|_2^2 \\ &= \left\| \begin{pmatrix} \mathbf{c}_1 - \Sigma_1\mathbf{y}_1 \\ \mathbf{c}_2 \end{pmatrix} \right\|_2^2 \\ &= \|\mathbf{c}_1 - \Sigma_1\mathbf{y}_1\|_2^2 + \|\mathbf{c}_2\|_2^2 \end{aligned}$$

Since $\mathbf{c}_2$ is independent of $\mathbf{x}$, it follows that $\|\mathbf{b} - A\mathbf{x}\|^2$ will be minimal if and only if

$$\|\mathbf{c}_1 - \Sigma_1 \mathbf{y}_1\| = 0$$

Thus $\mathbf{x}$ is a solution to the least squares problem if and only if $\mathbf{x} = V\mathbf{y}$, where $\mathbf{y}$ is a vector of the form

$$\begin{pmatrix} \Sigma_1^{-1}\mathbf{c}_1 \\ \mathbf{y}_2 \end{pmatrix}$$

In particular,

$$\begin{aligned} \mathbf{x} &= V\begin{pmatrix} \Sigma_1^{-1}\mathbf{c}_1 \\ \mathbf{0} \end{pmatrix} \\ &= V\begin{pmatrix} \Sigma_1^{-1} & O \\ O & O \end{pmatrix}\begin{pmatrix} \mathbf{c}_1 \\ \mathbf{c}_2 \end{pmatrix} \\ &= V\Sigma^+ U^T\mathbf{b} \\ &= A^+\mathbf{b} \end{aligned}$$

is a solution. If $\mathbf{z}$ is any other solution, $\mathbf{z}$ must be of the form

$$V\mathbf{y} = V\begin{pmatrix} \Sigma_1^{-1}\mathbf{c}_1 \\ \mathbf{y}_2 \end{pmatrix}$$

with $\mathbf{y}_2 \neq \mathbf{0}$. It follows then that

$$\|\mathbf{z}\|^2 = \|\mathbf{y}\|^2 = \|\Sigma_1^{-1}\mathbf{c}_1\|^2 + \|\mathbf{y}_2\|^2 > \|\Sigma^{-1}\mathbf{c}_1\|^2 = \|\mathbf{x}\|^2$$

If the singular value decomposition $U\Sigma V^T$ of A is known, it is a simple matter to compute the solution to the least squares problem. If $U = (\mathbf{u}_1, \ldots, \mathbf{u}_m)$ and $V = (\mathbf{v}_1, \ldots, \mathbf{v}_n)$, then defining $\mathbf{y} = \Sigma^+ U^T\mathbf{b}$, we have

$$\begin{aligned} y_i &= \frac{1}{\sigma_i}\mathbf{u}_i^T\mathbf{b} \qquad & i &= 1, \ldots, r \qquad (r = \text{rank of } A) \\ y_i &= 0 & i &= r+1, \ldots, n \end{aligned}$$

and hence

$$\begin{aligned} A^+\mathbf{b} = V\mathbf{y} &= \begin{bmatrix} v_{11}y_1 + v_{12}y_2 + \cdots + v_{1r}y_r \\ v_{21}y_1 + v_{22}y_2 + \cdots + v_{2r}y_r \\ \vdots \\ v_{n1}y_1 + v_{n2}y_2 + \cdots + v_{nr}y_r \end{bmatrix} \\ &= y_1\mathbf{v}_1 + y_2\mathbf{v}_2 + \cdots + y_r\mathbf{v}_r \end{aligned}$$

Thus the solution $\mathbf{x} = A^+\mathbf{b}$ can be computed in two steps:

1. Set $y_i = (1/\sigma_i)\mathbf{u}_i^T\mathbf{b}$ for $i = 1, \ldots, r$.
2. Let $\mathbf{x} = y_1\mathbf{v}_1 + \cdots + y_r\mathbf{v}_r$.

We conclude this section by outlining a method for computing the singular values of a matrix. Observe first that if A has singular value decomposition $U\Sigma V^T$ and $B = HAP^T$, where H is an $m \times m$ orthogonal matrix and P is an $n \times n$ orthogonal matrix, then B has singular value decomposition $(HU)\Sigma(PV)^T$. Thus the problem of finding the singular values of A can be simplified by applying orthogonal transformations to A to obtain a simpler matrix B with the same singular values. Golub and Kahan have shown that A can be reduced to upper bidiagonal form using Householder transformations.

Bidiagonalization

Let H_1 be a Householder transformation that annihilates all the elements below the diagonal in the first column of A. Let P_1 be a Householder transformation such that postmultiplication of H_1A by P_1 annihilates the last $n-2$ entries of the first row of H_1A while leaving the first column unchanged.

$$H_1AP_1 = \begin{bmatrix} \times & \times & 0 \cdots 0 \\ 0 & \times & \times \cdots \times \\ \vdots & & \\ 0 & \times & \times \cdots \times \end{bmatrix}$$

The next step is to apply a Householder transformation H_2 which annihilates the elements below the diagonal in the second column of H_1AP_1 while leaving the first row and column unchanged.

$$H_2H_1AP_1 = \begin{bmatrix} \times & \times & 0 \cdots 0 \\ 0 & \times & \times \cdots \times \\ 0 & 0 & \times \cdots \times \\ \vdots & & \\ 0 & 0 & \times \cdots \times \end{bmatrix}$$

$H_2H_1AP_1$ is then postmultiplied by a Householder transformation P_2 which annihilates the last $n-3$ elements in the second row while leaving the first two columns and the first row unchanged.

$$H_2H_1AP_1P_2 = \begin{bmatrix} \times & \times & 0 & 0 \cdots 0 \\ 0 & \times & \times & 0 \cdots 0 \\ 0 & 0 & \times & \times \cdots \times \\ \vdots & & & \\ 0 & 0 & \times & \times \cdots \times \end{bmatrix}$$

We continue in this manner until we obtain a matrix

$$B = H_n \cdots H_1AP_1 \cdots P_{n-2}$$

of the form

$$\begin{bmatrix} \times & \times & & & & \\ & \times & \times & O & & \\ & & \ddots & \ddots & & \\ & & & \times & \times & \\ & & & & \times & \\ & O & & & & \end{bmatrix}$$

Since $H = H_n \cdots H_1$ and $P^T = P_1 \cdots P_{n-2}$ are orthogonal, it follows that B has the same singular values as A.

The problem has now been simplified to that of finding the singular values of an upper bidiagonal matrix B. One could at this point form the symmetric tridiagonal matrix B^TB and then compute its eigenvalues using the QR algorithm. The problem with this approach is that in forming B^TB, we would still be squaring the condition number, and consequently our computed solution would be much less reliable. The method we will outline produces a sequence of bidiagonal matrices $B_1, B_2, \ldots$ that converges to a diagonal matrix Σ. The method involves applying a sequence of Givens transformations to B alternately on the right- and left-hand sides.

The Golub–Reinsch Algorithm

Let

$$R_k = \begin{bmatrix} I_{k-1} & O & O \\ O & G(\theta_k) & O \\ O & O & I_{n-k-1} \end{bmatrix} \quad \text{and} \quad L_k = \begin{bmatrix} I_{k-1} & O & O \\ O & G(\varphi_k) & O \\ O & O & I_{n-k-1} \end{bmatrix}$$

where the 2×2 matrices $G(\theta_k)$ and $G(\varphi_k)$ are given by

$$G(\theta_k) = \begin{pmatrix} \cos\theta_k & \sin\theta_k \\ \sin\theta_k & -\cos\theta_k \end{pmatrix} \quad \text{and} \quad R(\varphi_k) = \begin{pmatrix} \cos\varphi_k & \sin\varphi_k \\ \sin\varphi_k & -\cos\varphi_k \end{pmatrix}$$

for some angles θ_k and φ_k. The matrix $B = B_1$ is first multiplied on the right by R_1. This will have the effect of filling in the (2, 1) position.

$$B_1R_1 = \begin{bmatrix} \times & \times & & O & \\ \times & \times & \times & & \\ & & \times & & \\ & & & \ddots & \\ & & & & \times \\ & & & & \times \\ & O & & & \end{bmatrix}$$

Next L_1 is chosen so as to annihilate the element filled in by R_1. It will also have the effect of filling in the (1, 3) position. Thus

$$L_1B_1R_1 = \begin{bmatrix} \times & \times & \times & & & \\ & \times & \times & & & O \\ & & & \ddots & \ddots & \\ & & & & & \times \\ & & & & & \times \\ & O & & & & \end{bmatrix}$$

R_2 is chosen so as to annihilate the (1, 3) entry. It will fill in the (3, 2) entry of $L_1B_1R_1$. Next L_2 annihilates the (3, 2) entry and fills in the (2, 4) entry, and so on.

$$\underset{L_1B_1R_1R_2}{\begin{bmatrix} \times & \times & & & & \\ & \times & \times & & & O \\ & \times & \times & \times & & \\ & & & \ddots & \ddots & \\ & & & & & \times \\ & & & & & \times \\ & O & & & & \end{bmatrix}} \qquad \underset{L_2L_1B_1R_1R_2}{\begin{bmatrix} \times & \times & & & & \\ & \times & \times & \times & & O \\ & & \times & \times & & \\ & & & \ddots & \ddots & \\ & & & & & \times \\ & & & & & \times \\ & O & & & & \end{bmatrix}}$$

We continue this process until we end up with a new bidiagonal matrix,

$$B_2 = L_{n-1} \cdots L_1B_1R_1 \cdots R_{n-1}$$

Why should we be any better off with B_2 than with B_1? It can be shown that if the first transformation R_1 is chosen correctly, $B_2^TB_2$ will be the matrix obtained from $B_1^TB_1$ by applying one iteration of the QR algorithm with shift. The same process can now be applied to B_2 to obtain a new bidiagonal matrix B_3 such that $B_3^TB_3$ would be the matrix obtained by applying two iterations of the QR algorithm with shifts to $B_1^TB_1$. Even though the $B_i^TB_i$'s are never computed, we know that with the proper choice of shifts, these matrices will converge rapidly to a diagonal matrix. The B_i's then must also converge to a diagonal matrix Σ. Since each of the B_i's has the same singular values as B, the diagonal elements of Σ will be the singular values of B. The matrices U and V^T can be determined by keeping track of all the orthogonal transformations.

Only a brief sketch of the algorithm has been given. To include more would have been beyond the scope of this text. For complete details of the algorithm and an ALGOL program, we refer the reader to the paper by Golub and Reinsch in [19], p. 135.

EXERCISES

1. Find the solution $\mathbf{x}$ to the least squares problem given that $A = QR$ in each of the following.

(a) $Q = \begin{bmatrix} \frac{1}{\sqrt{2}} & \frac{1}{\sqrt{2}} \\ \frac{1}{\sqrt{2}} & -\frac{1}{\sqrt{2}} \\ 0 & 0 \end{bmatrix}, \quad R = \begin{pmatrix} 1 & 1 \\ 0 & 1 \end{pmatrix}, \quad \mathbf{b} = \begin{bmatrix} 1 \\ 1 \\ 1 \end{bmatrix}$

(b) $Q = \begin{bmatrix} 1 & 0 & 0 \\ 0 & \frac{1}{\sqrt{2}} & -\frac{1}{\sqrt{2}} \\ 0 & \frac{1}{\sqrt{2}} & \frac{1}{\sqrt{2}} \\ 0 & 0 & 0 \end{bmatrix}, \quad R = \begin{bmatrix} 1 & 1 & 0 \\ 0 & 1 & 1 \\ 0 & 0 & 1 \end{bmatrix}, \quad \mathbf{b} = \begin{bmatrix} 1 \\ 3 \\ 1 \\ 2 \end{bmatrix}$

(c) $Q = \begin{bmatrix} 1 & 0 & 0 \\ 0 & \frac{1}{\sqrt{2}} & -\frac{1}{\sqrt{2}} \\ 0 & \frac{1}{\sqrt{2}} & \frac{1}{\sqrt{2}} \end{bmatrix}, \quad R = \begin{bmatrix} 1 & 1 \\ 0 & 1 \\ 0 & 0 \end{bmatrix}, \quad \mathbf{b} = \begin{bmatrix} 1 \\ \sqrt{2} \\ -\sqrt{2} \end{bmatrix}$

(d) $Q = \begin{bmatrix} \frac{1}{2} & \frac{1}{\sqrt{2}} & 0 & \frac{1}{2} \\ \frac{1}{2} & 0 & \frac{1}{\sqrt{2}} & -\frac{1}{2} \\ \frac{1}{2} & 0 & -\frac{1}{\sqrt{2}} & -\frac{1}{2} \\ \frac{1}{2} & -\frac{1}{\sqrt{2}} & 0 & \frac{1}{2} \end{bmatrix}, \quad R = \begin{bmatrix} 1 & 1 & 0 \\ 0 & 1 & 1 \\ 0 & 0 & 1 \\ 0 & 0 & 0 \end{bmatrix}, \quad \mathbf{b} = \begin{bmatrix} 2 \\ -2 \\ 0 \\ 2 \end{bmatrix}$

2. Let

$$A = \begin{pmatrix} D \\ E \end{pmatrix} = \left[\begin{array}{cccc} d_1 & & & O \\ & d_2 & & \\ & & \ddots & \\ O & & & d_n \\ \hline e_1 & & & O \\ & e_2 & & \\ & & \ddots & \\ O & & & e_n \end{array}\right] \quad \text{and} \quad \mathbf{b} = \begin{bmatrix} b_1 \\ b_2 \\ \vdots \\ b_{2n} \end{bmatrix}$$

Use the normal equations to find the solution $\mathbf{x}$ to the least squares problem.

3. Let

$$A = \begin{bmatrix} 1 & 1 \\ \epsilon & 0 \\ 0 & \epsilon \end{bmatrix}$$

where ϵ is a small scalar.

(a) Determine the singular values of A exactly.

(b) Suppose that ϵ is sufficiently small that $1 + \epsilon^2$ gets rounded off to 1 on your calculator. Determine the eigenvalues of the calculated A^TA and compare the square roots of these eigenvalues to your answers in part (a).

4. Show that the pseudoinverse A^+ satisfies the four Penrose conditions.

5. Let B be any matrix that satisfies Penrose conditions (1) and (3) and let $\mathbf{x} = B\mathbf{b}$. Show that $\mathbf{x}$ is a solution to the normal equations $A^TA\mathbf{x} = A^T\mathbf{b}$.

6. Show each of the following.

(a) $(A^+)^+ = A$

(b) $(AA^+)^2 = AA^+$

(c) $(A^+A)^2 = A^+A$

7. Let $A_1 = U\Sigma_1 V^T$ and $A_2 = U\Sigma_2 V^T$, where

$$\Sigma_1 = \begin{bmatrix} \sigma_1 & & & & & \\ & \ddots & & & O & \\ & & \sigma_{r-1} & & & \\ & & & 0 & & \\ & & & & \ddots & \\ & O & & & & 0 \\ & & & & & \end{bmatrix},$$

and

$$\Sigma_2 = \begin{bmatrix} \sigma_1 & & & & & \\ & \ddots & & & O & \\ & & \sigma_{r-1} & & & \\ & & & \sigma_r & & \\ & & & & 0 & \\ & & & & & \ddots \\ & & & & & & 0 \\ & O & & & & & \\ & & & & & & \end{bmatrix}$$

and $\sigma_r = \epsilon > 0$. What are the values $\|A_1 - A_2\|_F$ and $\|A_1^+ - A_2^+\|_F$? What happens to these values as we let $\epsilon \to 0$?

8. Let $A = XY^T$, where X is an $m \times r$ matrix, Y^T is an $r \times n$ matrix, and X^TX and Y^TY are both nonsingular. Show that the matrix

$$B = \left[\, Y(Y^TY)^{-1}(X^TX)^{-1}X^T \right]$$

satisfies the Penrose conditions and hence must equal A^+. Thus A^+ can be determined from any factorization of this form.

Bibliography

A. LINEAR ALGEBRA

1. Anton, Howard. *Elementary Linear Algebra*, 2nd ed. New York: John Wiley & Sons, Inc., 1977.
2. Gantmacher, F. R. *The Theory of Matrices*, 2 vols. New York: Chelsea Publishing Company, Inc., 1960.
3. Kolman, Bernard. *Elementary Linear Algebra*, 2nd ed. New York: Macmillan Publishing Co., Inc., 1977.
4. Lancaster, Peter. *Theory of Matrices*. New York: Academic Press, Inc., 1969.
5. Shields, Paul C. *Elementary Linear Algebra*. Englewood Cliffs, N.J.: Prentice-Hall, Inc., 1977.

B. APPLIED LINEAR ALGEBRA

6. Bellman, Richard. *Introduction to Matrix Analysis*, 2nd ed. New York: McGraw-Hill Book Company, 1970.
7. Fletcher, T. J. *Linear Algebra Through Its Applications*. New York: Van Nostrand Reinhold Company, 1972.
8. Noble, Ben, and James W. Daniel. *Applied Linear Algebra*. Englewood Cliffs, N.J: Prentice-Hall, Inc., 1977.

9. Strang, Gilbert. *Linear Algebra and Its Applications*. New York: Academic Press, Inc., 1976.
10. Williams, Gareth. *Computational Linear Algebra with Models*. Boston: Allyn & Bacon, Inc., 1975.

C. NUMERICAL LINEAR ALGEBRA

11. Atkinson, Kendall E. *An Introduction to Numerical Analysis*. New York: John Wiley & Sons, Inc., 1978.
12. Dahlquist, G., and A. Bjorck. *Numerical Methods*. Englewood Cliffs, N.J.: Prentice-Hall, Inc., 1974.
13. Forsythe, G. E., and C. B. Moler. *Computer Solution of Linear Algebraic Systems*. Englewood Cliffs, N.J.: Prentice-Hall, Inc., 1967.
14. Forsythe, G. E., M. Malcom, and C. B. Moler. *Computer Methods for Mathematical Computations*. Englewood Cliffs, N.J.: Prentice-Hall, Inc., 1977.
15. Lawson, Charles L., and Richard J. Hanson. *Solving Least Squares Problems*. Englewood Cliffs, N.J.: Prentice-Hall, Inc., 1974.
16. Stewart, G. W. *Introduction to Matrix Computations*. New York: Academic Press, Inc., 1973.
17. Wilkinson, J. H. *Rounding Errors in Algebraic Processes*. Englewood Cliffs, N.J.: Prentice-Hall, Inc., 1963.
18. Wilkinson, J. H. *The Algebraic Eigenvalue Problem*. Oxford: Clarendon Press, 1965.
19. Wilkinson, J. H., and C. Reinsch. *Handbook for Automatic Computation*, vol. II, *Linear Algebra*. New York: Springer-Verlag, 1971.

D. BOOKS OF RELATED INTEREST

20. Cheney, E. W. *Introduction to Approximation Theory*. New York: McGraw-Hill Book Company, 1966.
21. Chiang, Alpha C. *Fundamental Methods of Mathematical Economics*. New York: McGraw-Hill Book Company, 1967.
22. Courant, R., and D. Hilbert. *Methods of Mathematical Physics*, vol. I. New York: Wiley–Interscience, 1953.
23. Kreider, D. L., R. G. Kuller, and D. R. Ostberg. *Elementary Differential Equations*. Reading, Mass: Addison-Wesley Publishing Company, Inc., 1968.
24. Rivlin, T. J. *The Chebyshev Polynomials*. New York: Wiley–Interscience, 1974.

References [1], [3], and [5] are elementary linear algebra texts. References [2] and [4] are written on a more advanced level. References [6] through [10] all have good selections of applications. References [13] and [19] contain ALGOL programs, while references [13], [14], and [15] contain FORTRAN programs. Extended bibliographies are included in the following references: [2], [8], [12], [14], [15], [16], and [18].

Answers to Selected Exercises

Chapter 1

SECTION 1

1. (a) (11, 3); (b) (4, 1, 3); (c) $(-2, 0, 3, 1)$; (d) $(-2, 3, 0, 3, 1)$
2. (a) $\begin{pmatrix} 1 & -3 \\ 0 & 2 \end{pmatrix}$ (b) $\begin{bmatrix} 1 & 1 & 1 \\ 0 & 2 & 1 \\ 0 & 0 & 3 \end{bmatrix}$ (c) $\begin{bmatrix} 1 & 2 & 2 & 1 \\ 0 & 3 & 1 & -2 \\ 0 & 0 & -1 & 2 \\ 0 & 0 & 0 & 4 \end{bmatrix}$
3. (a) $\{(-1, \alpha, 1-2\alpha) \mid \alpha \text{ is a real number}\}$; (b) $\varnothing$; (c) $\{(0, 0)\}$; (d) $\varnothing$
4. (a) and (c) are consistent; (a) is dependent while (c) is independent.
5. (a) $(\frac{1}{2}, \frac{2}{3})$; (b) (1, 1, 2); (c) $(1, 1, -1)$; (d) $(4, -3, 1, 2)$

SECTION 2

1. Row echelon form: (a), (c), (g), and (h). Reduced row echelon form: (c) and (g).

2. (a) Consistent and independent, $\{(5, 1)\}$; (b) inconsistent, $\varnothing$; (c) consistent and independent, $\{(0, 0)\}$; (d) consistent and dependent,

$$\left\{\left(\frac{5-\alpha}{4}, \frac{1+7\alpha}{8}, \alpha\right) \middle| \alpha \text{ is a real number}\right\}$$

(e) consistent and dependent, $\{(8-2\alpha, \alpha-5, \alpha)\}$; (f) inconsistent, $\varnothing$; (g) inconsistent, $\varnothing$; (h) inconsistent, $\varnothing$; (i) consistent and independent, $\{(0, \frac{3}{2}, 1)\}$; (j), (k), and (l) are consistent and dependent.

3. (a) $(0, -1)$; (b) $\{(\frac{3}{4}-\frac{5}{8}\alpha, -\frac{1}{4}-\frac{1}{8}\alpha, \alpha, 3) | \alpha \text{ is real}\}$; (c) $\{(0, \alpha, -\alpha)\}$; (d) $\{\alpha(-\frac{2}{3}, 0, \frac{1}{3}, 1)\}$

4. $(2, 0, -2, -2, 0, 2)$

SECTION 3

1. (a) $\begin{bmatrix} 6 & 2 & 8 \\ -4 & 0 & 2 \\ 2 & 4 & 4 \end{bmatrix}$ (b) $\begin{bmatrix} 4 & 1 & 6 \\ -5 & 1 & 2 \\ 3 & -2 & 3 \end{bmatrix}$ (c) $\begin{bmatrix} 3 & 2 & 2 \\ 5 & -3 & -1 \\ -4 & 16 & 1 \end{bmatrix}$

(d) $\begin{bmatrix} 3 & 5 & -4 \\ 2 & -3 & 16 \\ 2 & -1 & 1 \end{bmatrix}$ (f) $\begin{bmatrix} 5 & 5 & 8 \\ -10 & -1 & -9 \\ 15 & 4 & 6 \end{bmatrix}$ (h) $\begin{bmatrix} 5 & -10 & 15 \\ 5 & -1 & 4 \\ 8 & -9 & 6 \end{bmatrix}$

2. (a) $\begin{pmatrix} 15 & 19 \\ 4 & 0 \end{pmatrix}$ (c) $\begin{bmatrix} 19 & 21 \\ 17 & 21 \\ 8 & 10 \end{bmatrix}$ (d) $\begin{pmatrix} 36 & 10 & 56 \\ 10 & 3 & 16 \end{pmatrix}$

(b) and (e) are not possible.

3. (a) 3×3; (b) 1×2

8. $A = A^2 = A^3 = A^n$

9. $A^{2n} = I$, $A^{2n+1} = A$

11. Monday 1275, Tuesday 936, Wednesday 457.8, Thursday 1105, Friday 457.8

12. 4500 married, 5500 single

SECTION 4

1. (a) $\begin{pmatrix} 3 & 2 \\ 2 & -3 \end{pmatrix}\begin{pmatrix} x_1 \\ x_2 \end{pmatrix} = \begin{pmatrix} 1 \\ 5 \end{pmatrix}$ (c) $\begin{bmatrix} 2 & 1 & 1 \\ 1 & -1 & 2 \\ 3 & -2 & -1 \end{bmatrix}\begin{bmatrix} x_1 \\ x_2 \\ x_3 \end{bmatrix} = \begin{bmatrix} 4 \\ 2 \\ 0 \end{bmatrix}$

4. (a) $\begin{pmatrix} -2 & 0 \\ 0 & 1 \end{pmatrix}$ (b) $\begin{bmatrix} 1 & 0 & 0 \\ 0 & 0 & 1 \\ 0 & 1 & 0 \end{bmatrix}$ (c) $\begin{bmatrix} 1 & 0 & 0 \\ 0 & 1 & 0 \\ 0 & 2 & 1 \end{bmatrix}$

5. (a) $\begin{bmatrix} 0 & 0 & 1 \\ 0 & 1 & 0 \\ 1 & 0 & 0 \end{bmatrix}$ (b) $\begin{pmatrix} 1 & -3 \\ 0 & 1 \end{pmatrix}$ (c) $\begin{bmatrix} \frac{1}{2} & 0 & 0 \\ 0 & 1 & 0 \\ 0 & 0 & 1 \end{bmatrix}$

8. (a) $\frac{1}{13}\begin{pmatrix} 5 & -2 \\ -1 & 3 \end{pmatrix}$ (c) $\frac{1}{7}\begin{bmatrix} -10 & -33 & 21 \\ 3 & 12 & -7 \\ 2 & 1 & 0 \end{bmatrix}$

SECTION 5

1. (a) $(I \quad A^{-1})$ (b) $\begin{pmatrix} I \\ A^{-1} \end{pmatrix}$ (c) $\begin{pmatrix} A^2 & A \\ A & I \end{pmatrix}$

(d) $A^2 + I$ (e) $\begin{pmatrix} I & A^{-1} \\ A & I \end{pmatrix}$

3. (i) $A\mathbf{b}_1 = \begin{pmatrix} 3 \\ 3 \end{pmatrix}$, $A\mathbf{b}_2 = \begin{pmatrix} 4 \\ -1 \end{pmatrix}$

(ii) $(1 \quad 1)B = (3 \quad 4)$, $(2 \quad -1)B = (3 \quad -1)$

(iii) $AB = \begin{pmatrix} 3 & 4 \\ 3 & -1 \end{pmatrix}$

4. (a) $\left[\begin{array}{cc|cc} 3 & 1 & 1 & 1 \\ 3 & 2 & 1 & 2 \\ \hline 1 & 1 & 1 & 1 \\ 1 & 2 & 1 & 1 \end{array}\right]$ (b) $\left[\begin{array}{cc|cc} 1 & 1 & 1 & 1 \\ 0 & 1 & 0 & 0 \\ \hline 3 & 1 & 1 & 1 \\ 0 & 1 & 0 & 1 \end{array}\right]$

(c) $\left[\begin{array}{cc|cc} 2 & 2 & 2 & 2 \\ 2 & 4 & 2 & 2 \\ \hline 3 & 1 & 1 & 1 \\ 3 & 2 & 1 & 2 \end{array}\right]$ (d) $\left[\begin{array}{cc|cc} 1 & 1 & 1 & 1 \\ 1 & 2 & 1 & 1 \\ \hline 3 & 2 & 1 & 2 \\ 3 & 1 & 1 & 1 \end{array}\right]$

5. (b) $\left[\begin{array}{ccc|c} 0 & 2 & 0 & -2 \\ 8 & 5 & 8 & -5 \\ 3 & 2 & 3 & -2 \\ \hline 5 & 3 & 5 & -3 \end{array}\right]$ (d) $\left[\begin{array}{cc} 3 & -3 \\ 2 & -2 \\ 1 & -1 \\ \hline 5 & -5 \\ 4 & -4 \end{array}\right]$

Chapter 2

SECTION 1

1. (a) Even; (b) odd; (c) even; (d) even; (e) odd
3. (a) -39; (b) 0; (c) 8; (d) 20
4. (a) 2; (b) -4; (c) 0; (d) 0

SECTION 2

1. (a) -24; (b) 30; (c) -1
2. (a) 10; (b) 20
3. (a), (e), and (f) are singular while (b), (c), and (d) are nonsingular.

SECTION 3

1. (a) $|A| = -7$, adj $A = \begin{pmatrix} -1 & -2 \\ -3 & 1 \end{pmatrix}$, $A^{-1} = \begin{bmatrix} \frac{1}{7} & \frac{2}{7} \\ \frac{3}{7} & -\frac{1}{7} \end{bmatrix}$

(c) $|A| = 3$, adj $A = \begin{bmatrix} -3 & 5 & 2 \\ 0 & 1 & 1 \\ 6 & -8 & -5 \end{bmatrix}$, $A^{-1} = \frac{1}{3}$ adj A

2. (a) $(\frac{5}{7}, \frac{8}{7})$; (b) $(2, -1, 2)$; (c) $(-\frac{2}{3}, \frac{2}{3}, \frac{1}{3}, 0)$

Chapter 3

SECTION 1

2. V is not a vector space. Axiom 6 does not hold.

6. If $\mathbf{x} + \mathbf{y} = \mathbf{x}$ for all $\mathbf{x}$ in the vector space, then $\mathbf{0} = \mathbf{0} + \mathbf{y} = \mathbf{y}$.

SECTION 2

1. (b) and (c) are subspaces; (a) and (d) are not.

2. (a), (b), (d), and (e) are subspaces; (c) and (f) are not.

3. (a) $\{(0, 0)^T\}$; (b) $\mathbb{S}\{(-2, 1, 0, 0)^T, (3, 0, 1, 0)^T\}$; (c) $\mathbb{S}\{(1, 1, 1)^T\}$; (d) $\mathbb{S}\{(-5, 0, -3, 1)^T, (1, -1, 0, 0)^T\}$

5. Only the set in part (c) is a subspace of P_4.

6. (a), (b), and (d) are subspaces.

7. (a), (c), and (e) are spanning sets.

8. (a) and (b) are spanning sets.

9. (b) and (c)

SECTION 3

1. (a) and (e) are linearly independent; (b), (c), and (d) are linearly dependent.

2. (a) and (e) are linearly independent; (b), (c), and (e) are not.

3. (a) and (b) are linearly dependent while (c) and (d) are linearly independent.

SECTION 4

1. (a) 2, basis; (b) 1; (c) 2; (d) 1; (e) 2, basis

2. (a) 3, basis; (b) 3; (c) 2; (d) 1; (e) 2

3. (b) $\{(1, 1, 1)^T\}$, dimension 1; (c) $\{(1, 0, 1)^T, (0, 1, 1)^T\}$, dimension 2

4. (a) $\{E_{11}, E_{12}\}$; (b) $\{E_{11}, E_{21}, E_{22}\}$; (d) $\{E_{12}, E_{21}, E_{22}\}$; (e) $\{E_{11}, E_{22}, E_{21} + E_{12}\}$

5. 2

6. (a) 3; (b) 3; (c) 2; (d) 2

7. (a) $\{x, x^2\}$; (b) $\{x - 1, (x - 1)^2\}$; (c) $\{x(x - 1)\}$

SECTION 5

1. (a) The row space and column space have dimension 2. There are many possible bases for each of these spaces. The bases derived from the reduced row echelon form and the reduced column echelon form are

(i) $\{(1, 0, 2), (0, 1, 0)\}$ (row space)

(ii) $\{(1, 0, 2)^T, (0, 1, 1)^T\}$ (column space)

The set $\{(-2, 0, 1)^T\}$ is a basis for the nullspace.

2. The following examples are all in reduced row echelon form:

(a) $\begin{bmatrix} 0 & 0 & 0 \\ 0 & 0 & 0 \\ 0 & 0 & 0 \end{bmatrix}$ (b) $\begin{bmatrix} 1 & 0 & 0 \\ 0 & 0 & 0 \\ 0 & 0 & 0 \end{bmatrix}$ (c) $\begin{bmatrix} 1 & 0 & 0 \\ 0 & 1 & 0 \\ 0 & 0 & 0 \end{bmatrix}$

(d) $\begin{bmatrix} 1 & 0 & 0 \\ 0 & 1 & 0 \\ 0 & 0 & 1 \end{bmatrix}$

5. If $\mathbf{x}_j$ is a solution to $A\mathbf{x} = \mathbf{e}_j$ for $j = 1, \ldots, m$ and $X = (\mathbf{x}_1, \mathbf{x}_2, \ldots, \mathbf{x}_m)$, then $AX = I_m$.

Chapter 4

SECTION 1

1. (a) Reflection about x_2 axis; (b) reflection about the origin; (c) reflection about the line $x_2 = x_1$; (d) rotation by an angle θ
2. All except (c) are linear transformations from R^3 into R^2.
3. (b) and (c) are linear transformations from R^2 into R^3.
4. $L(e^x) = e^x - 1$ and $L(x^2) = x^3/3$
5. (a) and (c) are linear transformations from $C[0, 1]$ into R^1.
6. (a), (b), and (d) are linear transformations on $M_{n,n}$.
7. (a) and (c) are linear transformations from P_2 into P_3.

SECTION 2

1. (a) $\begin{pmatrix} 1 & 1 & 0 \\ 0 & 0 & 0 \end{pmatrix}$ (b) $\begin{pmatrix} 1 & 0 & 0 \\ 0 & 1 & 0 \end{pmatrix}$ (c) $\begin{pmatrix} -1 & 1 & 0 \\ 0 & -1 & 1 \end{pmatrix}$

2. (a) $\begin{bmatrix} 0 & 0 & 1 \\ 0 & 1 & 0 \\ 1 & 0 & 0 \end{bmatrix}$ (b) $\begin{bmatrix} 1 & 0 & 0 \\ 1 & 1 & 0 \\ 1 & 1 & 1 \end{bmatrix}$ (c) $\begin{bmatrix} 0 & 0 & 2 \\ 3 & 1 & 0 \\ 2 & 0 & -1 \end{bmatrix}$

3. $\begin{pmatrix} 1 & \frac{1}{2} \\ 1 & 0 \end{pmatrix}$

4. $\begin{pmatrix} 1 & \frac{1}{2} & \frac{1}{2} \\ -2 & 0 & 0 \end{pmatrix}$ (a) $\begin{pmatrix} \frac{1}{2} \\ -2 \end{pmatrix}$ (d) $\begin{pmatrix} 5 \\ -8 \end{pmatrix}$

5. (b) $\begin{bmatrix} 0 & 0 & 1 \\ 0 & 1 & -1 \\ 1 & -1 & 0 \end{bmatrix}$

6. $\begin{bmatrix} 1 & 1 & 0 \\ 0 & 1 & 2 \\ 0 & 0 & 1 \end{bmatrix}$

SECTION 3

1. $\begin{bmatrix} 1 & -1 & 0 \\ 0 & 1 & -1 \\ 0 & 0 & 1 \end{bmatrix}$

2. (c) $\begin{bmatrix} 1 & 0 & 1 \\ 0 & 1 & 0 \\ 0 & 0 & 1 \end{bmatrix}$

3. (a) $\begin{bmatrix} 1 & 0 & 0 \\ 0 & 1 & 1 \\ 0 & 1 & -1 \end{bmatrix}$ (c) $\begin{bmatrix} 0 & 0 & 0 \\ 0 & 1 & 0 \\ 0 & 0 & -1 \end{bmatrix}$

Chapter 5

SECTION 1

1. (a) $0°$; (b) $90°$

2. (a) $\sqrt{14}$ (scalar projection), $(2, 1, 3)^T$ (vector projection);
(b) 0, $\mathbf{0}$; (c) $\dfrac{14\sqrt{13}}{13}$, $(\frac{42}{13}, \frac{28}{13})^T$; (d) $\dfrac{8\sqrt{21}}{21}$, $(\frac{8}{21}, \frac{16}{21}, \frac{32}{21})^T$

3. (1.8, 3.6)

4. (1.4, 3.8)

5. $\frac{2}{3}$

SECTION 2

1. (a) $\{(3, 4)^T\}$ basis for $R(A^T)$, $\{(-4, 3)^T\}$ basis for $N(A)$, $\{(1, 2)^T\}$ basis for $R(A)$, $\{(-2, 1)^T\}$ basis for $N(A^T)$
(d) Basis for $R(A^T)$: $\{(1, 0, 0, 0)^T, (0, 1, 0, 0)^T, (0, 0, 1, 1)^T\}$
Basis for $N(A)$: $\{(0, 0, -1, 1)^T\}$
Basis for $R(A)$: $\{(1, 0, 0, 1)^T, (0, 1, 0, 1)^T, (0, 0, 1, 1)^T\}$
Basis for $N(A^T)$: $\{(1, 1, 1, -1)^T\}$

3. (b) The orthogonal complement is spanned by $(-5, 1, 3)^T$

6. $\dim N(A) = n - r$, $\dim N(A^T) = m - r$

SECTION 3

2. (a) 1; (b) $1/\pi$; (c) $\frac{1}{6}$

8. (a) $\|\mathbf{x}\|_1 = 7$, $\|\mathbf{x}\|_2 = 5$, $\|\mathbf{x}\|_\infty = 4$
(b) $\|\mathbf{x}\|_1 = 4$, $\|\mathbf{x}\|_2 = \sqrt{6}$, $\|\mathbf{x}\|_\infty = 2$
(c) $\|\mathbf{x}\|_1 = 3$, $\|\mathbf{x}\|_2 = \sqrt{3}$, $\|\mathbf{x}\|_\infty = 1$

9. (a) $\sqrt{\pi}$; (b) $\sqrt{\pi}$; (c) $\sqrt{2\pi}$

10. (a) $\dfrac{\sqrt{34}}{2}$; (b) $\dfrac{\sqrt{514}}{4}$

15. (a) Not a norm; (b) norm; (c) norm

SECTION 4

1. (a) $\|A\|_F = \sqrt{2}$, $\|A\|_\infty = 1$, $\|A\|_1 = 1$
(b) $\|A\|_F = 5$, $\|A\|_\infty = 5$, $\|A\|_1 = 6$
(c) $\|A\|_F = \|A\|_\infty = \|A\|_1 = 1$
(d) $\|A\|_F = 7$, $\|A\|_\infty = 6$, $\|A\|_1 = 10$
(e) $\|A\|_F = 9$, $\|A\|_\infty = 10$, $\|A\|_1 = 12$

7. (a) $\|A\mathbf{x}\|_\infty \leqslant \|A\mathbf{x}\|_2 \leqslant \|A\|_2\|\mathbf{x}\|_2 \leqslant \sqrt{n}\,\|A\|_2\|\mathbf{x}\|_\infty$

SECTION 5

2. (a) $(2, 1)^T$; (c) $(1.6, 0.6, 1.2)^T$

3. (a) $(0, 0, 2)^T$; (c) $(0.6, -0.2, 0.4, -0.8)^T$

4. (a) $y = 1.8 + 2.9x$

SECTION 6

1. (a) and (d)

2. (b) $\mathbf{x} = -\dfrac{\sqrt{2}}{3}\mathbf{x}_1 + \dfrac{5}{3}\mathbf{x}_2$; $\|\mathbf{x}\| = \left[\left(-\dfrac{\sqrt{2}}{3}\right)^2 + \left(\dfrac{5}{3}\right)^2\right]^{1/2} = \sqrt{3}$

3. $\mathbf{p} = (\frac{23}{18}, \frac{41}{18}, \frac{8}{9})^T$; $\mathbf{p} - \mathbf{x} = (\frac{5}{18}, \frac{5}{18}, -\frac{10}{9})^T$

6. (b)(i) $(\sqrt{3}, \sqrt{6})^T$; (ii) $\frac{1}{2}(4\sqrt{3}, -\sqrt{6})^T$; (iii) $\frac{1}{6}(8\sqrt{3}, -\sqrt{6})^T$

SECTION 7

1. $u_1(x) = \dfrac{1}{\sqrt{2}}$, $u_2(x) = \dfrac{\sqrt{6}}{2}x$, $u_3(x) = \dfrac{3\sqrt{10}}{4}\left(x^2 - \dfrac{1}{3}\right)$

2. (a) $\left\{\dfrac{1}{3}(2, 1, 2)^T, \dfrac{\sqrt{2}}{6}(-1, 4, 1)^T\right\}$

(b) $Q = \begin{bmatrix} \frac{2}{3} & -\frac{\sqrt{2}}{6} \\ \frac{1}{3} & \frac{2\sqrt{2}}{3} \\ \frac{2}{3} & -\frac{\sqrt{2}}{6} \end{bmatrix}$ $\quad R = \begin{bmatrix} 3 & \frac{5}{3} \\ 0 & \frac{\sqrt{2}}{3} \end{bmatrix}$

(c) $\mathbf{x} = \begin{pmatrix} 9 \\ -3 \end{pmatrix}$

6. (b) $Q\mathbf{x}$ can be obtained by a rotation of an angle θ followed by a reflection about the x_2 axis.

SECTION 8

1. (a) $T_4 = 8x^4 - 8x^2 + 1$, $T_5 = 16x^5 - 20x^3 + 5x$

(b) $H_4 = 16x^4 - 48x^2 + 12$, $H_5 = 32x^5 - 160x^3 + 120x$

2. $p_1(x) = x$, $p_2(x) = x^2 - \dfrac{4}{\pi} + 1$

4. $p(x) = (\sinh 1)P_0(x) + \dfrac{3}{e}P_1(x) + 5\left(\sinh 1 - \dfrac{3}{e}\right)P_2(x)$

$p(x) \approx 0.9963 + 1.1036x + 0.5367x^2$

6. (a) $U_0 = 1$, $U_1 = 2x$, $U_2 = 4x^2 - 1$

Chapter 6

SECTION 1

1. (a) $\lambda_1 = 5$, the eigenspace is spanned by $(1, 1)^T$

$\lambda_2 = -1$, the eigenspace is spanned by $(1, -2)^T$

(b) $\lambda_1 = 2 + i$, the eigenspace is spanned by $(1, 1+i)^T$
$\lambda_2 = 2 - i$, the eigenspace is spanned by $(1, 1-i)^T$
(c) $\lambda_1 = 2$, the eigenspace is spanned by $(1, 1, 0)^T$
$\lambda_2 = 1$, the eigenspace is spanned by $(1, 0, 0)^T$, $(0, 1, -1)^T$
(d) $\lambda_1 = 2$, the eigenspace is spanned by $(7, 3, 1)^T$
$\lambda_2 = 1$, the eigenspace is spanned by $(3, 2, 1)^T$
$\lambda_3 = 0$, the eigenspace is spanned by $(1, 1, 1)^T$

4. β is an eigenvalue of B if and only if $\beta = \lambda - \alpha$ for some eigenvalue λ of A.

6. $\lambda_1 \mathbf{x}^T\mathbf{y} = (A\mathbf{x})^T\mathbf{y} = \mathbf{x}^T A^T \mathbf{y} = \lambda_2 \mathbf{x}^T\mathbf{y}$

15. If $A\mathbf{z} = \lambda\mathbf{z}$, then $A\bar{\mathbf{z}} = \bar{A}\bar{\mathbf{z}} = \overline{A\mathbf{z}} = \overline{\lambda\mathbf{z}} = \bar{\lambda}\bar{\mathbf{z}}$

SECTION 2

1. (a) $\mathbf{Y} = \begin{pmatrix} c_1e^{2t} + c_2e^{3t} \\ c_1e^{2t} + 2c_2e^{3t} \end{pmatrix}$

(b) $\mathbf{Y} = \begin{pmatrix} c_1e^t \cos t + c_2e^t \sin t \\ c_1e^t \sin t - c_2e^t \cos t \end{pmatrix}$

(c) $\mathbf{Y} = \begin{bmatrix} -2c_1e^t - 2c_2e^{-t} + c_3e^{\sqrt{2}\,t} + c_4e^{-\sqrt{2}\,t} \\ c_1e^t + c_2e^{-t} - c_3e^{\sqrt{2}\,t} - c_4e^{-\sqrt{2}\,t} \end{bmatrix}$

2. (a) $\mathbf{Y} = \begin{pmatrix} e^{-3t} + 2e^t \\ -e^{-3t} + 2e^t \end{pmatrix}$

(b) $\mathbf{Y} = \begin{pmatrix} e^t \cos t + 2e^t \sin 2t \\ e^t \sin 2t - 2e^t \cos 2t \end{pmatrix}$

(c) $\mathbf{Y} = \begin{bmatrix} -6e^t + 2e^{-t} + 6 \\ -3e^t + e^{-t} + 4 \\ -e^t + e^{-t} + 2 \end{bmatrix}$

4. $x_1(t) = \cos t + 3 \sin t + \dfrac{1}{\sqrt{3}} \sin \sqrt{3}\, t$

$x_2(t) = \cos t + 3 \sin t - \dfrac{1}{\sqrt{3}} \sin \sqrt{3}\, t$

6. (a) $m_1x_1''(t) = -kx_1 + k(x_2 - x_1)$
$m_2x_2''(t) = -k(x_2 - x_1) + k(x_3 - x_2)$
$m_3x_2''(t) = -k(x_3 - x_2) - kx_3$

(b) $x_1(t) = x_3(t) = \dfrac{\beta}{4} \cos \sqrt{\alpha}\, t + \dfrac{\alpha}{4} \cos \sqrt{\beta}\, t$

$x_2(t) = \dfrac{\sqrt{2}\,\beta}{4} \cos \sqrt{\alpha}\, t - \dfrac{\sqrt{2}\,\alpha}{4} \cos \sqrt{\beta}\, t$

where $\alpha = 2 - \sqrt{2}$, $\beta = 2 + \sqrt{2}$

7. $p(\lambda) = (-1)^n(\lambda^n - a_{n-1}\lambda^{n-1} - \cdots - a_1\lambda - a_0)$

8. $y_1(t) = -e^{2t} + e^{-2t} + e^t$, $y_2(t) = -e^{2t} - e^{-2t} + 2e^t$

SECTION 3

5. $|B - \lambda I| = |S^{-1}AS - \lambda S^{-1}S| = |S^{-1}(A - \lambda I)S| = |A - \lambda I|$
Since B and A have the same characteristic polynomial, they must have the same eigenvalues. Thus similar matrices have the same eigenvalues.

8. (a) $A = \begin{bmatrix} 0.70 & 0.20 & 0.10 \\ 0.20 & 0.70 & 0.10 \\ 0.10 & 0.10 & 0.80 \end{bmatrix}$

(c) The membership of all three groups will approach 100,000 as n gets large.

SECTION 4

1. (b) and (f) are Hermitian while (b), (c), (e), and (f) are normal.
8. (b) $\|U\mathbf{x}\|^2 = (U\mathbf{x})^H U\mathbf{x} = \mathbf{x}^H U^H U\mathbf{x} = \mathbf{x}^H\mathbf{x} = \|\mathbf{x}\|^2$
9. U is unitary, since
$U^H U = (I - 2\mathbf{u}\mathbf{u}^H)^2 = I - 4\mathbf{u}\mathbf{u}^H + 4\mathbf{u}(\mathbf{u}^H\mathbf{u})\mathbf{u}^H = I$

SECTION 5

1. (a) $\begin{bmatrix} 3 & -\frac{5}{2} \\ -\frac{5}{2} & 1 \end{bmatrix}$ (b) $\begin{bmatrix} 2 & \frac{1}{2} & -1 \\ \frac{1}{2} & 3 & \frac{3}{2} \\ -1 & \frac{3}{2} & 1 \end{bmatrix}$

2. (a) $Q = \dfrac{1}{\sqrt{2}}\begin{pmatrix} 1 & 1 \\ 1 & -1 \end{pmatrix}$, $\dfrac{(x')^2}{4} + \dfrac{(y')^2}{12} = 1$, ellipse

(d) $Q = \dfrac{1}{\sqrt{2}}\begin{pmatrix} 1 & 1 \\ -1 & 1 \end{pmatrix}$, $\left(y' + \dfrac{\sqrt{2}}{2}\right)^2 = -\dfrac{\sqrt{2}}{2}(x' - \sqrt{2})$ or

$(y'')^2 = -\dfrac{\sqrt{2}}{2}x''$

3. (a) Ellipsoid; (c) paraboloid; (d) paraboloid;
(f) hyperboloid (1 sheet); (g) cone; (h) hyperboloid (2 sheets)

SECTION 6

1. (a) Positive definite; (b) indefinite; (c) negative definite;
(d) indefinite;
2. (a) Minimum; (b) saddle point; (c) saddle point;
(d) local maximum

SECTION 7

1. $x_1 = 70{,}000$; $x_2 = 56{,}000$; $x_3 = 44{,}000$
2. $x_1 = x_2 = x_3$
3. $(I - A)^{-1} = I + A + \cdots + A^{m-1}$

4. (a) $(I-A)^{-1} = \begin{bmatrix} 1 & -1 & 3 \\ 0 & 0 & 1 \\ 0 & -1 & 2 \end{bmatrix}$

(b) $A^2 = \begin{bmatrix} 0 & -2 & 2 \\ 0 & 0 & 0 \\ 0 & 0 & 0 \end{bmatrix}$, $A^3 = \begin{bmatrix} 0 & 0 & 0 \\ 0 & 0 & 0 \\ 0 & 0 & 0 \end{bmatrix}$

5. (b) and (c) are reducible.

Chapter 7

SECTION 1

1. (a) 0.231×10^4; (b) 0.362×10^2; (c) 0.128×10^{-1};
(d) 0.824×10^5

2. (a) $\epsilon = -2$; $\delta \approx -8.7 \times 10^{-4}$; (b) $\epsilon = 0.04$; $\delta \approx 1.2 \times 10^{-3}$;
(c) $\epsilon = 3.0 \times 10^{-5}$; $\delta \approx 2.3 \times 10^{-3}$; (d) $\epsilon = -31$; $\delta \approx -3.8 \times 10^{-4}$

3. (a) 10,420, $\epsilon = -0.0018$, $\delta \approx -1.7 \times 10^{-7}$;
(b) 0, $\epsilon = -8$, $\delta = -1$;
(c) 1×10^{-4}, $\epsilon = 5 \times 10^{-5}$, $\delta = 1$;
(d) 82,190, $\epsilon = 25.7504$, $\delta \approx 3.1 \times 10^{-4}$

4. (a) 0.1043×10^6; (b) 0.1045×10^6; (c) 0.1045×10^6

5. 23

SECTION 2

1. $A = \begin{bmatrix} 1 & 0 & 0 \\ 2 & 1 & 0 \\ -3 & 2 & 1 \end{bmatrix} \begin{bmatrix} 1 & 1 & 1 \\ 0 & 2 & -1 \\ 0 & 0 & 3 \end{bmatrix}$

2. (a) $(2, -1, 3)^T$; (b) $(1, -1, 3)^T$; (c) $(1, 5, 1)^T$

3. (a) n^2 multiplications and $n(n-1)$ additions
(b) n^3 multiplications and $n^2(n-1)$ additions

6. $5n-4$ multiplications/divisions, $3n-3$ additions/subtractions

7. (a) $[(n-j)(n-j+1)]/2$ multiplications; $[(n-j-1)(n-j)]/2$ additions
(c) It requires on the order of $\frac{2}{3}n^3$ additional multiplications/divisions to compute A^{-1} given the LU factorization

SECTION 3

1. (a) $(1, 1, -2)$; (b) $\begin{bmatrix} 0 & 0 & 1 \\ 1 & 0 & 0 \\ 0 & 1 & 0 \end{bmatrix} \begin{bmatrix} 1 & 0 & 0 \\ 2 & 1 & 0 \\ 0 & 3 & 1 \end{bmatrix} \begin{bmatrix} 1 & 2 & -2 \\ 0 & 1 & 8 \\ 0 & 0 & -23 \end{bmatrix}$

2. (a) $(1, 2, 2)$; (b) $(4, -3, 0)$; (c) $(1, 1, 1)$

3. Error $\dfrac{-2000\epsilon}{0.6} \approx -3333\epsilon$. If $\epsilon = 0.001$, then $\delta = -\frac{2}{3}$

4. (1.667, 1.001)

5. (5.002, 1.000)

6. (5.001, 1.001)

SECTION 4

1. $\text{cond}_\infty A = (4.0005)(4000.5) = 16{,}004.00025$
2. (a) $(-0.48, 0.8)$; (b) $(-2.902, 2.0)$
3. (i) $A_n^{-1} = \begin{pmatrix} 1-n & n \\ n & -n \end{pmatrix}$, (ii) $\text{cond}_\infty A_n = 4n$, (iii) $\lim_{n\to\infty} \text{cond}_\infty A_n = \infty$
5. $\text{cond}_\infty (A) = 28$
6. (a) $\mathbf{r} = (-0.1, -0.9, -0.6)^T$, $\|\mathbf{r}\|_\infty = 0.9$ (b) 3.6; (c) $\mathbf{x} = (1, -2, 3)^T$, $\delta = 0.1$
7. $\text{cond}_1 (A) = 6$
8. 1.4

SECTION 5

1. It takes three multiplications, two additions, and one square root to determine H. It takes four multiplications/divisions, one addition, and one square root to determine G. The calculation of GA requires $4n$ multiplications and $2n$ additions while the calculation of HA requires $3n$ multiplications/divisions and $3n$ additions.
2. (a) $n - k + 1$ multiplications/divisions, $2n - 2k + 1$ additions (b) $n(n - k + 1)$ multiplications/divisions, $n(2n - 2k + 1)$ additions
3. (a) $4(n - k)$ multiplications, $2(n - k)$ additions (b) $4n(n - k)$ multiplications, $2n(n - k)$ additions
4. (a) rotation; (b) rotation; (c) Givens transformation; (d) Givens transformation

SECTION 6

1. $\mathbf{x}^{(1)} = (3, 2, 1)^T$, $\mathbf{x}^{(2)} = (0, 1, 1)^T$, $\mathbf{x}^{(3)} = (1, 1, 1)^T$, $\mathbf{x}^{(4)} = (1, 1, 1)^T$
2. (a) $\mathbf{x}^{(1)} = (1.2, 1.1, 1.1)^T$; (b) $\mathbf{x}^{(1)} = (1.20, 1.08, 0.972)^T$
3. The iteration scheme will converge for the matrices in parts (b), (c), (d), and (e).
6. (a) $B = \left(1 - \dfrac{1}{\alpha}\right)I + \dfrac{1}{\alpha}D^{-1}(L + U)$, $\mathbf{c} = \dfrac{1}{\alpha}D^{-1}\mathbf{b}$ (b) $B = (\alpha D - L)^{-1}((\alpha - 1)D + U)$, $\mathbf{c} = (\alpha D - L)^{-1}\mathbf{b}$

SECTION 7

2. (a) $\sigma_1 = \sqrt{10}$, $\sigma_2 = 0$; (b) $\sigma_1 = 3$, $\sigma_2 = 2$; (c) $\sigma_1 = 4$, $\sigma_2 = 2$; (d) $\sigma_1 = 3$, $\sigma_2 = 2$, $\sigma_3 = 1$. The matrices U and V are not unique. The reader may check his or her answers by multiplying out $U\Sigma V^T$.
3. (b) Rank of $A = 2$, $\|A\|_2 = 3$, $A' = \begin{pmatrix} 1.2 & -2.4 \\ -0.6 & 1.2 \end{pmatrix}$

SECTION 8

1. (a) $\mathbf{u}_1 = \begin{pmatrix} 1 \\ 1 \end{pmatrix}$; (b) $A_2 = \begin{pmatrix} 2 & 0 \\ 0 & 0 \end{pmatrix}$; (c) $\lambda_1 = 2$, $\lambda_2 = 0$ The eigenspace corresponding to λ_1 is spanned by $\mathbf{u}_1$.

2. (a) $\mathbf{v}_1 = \begin{bmatrix} 3 \\ 5 \\ 3 \end{bmatrix}$, $\mathbf{u}_1 = \begin{bmatrix} 0.6 \\ 1.0 \\ 0.6 \end{bmatrix}$, $\mathbf{v}_2 = \begin{bmatrix} 2.2 \\ 4.2 \\ 2.2 \end{bmatrix}$, $\mathbf{u}_2 = \begin{bmatrix} 0.52 \\ 1.00 \\ 0.52 \end{bmatrix}$, $\mathbf{v}_3 \approx \begin{bmatrix} 2.05 \\ 4.05 \\ 2.05 \end{bmatrix}$

(b) $\lambda_1' = 4.05$; (c) $\lambda_1 = 4$, $\delta = 0.0125$

3. (b) A has no dominant eigenvalue.

4. $A_2 = \begin{pmatrix} 3 & -1 \\ -1 & 1 \end{pmatrix}$, $A_3 = \begin{pmatrix} 3.4 & 0.2 \\ 0.2 & 0.6 \end{pmatrix}$

$\lambda_1 = 2 + \sqrt{2} \approx 3.414$, $\lambda_2 = 2 - \sqrt{2} \approx 0.586$

SECTION 9

1. (a) $(\sqrt{2}, 0)^T$; (b) $(1 - 3\sqrt{2}, 3\sqrt{2}, -\sqrt{2})^T$; (c) $(1, 0)^T$; (d) $(1 - \sqrt{2}, \sqrt{2}, -\sqrt{2})^T$

2. $\hat{x}_i = \dfrac{d_i b_i + e_i b_{n+i}}{d_i^2 + e_i^2}$, $i = 1, \ldots, n$

3. (a) $\sigma_1 = \sqrt{2 + \epsilon^2}$, $\sigma_2 = \epsilon$; (b) $\lambda_1' = 2$, $\lambda_2' = 0$, $\sigma_1' = \sqrt{2}$, $\sigma_2' = 0$

7. $\|A_1 - A_2\|_F = \epsilon$, $\|A_1^+ - A_2^+\|_F = 1/\epsilon$. As $\epsilon \to 0$, $\|A_1 - A_2\|_F \to 0$ and $\|A_1^+ - A_2^+\|_F \to \infty$.

Index